17th National Conference of the Australian Meteorological and Oceanographic Society 2010

AA002917

IOP Conference Series: Earth and Environmental Science
Volume 11

Canberra, Australia
27 - 29 January 2010

Editors:

Barbara A. Burns
Clem Davis
Andrew E. Kiss
John R. Taylor

ISBN: 978-1-5108-0200-1
ISSN: 1755-1307

Printed from e-media with permission by:

Curran Associates, Inc.
57 Morehouse Lane
Red Hook, NY 12571

Some format issues inherent in the e-media version may also appear in this print version.

Printed by Curran Associates, Inc. (2015)

For permission requests, please contact the Institute of Physics at the address below.

Institute of Physics
Dirac House, Temple Back
Bristol BS1 6BE UK

Phone: 44 1 17 929 7481
Fax: 44 1 17 920 0979

techtracking@iop.org

Additional copies of this publication are available from:

Curran Associates, Inc.
57 Morehouse Lane
Red Hook, NY 12571 USA
Phone: 845-758-0400
Fax: 845-758-2634
Email: curran@proceedings.com
Web: www.proceedings.com

Table of contents

Volume 11

17th National Conference of the Australian Meteorological and Oceanographic Society
27–29 January, 2010, Canberra, Australia

Accepted papers received: 26 July 2010
Published online: 16 August2010

Antarctic weather,ocean and climate systems

Weather, ocean and climate forecasting

Observing and modelling the integrated earth system

Ocean systems and dynamics

Climate trends, variability and extremes: past present and future

Weather and forecasting at Wilkins ice runway, Antarctica

Scott Carpentier

Bureau of Meteorology, 111 Macquarie Street, Hobart, Tasmania, 7001

Email: s.carpentier@bom.gov.au

Abstract. Aviation forecasts for Wilkins ice runway in East Antarctica are developed within the conceptual framework of flow against a single dome shaped hill. Forecast challenges include the sudden onset of blizzards associated with the formation of an internal gravity wave; frontal weather; transient wake vortices and mesoscale lows; temperature limitations on runway use; and snow and fog events. These key weather aspects are presented within the context of synoptic to local scale climatologies and numerical weather prediction models.

1. Introduction

In 2008 the Australian Antarctic Division (AAD) introduced an air service between Hobart in Tasmania and Casey in East Antarctica. The weekly service is provided by an Airbus A319 jet using the Wilkins ice runway. Wilkins runway (66° 41´S, 111° 30´ E) is located at approximately 750 m above sea level (ASL), 63 km southeast of Casey station, on the western slope of Law Dome, which rises to 1395m ASL and is about 150km in diameter near sea level (figure 1).

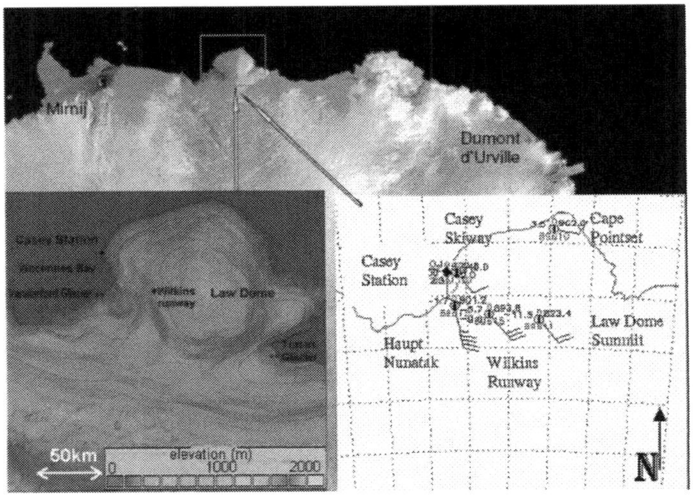

Figure 1. Antarctic Digital Elevation Model (DEM, top), local topography with dark lines at 100m contours (bottom left) and AWS network with full wind barbs in knots (bottom right). DEM image from USGS website: http://terraweb.wr.usgs.gov/projects/Antarctica/antdemsr.html#nuvu

Law Dome is an ice covered dome shaped hill protruding out from the Antarctic continent and its smooth near symmetrical shape lends itself to comparisons with idealized numerical studies of wind flow against a solitary dome shaped hill, such as by Ding and Street [1] and Epifanio and Rotunno [2]. In addition to the deflection of wind and cloud by Law Dome, the weather at Wilkins runway is complicated by interactions between the strongly contrasting airmasses of cold/dry "continental" and warm/moist "maritime" that meet in the area. With these ideas, a number of conceptual models have been developed to aid the aviation forecaster in making sense of the vagaries in the local weather.

2. Aviation forecast resources

In support of air operations, terminal aerodrome forecasts (TAFs) of expected meteorological conditions for the next 24 hours at the runway are routinely prepared. These forecasts are augmented by: a network of automatic weather stations (AWS); polar orbiting and geostationary satellite images received directly via a High Resolution Picture Transmission (HRPT) receiver at Davis and an x-band receiver at Casey; a twice daily upper air program at Casey; and numerical weather prediction models (NWP) including the National Centre for Environmental Prediction Global Forecast System (NCEP GFS) with a horizontal resolution of around 56km, the limited area Antarctic Mesoscale Prediction System at 15 km resolution (AMPS WRF, http://www.mmm.ucar.edu/rt/amps/), and the Polar Limited Area Prediction System at around 28 km resolution (polarLAPS, [3]). Both AMPS WRF and polarLAPS are nested within the NCEP GFS, but have more sophisticated sea ice analysis and physics parameterisations that are better suited to the polar environment.

3. Large scale winds

Antarctic topography drives the low level anticyclonic flow over the continent [4-5]). From a 1 year archive of the AMPS near surface winds over the continent, Parish and Bromwich [5] show that the southeasterly outflow off the continent falling into the Law Dome region undergoes some confluence and therefore acceleration as it is deflected around the southern slope of Law Dome and funneled into the Vanderford Glacier, terminating in Vincennes Bay. However, katabatic flow is limited to a few hundred metres above the surface [6]; consequently katabatic outflow off the continent rarely reaches the elevation of Wilkins. This is supported by comparing wind roses from Haupt Nunatak, located near the coast along the Vanderford Glacier and Wilkins runway, which is located at approximately 700m ASL directly upslope to the east. Southeasterly winds are the most common direction at Haupt Nunatak, whereas this direction is quite rare at Wilkins runway (figure 2). When southeasterlies occur at Wilkins, they are synoptically supported as opposed to being part of the background katabatic flow.

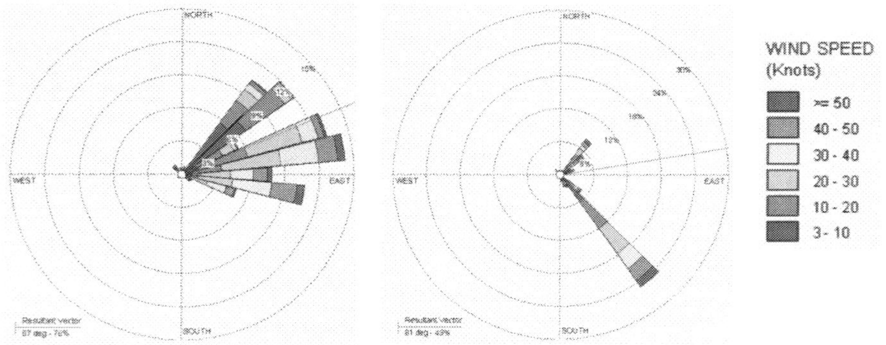

Figure 2. Wind roses at Wilkins runway (left) and Haupt Nunatak on the Vanderford Glacier (right). 36 points of compass. Derived from all hourly observations since 2007 and 2003 to 2010 respectively.

Over the summer months, the circumpolar trough typically lies offshore and combines with the Antarctic surface high to set up an easterly wind regime along the Antarctic coast (for example [7]). Consequently, Wilkins runway typically lies in the wake of Law Dome.

4. Forecasting the wind at Wilkins

All winds at Wilkins runway are perturbed by Law Dome. When there is little synoptic scale pressure gradient, a light east to northeast katabatic wind from the Law Dome summit is observed. When a pressure gradient does exist, one first needs to deduce whether the upstream wind flow is being blocked and forced around Law Dome (subcritical flow), or whether it flows over Law Dome (supercritical). This can be forecast with some success by considering the Froude number of the upstream layer stretching from sea level to the height of Law Dome, which is near the 850hPa level. The Froude number is a dimensionless ratio of the kinetic energy of the flow versus the potential energy required to lift that same layer over the obstacle, and can be defined as:

$$F = \frac{u}{Nh} \qquad \text{where} \qquad N^2 = \frac{g}{\theta_0}\frac{d\theta}{dz}$$

and where u is the average wind speed of the surface to 850hPa layer, h the height of the obstacle (ie. 850hPa level), N the Brunt-Väisälä frequency, g the gravitational acceleration, θ the potential temperature and therefore $\frac{d\theta}{dz}$ the vertical gradient of potential temperature, and θ_0 the average potential temperature of the layer. The switch from subcritical to supercritical flow typically occurs when F rises above unity ($F \geq 1$). Experience suggests that when $0.7<F<1$ the upstream flow is partially blocked by Law Dome and therefore the wind in the lee of Law Dome may exhibit combinations of minor downslope accelerations on the upper reaches of the lee slope (section 4.1.1) and wake vortices in the lower reaches (sections 4.1.2 and 4.1.3).

Unfortunately, the next available upper air program upstream (east) of Law Dome is at the coastal station of Dumont d'Urville (140E), over 1000km away. NWP models are therefore used to calculate the Froude number. In addition to giving a qualitative idea of the structure of the flow around Law Dome, the Froude number allows the forecaster to assess what variations in the upstream flow are required to shift from blocked to overshooting regimes, which could be as markedly different as sunny with light winds versus blizzard conditions (see 4.1.1 and figure 3). It should be noted that, as Law Dome has a diameter of approximately 150km near sea level, even the global models resolve Law Domes perturbation on the wider wind field to a useful degree.

4.1.1. Downslope wind storms

The passage of transient low pressure systems in the midlatitudes or along the circumpolar trough, and their associated warm fronts typically affect the area on a weekly to daily basis. These systems are mostly responsible for strengthening the easterly flow enough to tip the balance from blocked flow to flow-over regimes. Flow over regimes typically set up a vertically propagating internal gravity wave, which leads to the bulk of strong wind events at Wilkins runway and at Casey. The mechanism for this is well described by Turner *et al.* [8].

It is common for a hydraulic jump to occur at the termination of the wave, and this region is associated with intense small-scale turbulent motions [2]. Turbulence in the wake of Law Dome is a real concern for both forecaster and pilot. The 23z and 11z sonde flights from Casey can highlight certain layers where turbulence is most likely, through vertical velocities and indices such as the Froude number, Brunt-Väisälä frequency and Richardson number [9].

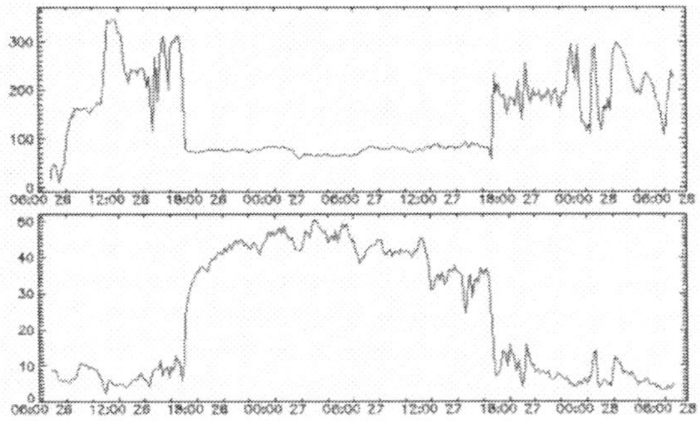

Figure 3. Variable winds associated with a hydraulic jump before and at the termination of a downslope windstorm at Casey. The gale lasted 24 hours from 18z on the 26th to 18z on the 27th. Note the stable and near constant wind direction once an internal gravity wave sets up. Abscissa: Hour (UTC) and day. Top panel ordinate: wind direction (true). Bottom panel ordinate: wind speed (knots).

4.1.2. Vincennes Bay mesolow

Figure 4 (left) shows a diffluent easterly (upstream) producing twin wake vortices that combine into a westerly "counterflow" of 15 knots at Casey. Though in this case the AMPS captures the occurrence of a counter flow in the wake, these situations are invariably poorly forecast. In this event, Casey recorded 25 knot westerlies for a few hours, 6 hours before the AMPS WRF model.

Westerlies occasionally occur in Vincennes Bay and Casey during the passage of a strong midlatitude ridge, but these again are more often deflected into northwesterlies at Wilkins due to blocking of the flow by Law Dome. Figure 4 (middle) shows an approaching midlatitude ridge and strong outflow off the continent combining to produce a mesolow (a mesocyclone of diameter less than 1000km and lifetime of generally less than 24 hours) in Vincennes Bay. On rare occasions, the Vincennes Bay mesolow can spin up to near gale force.

Forced ascent of the westerly against the slope of Law Dome invariably leads to the formation of low cloud or fog through adiabatic cooling and cooling through contact with the icy surface. In the situation of an approaching midlatitude ridge, the mesolow and moisture is typically shallow, with clear dry air occurring above the inversion, due to subsidence into the surface ridge. When flight forecasting in these scenarios, gauging the height of the subsidence inversion is crucial to forecasting whether Wilkins runway will be situated in a rising moist boundary layer, or in the clear dry subsiding air above the inversion. In figure 4 (right), Wilkins runway is in the cloud below the inversion.

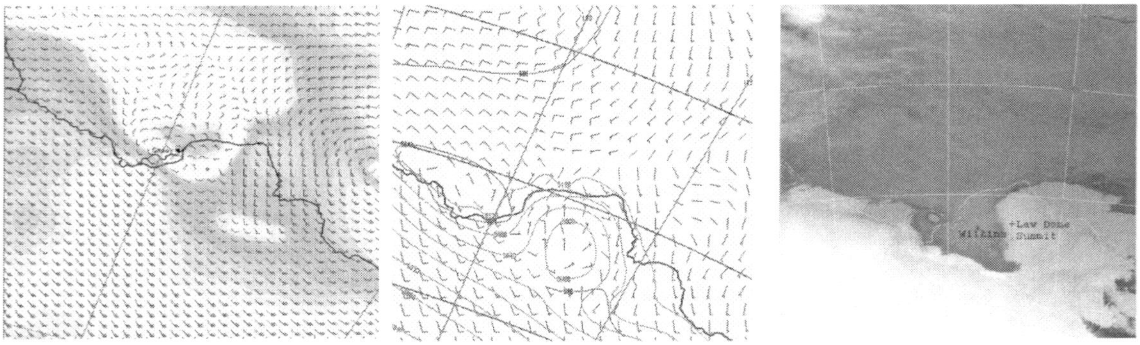

Figure 4. Refer to text in 4.1.2. Left: AMPS-WRF surface winds 20/10/08 (image courtesy of the National Science Foundation and National Center for Atmospheric Research). Centre: polarLAPS surface winds 21/12/2006 (with permission N.Adams): both models with full wind barbs 10 knots. Right: NOAA 15 band 3 on 9/12/2003, low cloud in grey and clear air in white.

4.1.3. Steaming fog

The Vincennes Bay mesolow has been observed to wrap steaming fog from a light and relatively colder southerly off the Vanderford Glacier back onto Casey station. For example on the 21st December 2005, fog was observed at Casey along with a wind shift from NNE to NNW (figure 5). The fog persisted for an hour at the station before dissipating into scattered stratus, as the wind shifted back NNE. This event coincided with the passage of a strong midlatitude ridge.

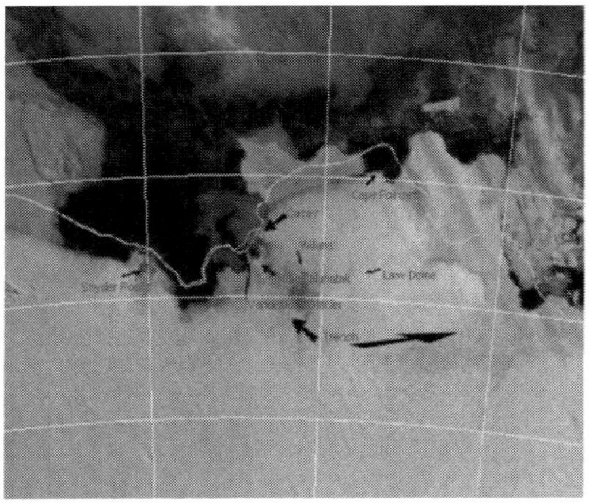

Figure 5. NOAA 17 combined bands: 1-2-3. Fog appears as a swirl of flat grey/blue. Higher cloud at the top of the picture is more textured. Open water appears black and the icy continent and sea ice appear yellow.

4.1.4. Synoptically enhanced southeasterly flow off the continent

It was argued in section 3 that the katabatic outflow off the continent is too shallow to reach the runway. The bottom right panel of figure 1 shows AWS observations over the area whilst a deep low is situated to the northeast of Law Dome. This sets up a strong southeasterly surging off the continent. In this instance, we observe channeling of the flow and probably an additional katabatic component into the Vanderford Glacier (Haupt Nunatak) at 45kts, 30kts at Wilkins (750m ASL), and 20 knots atop Law Dome (near 1400m ASL). However, very little of the southeasterly is reaching Casey and the primary skiway. Also, there is only 10 knots of easterly wrapping around the northern flank of Law Dome at Cape Pointset.

4.1.5. Other wind directions

The Wilkins runway windrose (figure 2, left panel) shows that the prevailing wind is from the northeast through east to east-southeast. Apart from rare northwesterlies associated with the passage of a midlatidude ridge and/or the development of a Vincennes Bay mesolow, other wind directions are practically non-existent. This is likely the result of the strong forcing that the Antarctic continent plays on the low level flow [4-5].

5. Wilkins runway temperature

A significant melt occurs along the coastal margin of Law Dome in the peak of summer, which can hamper travel between the station and the runway. A cold environment is critical for the strength and integrity of the ice runway and a temperature at or below -7°C at the runway is necessary to achieve the surface friction constraints imposed by the Australian Civil Aviation Authority. Therefore flights typically leave Hobart at 5am for a 9:30am (2230UTC) arrival at Wilkins, coinciding with the start of warming of the runway surface (figure 6). Incidentally, temperature was a limiting factor when the runway location and proximity to Casey station was first being considered.

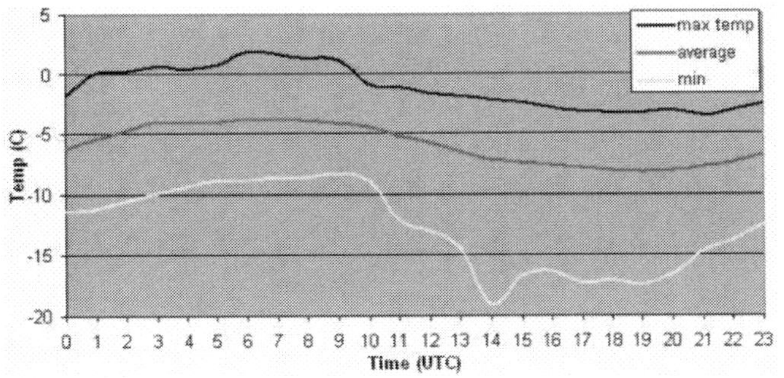

Figure 6 Wilkins average January temperatures from all observations 2007-2010. Blue is the highest maxima, yellow is the lowest minima and purple is the mean.

6. Precipitation

Due to a very sparse ground observation network, and the difficulties in measuring actual precipitation that is uncontaminated by snow lifted from the surface, direct measurement based high latitude precipitation climatologies remain error prone. Remote sensing techniques, glaciological approaches and atmospheric models have all added to the precipitation dataset, but these techniques also have their shortcomings [10]. The Antarctic mean annual precipitation from an 11 year climatology (1998-2008) of polarLAPS [11], nested in the NCEP2 reanalysis [12] is presented in figure 7.

From figure 7 we see dry continental air flowing out from the main Antarctic basins and ice shelves. The higher precipitation along the coast suggests that maritime moisture rarely penetrates deep into the continent. The eastern side of Law Dome is modelled as an area of very high precipitation by Antarctic standards (>2.5m of equivalent water), whilst the western slope, including Wilkins runway, is situated in a "rain shadow".

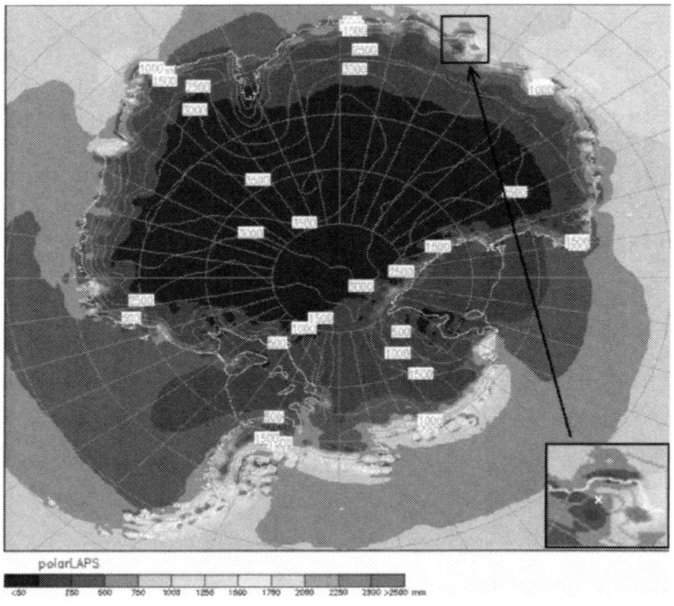

Figure 7. Mean annual precipitation (mm) over Antarctica from an 11 year climatology (1998-2008) of polarLAPS. "X" in inset marks Wilkins runway. White topographic contour interval is 500m.

7. Conclusions

Forecasting the weather at Wilkins runway is critical for safe and efficient air and overland operations. The weather at Wilkins runway is complicated by the deflection of wind and cloud by Law Dome, as well as interactions between the strongly contrasting cold and dry airmass off the continent with the relatively more warm and moist maritime Southern Ocean airmass that meet in the area. Subtle variations in characteristics of the upstream airmass, such as stability, speed, direction, diffluence etc. can lead to widely different conditions at the runway. Some success in forecasting has been found over the years by interpreting the local observations within the framework of conceptual models based on numerical studies of flow against a single dome shaped hill. Direct observations of the airmass to the east, within closer range than Dumont d'Urville, could only improve the chances of a successful forecast. Unfortunately, the logistics in deploying and maintaining an AWS in such a remote and harsh environment make this extremely difficult and costly.

References

[1] Ding L and Street R L 2003 *J. Atm. Sci* . **60** 1678-90
[2] Epifanio C C and Rotunno R 2005 *J. Atm. Sci.* **62** 3127-50
[3] Adams N 2006 Continued Developments in Numerical Weather Prediction and Weather Forecasting in Support of the Australian Antarctic Program. The Antarctic Meteorological Observations, Modeling and Forecasting Workshop, Boulder CO., USA June 13-15 2006
[4] van den Broeke M R and van Lipzig N P M 2003 *Mon. Wea. Rev.* **131** 733-43
[5] Parish T R and Bromwich D H 2007 *Mon.Wea. Rev.* **135** 1961-73
[6] Phillpot H R and Zillman J W 1970 *J. Geophys. Res.* **75** 4161-69
[7] Simmonds I, Keay K and Lim E P 2002 *Mon. Wea. Rev.* **131** 272-88
[8] Turner J, Lachlan-Cope T A, Marshall G J, Pendlebury S and Adams N 2001 *J. Geophys. Res.* **106** 7291-7311
[9] Adams N 1996 *Aust. Met. Mag.* **45** 219-232
[10] Bromwich D H, Guo Z, Bai L and Chen Q S 2004 *J. Climate* **17** 427-47
[11] Adams N 2009 Output from an eleven year climate run of a regional Antarctic Numerical Weather prediction System. MOCA-09: IAMAS - IAPSO - IACS Joint Assembly, Montreal, July 19 - 29 2009
[12] Kalnay E *et al.* 1996 *Bull. Amer. Meteor. Soc.* **77** 437-70

Tides, the PIG, and "warm" water

Robin Robertson

School of Physical, Environmental, and Mathematical Sciences,
University of New South Wales @ Australian Defence Force Academy
Canberra, ACT 2600 Australia

Email: r.robertson@adfa.edu.au

Abstract. The present rapid melting of the Pine Island Glacier (PIG) has been attributed to basal melting driven by the ocean. Specifically, this ocean melting is attributed to currents and tides pumping "warm" Circumpolar Deep Water (CDW) into the ice shelf cavity. To identify tidal activity in the region, an observational time series of yo-yo CTD (Conductivity, Temperature, and Depth) data collected in the PIG outflow region was analyzed. The water column in front of the PIG consisted of two primary layers, a meltwater layer exiting the ice shelf cavity over a layer of CDW. Semidiurnal tides were present in both layers, with both the strength and direction of the tides differing between the two layers. The upper layer tides were stronger and directed in and out of the cavity, while the lower layer tides were primarily directed along the front of the cavity. Energy was found to be transferred from the semidiurnal tide to other frequencies and to be reflected by the ice shelf front. These mechanisms were most prominent at the interfaces between layers and indicate potential mixing between the layers. In conclusion, tides were found to contribute to the circulation into the ice shelf cavity and also to mixing of the exiting water, which influences pumping of the CDW into the ice shelf cavity and melting of the PIG.

1. Introduction

The Pine Island Glacier (PIG) is melting 100 times faster than other Antarctic ice shelves as determined by satellite observations [1]. Melting of the PIG is significant both for the potential global sea level rise, which has been estimated from 0.2-0.3 m [2] to 1.0-1.9 m [3], and the freshening of the Ross and Amundsen Seas [4], which has been attributed to melting of the ice shelves [5]. Both the increase in sea level and the freshening have global climate implications, since the global deep water of the thermohaline circulation is formed in the Antarctic Seas. The freshening is expected to reduce deep water formation, since as water freshens it becomes less dense, sinks less, and less qualifies as deep water.

Melting of the PIG is believed to be driven by the ocean, with "warm" Circumpolar Deep Water (CDW) entering the ice shelf cavity and fuelling the melting [5]. Within the ice shelf cavity, "warm", salty CDW enters the cavity in the bottom portion of the water column. Coming in contact with the ice shelf near the grounding line, the "warm" water melts the ice, becoming cooler and fresher as the new melt water mixes with the former CDW. This meltwater mix is denoted Ice Shelf Water (ISW) and is fresher, cooler, and less dense than the CDW. It flows out of the ice shelf cavity in the upper water column adjacent to the ice shelf. The meltwater exits the ice shelf cavity going to the appropriate depth for its density. When this density level is further up in the water column than the bottom of the ice shelf, the exiting ISW undergoes mixing as it flows out, similar to flow over a sill, except upside down. Furthermore, as the CDW warms, melting of the ice shelf and freshening is expected to increase. If the process accelerates, the PIG may be in "hot" rather than "warm" water.

In early 2009, Stan Jacobs and Adrian Jenkins led a research cruise, NBP0901, to investigate the hydrography and circulation of the eastern Amundsen Sea with the goal of better understanding the effect of the ocean forcing on the ice shelves. This study examines a small portion of the data collected during this cruise, specifically a time series of profiles collected at the outflow in the southern portion of PIG. My goals are to characterize the outflow water of the PIG ice shelf cavity and determine the water column structure and the role of tides in this outflow.

2. Observational Data

The observational data is composed primarily of hydrographic data from Conductivity, Temperature, and Depth (CTD) casts combined with Lowered Acoustic Doppler Current Profiler (LADCP) profiles. During NBP0901, the CTD/LADCP was operated in a yo-yo fashion in the southern notch of the PIG (Figure 1) in order to investigate temporal variability of the outflow. Twenty-four casts, casts 58-81, were made at one hour intervals in the southern notch of the PIG at the site of cast 16 (yellow star in Figure 1) with the first one starting its descent at 21:00 GMT Jan 24, 2009 and the last one at 20:00 Jan 25, 2009. The first five casts extended to within 10 m of the bottom and the remainder to 800 m. All casts, except the fourth, started their descents on the hour. The fourth cast was 15 min. late starting its descent. CTD data was collected with a Sea-Bird SBE 9/11 using dual sensors for potential temperature, θ, salinity, S, and dissolved oxygen, O_2. CTD data was processed and averaged to provide estimates of θ and S at 1 m intervals from the surface to the extent of the cast. Typical profiles of θ, S, and the potential density anomaly, *sigma-*θ are shown in Figure 2. A 300kHz RDI LADCP was attached on the CTD rosette frame looking downward, measuring horizontal velocities. The LADCP data was processed using Global Positioning System data and the Shipboard ADCP data to constrain the solution when combining the scans of the profile. Velocity profiles of the East-West, U, and North-South, V, velocities were generated with 20 m depth bins.

Figure 1. Bathymetry of the Pine Island Bay (provided by Frank Nitsche). The yo-yo location is indicated by a yellow star.

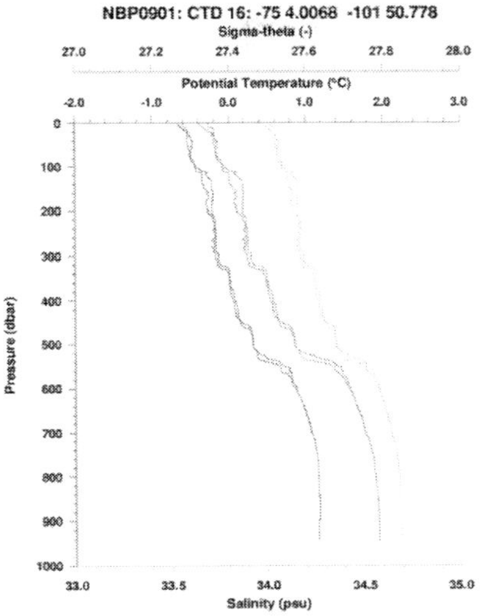

Figure 2. Profiles of potential temperature (blue), salinity (green) and potential density anomaly (sigma-θ, red) with pressure from a profile collected before the yo-yo.

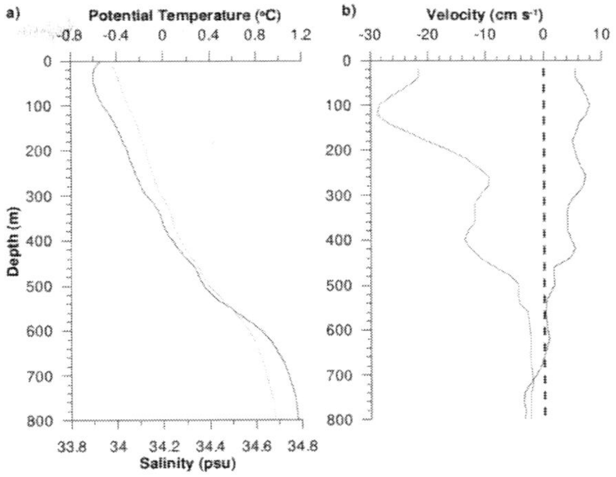

Figure 3. a) Profiles of average potential temperature (blue) and salinity (green) over the yo-yo. b) Profiles of average East-West (red) and North-South (blue) velocities during the yo-yo as determined from the LADCP.

3. Analysis

3.1. Water column structure

In the southern notch of the PIG, the mean θ and S gradually increase with depth (blue and green lines in Figure 3, respectively) from 50 m to 800 m. The gradual increases show marked changes in slope between 500 and 600 m. A more detailed look at the individual profiles shows distinct steps or layers as θ gradually increases with depth (Figure 2). Layer or "step" heights range from 20-120 m, although smaller layers or "steps" down to the data resolution of 1 m were also present. The layers are generally larger, ~100 m height, above 300m and smaller below this depth, 50 m or less in height. The lower water column, below 600 m, showed few changes in θ.

The mean velocities were also depth-dependent with both U (red line in Figure 3b) and V (blue line in Figure 3b) generally decreasing in magnitude with depth below their peaks of 28 cm s^{-1} and 8 cm s^{-1} at ~ 120 m and 100 m to 2.5 cm s^{-1} and ~0.2 cm s^{-1} at 550 m, respectively. Here U indicates the flow out of the notch and V the flow across the notch. Velocities below 550 m are small, < 3 cm s^{-1} and V changes direction at ~650 m. Substantial structure can be seen in the mean velocities, with U having two distinct peaks of ~28 cm s^{-1} and 14 cm s^{-1} at ~120 and 400 m and V three peaks of ~8 cm s^{-1}, 7 cm s^{-1} and 5 cm s^{-1},at ~100 , 260, and 420 m, respectively. The mean velocities suggest a north westward flow out of the notch at depths above 650 m, below which flow is southwest.

The water column at the southern notch of PIG is definitely layered, although the layers are not likely to be double diffusive-convective steps, since velocities are too strong for double-diffusion. The water column can be broken into layers based on the values and behaviour of θ, S, ρ, U, and V. The upper 50 m, surface layer is influenced by the wind, solar radiation, and melting sea ice and icebergs. This discussion will focus on the layers below the surface layer. There were two primary layers: an upper layer from ~100 m to 450 m and a lower layer from 550 m to the bottom, ~925 m (Figure 2). Transition layers existed between the upper layer and both the surface mixed layer and the lower layer. Within the transition layers, behaviour is influenced by processes in both of the adjacent layers.

3.1.1. *Upper layer.* The upper layer is composed of meltwater with temperatures below 0.0°C. It was very active with θ profiles often changing between down- and up-casts. Both "steps" and intrusions occurred frequently in the upper layer. Further analysis indicated that the upper layer should be divided into two layers, an upper upper layer 1 from 100 to 300 m and a lower upper layer 2 from 320 to 450 m, with a narrow transition region between them. The primary support for division of the upper layer stems from the variability of θ and the mean velocities. With the exception of the surface mixed layer, depths of high θ variability coincide with the layer depths and the interfaces, evident in the θ standard deviations (not shown). High variability at the interfaces is expected, since fluctuations of a layer of constant θ will not show variability, whereas fluctuation of an interface between two layers results in high variability. The data set is not extensive enough to identify whether the two upper layers are a result of mixing processes in the ice shelf cavity or mixing during the meltwater exit from the cavity.

3.1.2. *Lower layer.* The lower layer, with θs exceeding 0.6°C, was composed of CDW. The velocity out of the notch, U, was steady at ~2 cm s^{-1}. Flow across the notch, V, varied with depth from 0-3 cm s^{-1}. The resulting flow was slow, exiting the notch to the west southwest. This flow is restricted by the topography, since the notch location is in the southern part of a trench going from PIG under the ice shelf (Figure 1). CDW does not appear to be flowing into the cavity on the southern end, even though the deepest portion of the topography is nearby. However, CDW must flow into the ice cavity somewhere, so it is either deeper in the water column below the profiles, which is unlikely, or further north along the ice shelf front, more likely. θ profiles in the lower layer changed very little between casts (not shown). Unlike the upper layer, the lower layer was very stable during the time of these observations.

3.2. Signatures of tides and waves

The 24-hour time series were analyzed for tides using *Pawlowicz's* [6] routines for both horizontal velocities (Figure 4) and temperatures (not shown). The short time series, 24 hourly points, resulted in large uncertainties for the tidal constituents (error bars in Figure 4).

3.2.1. *Tides.* The strongest semidiurnal tides, ~ 11 cm s^{-1}, occurred in the upper 50 m, the surface mixed layer. In the upper layer, semidiurnal tides fluctuated between 2 and 8 cm s^{-1} with peaks at 150 and 310 m (Figure 4). Tides were weak in the lower layer, 1-2 cm s^{-1}. Semidiurnal tides are also apparent in the spectra with peaks at 2 cpd (cycles per day) in the upper layer (Figure 5). Semidiurnal tides are stronger in the East-West direction (upper panel in Figure 5) than they are in the North-South direction (lower panel in Figure 5). Energy is also apparent at the diurnal frequency again with the East-West direction (upper panel in Figure 5) stronger than the North-South direction (lower panel in Figure 5).

3.2.2. Waves. Energy is also apparent in the spectra at frequencies other than the tidal frequencies (Figure 5), particularly the 6 hr and 3 hr harmonics of the tides. This indicates the existence of internal waves at these frequencies. Bispectral analysis confirms that the higher frequency energy originated from the semidiurnal and diurnal tidal frequencies. Significantly more high frequency energy occurs in the upper layer, 100-300 m than in the lower layer. And more high frequency energy occurs in the East-West direction (Figure 5a) than in the North-South direction (Figure 5b). The notch where these profiles were collected is very close to the ice shelf, which extends below 300 m. The East-West velocities and tides are reflected by the ice shelf front. These reflected waves interact non-linearly with the tides producing more harmonics, which is reflected in the increase in high frequency energy in the upper layer.

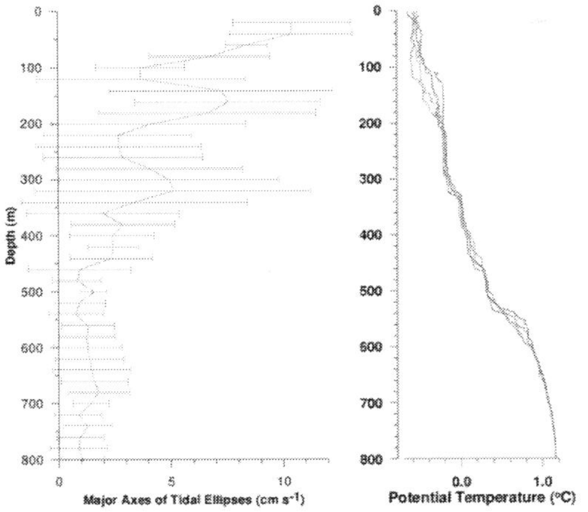

Figure 4. a) Profiles of the major axes of the semidiurnal tidal ellipses. The error bars indicate the observational uncertainty as determined by the tidal analysis. b) Potential temperature profiles for CTD cast 16 (blue) before the yo-yo and CTD 58 (red), the first profile of the yo-yo.

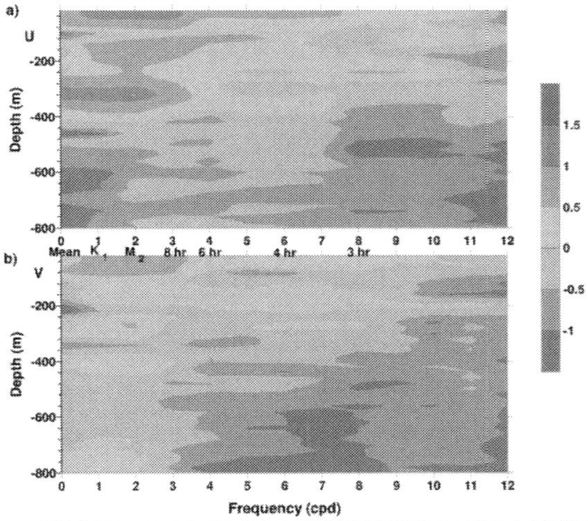

Figure 5. Spectra of the a) East-West (upper) and b) North-South (lower) velocities. Contour levels are the log of the spectral density ($\log_{10}$ [(cm $s^{-1})^2$/cpd).

4. Summary

A 24 hour time series of CTD and LADCP data was collected in the outflow region of the Pine Island Glacier. The water column was composed of two layers: an upper layer of meltwater from 100-450 m and a lower layer, below 550 m. There was a strong flow of the meltwater out of the ice shelf cavity and a weak outflow of the CDW. Inflowing CDW must flow in at another portion of the ice shelf front or in the 100 m below the extent of the CTD yo-yo. The upper layer was further split into two layers. Strong semidiurnal tides were found in the upper layer and weak ones in the lower layer. Diurnal tides were also evident in the upper layer and were stronger in the East-West direction than in the North-South direction. Internal waves with periods of 3 and 6 hours occurred and were also stronger in the East-West direction than in the North-South direction. This higher frequency energy originated at the tidal frequencies.

This is in no way a definitive treatment of the PIG outflow. A 24-hour time series is too short and barely sufficient to determine the semidiurnal tide. It barely spans one K_1 tidal cycle and does not come close to the 15 days needed to distinguish the four major tidal constituents from each other or cover a spring-neap tidal cycle. However until data from a mooring with instruments in the upper water column is placed either in this region or on the ice shelf for 15 days or longer, this is the best picture we have. Also such a mooring is unlikely to survive a long term deployment, due to iceberg activity in the region and the rapid growth of and calving from the ice shelf.

Acknowledgements

I would like to thank Stan Jacobs for inviting me on the cruise, the crew and support personnel on the RVIB Nathaniel B. Palmer during NPB0901, and the people who collected the CTD data: Katherine Leonard, Christopher Little, and Kenneth Mankoff. Thanks are also due to Pierre Dutrieux for providing the shipboard ADCP data.

References

[1] Rignot E, Bamber J L, Van Den Broeke M R, Davis C, Li Y, Van De Berg W, and Van Meijgaard E 2008 *Nat. Geosci.* **1** 106
[2] Barley S 2009 *Proc. Roy. Soc. A* doi:10.1098.rspa.2009.0434
[3] Pearce F 2007 *With Speed and Violence: Why scientists fear tipping points in climate change* (Boston, MA:Beacon Press)
[4] Jacobs S 2006 *Philos. T. R. Soc.* **365** doi: 10.1098/rsta.2006.1784 1657
[5] Shepard A, Wingham D, Rignot E 2004 *Geophys. Res. Lett.* **31** doi:10.1029/2004GL021106
[6] Pawlowicz, R, Beardsley R , and Lentz S 2002 *Comput. and Geosci.* **28** 929

Diagnosing indicators of large-scale forcing of east-coast cyclogenesis

Andrew J. Dowdy, Graham A. Mills and Bertrand Timbal

The Centre for Australian Weather and Climate Research, Box 1289 Melbourne VIC 3001 Australia.

E-mail: A.Dowdy@bom.gov.au

Abstract. Extra-tropical cyclones that develop near the east coast of Australia often have severe consequences such as flash flooding and damaging winds and seas, as well as beneficial consequences such as being responsible for heavy rainfall events that contribute significantly to total rainfall and runoff. There is subjective evidence that the development of most major events, commonly known as East Coast Lows, is associated with the movement of a high amplitude upper-tropospheric trough system over eastern Australia. This paper examines a number of upper-tropospheric diagnostic quantities that might provide a basis for preparing a climatology of the large-scale drivers of east-coast cyclogenesis. A preliminary climatology of these diagnostic quantities, based on ECMWF interim reanalyses, is compared with a database of observed East Coast Low events. The potential application of these diagnostics to global climate model simulations of past and future climates is also discussed.

1. Introduction

Cutoff lows in the region around the east coast of Australia are commonly referred to as East Coast Lows (ECLs). ECLs are perhaps the most significant meteorological phenomena affecting the Australian East Coast. The most intense ECLs can have similar consequences to those caused by tropical cyclones even though they are distinctly different phenomena. The ECL that occurred in early June 2007 near the city of Newcastle in New South Wales (NSW) led to nine deaths, extensive flooding, the grounding of the bulk coal carrying ship the Pasha Bulker, and insurance claims of around $1.6 billion, making it the 6th most costly natural disaster in Australia's history (Rob Webb, personal communication). The extreme winds sometimes generated by ECLs pose a major hazard to marine activities and on-shore infrastructure, while the ocean waves and storm-surge generated can cause severe coastal erosion. The extreme rainfall events associated with ECLs can cause significant flooding, as well as providing significant replenishment of reservoirs.

Accurate forecasts of the precise locations and intensities of weather associated with ECLs are essential for community preparedness for individual events. Additionally, knowledge of the climatological distribution of the environments in which they occur, including interannual variability and long-term trends, will be beneficial if changes in their occurrence are to be assessed under future climate change scenarios. This knowledge is essential for developing improved projections for eastern Australia, including projections of heavy rainfall events, damaging winds and coastal erosion.

The term East Coast Low (ECL) is generally used in Australia to refer to a system with closed cyclonic circulation at sea level that forms and/or intensifies in a maritime environment within the vicinity of the east coast of Australia, although it should also be noted that a range of secondary criteria are used in different studies. There is a relatively limited body of work on the climatology of ECLs. Hopkins and Holland [1] presented a climatology of Australian east-coast cyclones from 1958 to 1992 based on MSLP analyses and station-based surface wind and rainfall data, while Speer and Wiles [2] recently produced a climatology of maritime cyclones in the vicinity of the east coast of Australia based on a synoptic typology using data from 1970 to 2006. These climatologies are based either on observed coastal impacts (such as rainfall or wind) or on a partially subjective classification of MSLP analyses, and are thus purely surface-based. While some studies have qualitatively described the upper-tropospheric wave/jet structures of particular events [3, 4], hitherto no systematic studies of upper-tropospheric patterns associated with ECL occurrence have been presented.

An examination of the highest-impact ECLs during the last 40 years indicates that the cyclogenesis was preceded by the development of a marked upper-tropospheric cut-off low over eastern Australia. Examples of the 300 hPa geopotential height and wind fields are shown in Figure 1 for the 1974 "Sygna" storm and the 2007 "Pasha Bulker" storm.

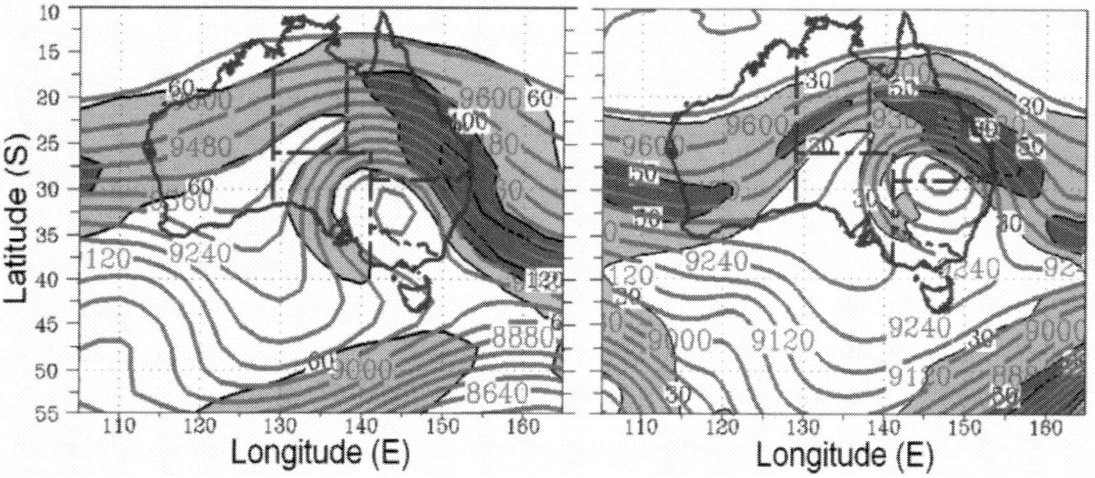

Figure 1. Geopotential height (in metres: red contours) and wind speed (in kilometres per hour: shaded gray-scale contours) at 300 hPa for the "Sygna" ECL on 24 May 1974 (left panel: data obtained from ERA-40 reanalysis [5]) and the "Pasha Bulker" ECL on 8 June 2007 (right panel: data obtained from LAPS numerical weather prediction [6]).

The patterns of the geopotential height and wind fields are strikingly similar between these two events, with marked closed lows over NSW and a major jet streak on the eastern flank. Strikingly similar patterns were also found to occur for a number of other highly publicised events which led to the following hypothesis regarding a typical upper-tropospheric evolution leading up to explosive ECL development:

- There is, 48-72 hours prior to surface intensification, an upper-tropospheric split-jet or blocking pattern, with a positively tilted trough or cut-off low, over eastern Australia longitudes.
- The upstream trough over the Indian Ocean amplifies, with subsequent amplification of the immediate upstream ridge south of Australia.
- The next stage of this energy-dispersion process is the development of a southerly jet streak on the western side of the positively tilted trough over eastern Australia.

- As this southerly jet streak propagates towards the apex of the trough (i.e. to lower latitudes), the trough/cut-off low deepens and begins to tilt towards the northeast and move towards the coast.
- The consequence of this evolution is a focussing of isentropic potential vorticity (IPV) advection on the northeast side of the cut-off low, which forces pressure falls at the surface.
- If low-level ingredients are suitable, explosive development (i.e. rapid intensification) of an ECL occurs.

Having formed this hypothesis, two questions arise if one wants to infer the impact of climate change on the risk of ECL formation. Firstly, how often does such an upper-tropospheric evolution occur without an ECL development (i.e. a 'false alarm'), and as a corollary, how often does an ECL form without such an upper-tropospheric sequence of events (i.e. a 'miss')? Secondly, if it can be shown that indeed the majority of ECL events are associated with these large-scale upper-tropospheric structures, is it possible to use them as a climatological indicator of ECL occurrence in Global Climate Model (GCM) simulations of future climate? Even though current GCMs may not have the resolution to be able to resolve the detailed intensity or impact of individual ECL events, they may have the ability to resolve the large-scale upper-tropospheric structures associated with ECL occurrence.

Based on the considerations described above, four distinct diagnostic quantities were selected: the Laplacian of geopotential height, the Blocking Index [7], IPV and a forcing term based on the advection by the geostrophic winds of the gradient of quasi-geostrophic potential vorticity [8]. A systematic examination of these diagnostics in relation to the ECL events in the Speer and Wiles [2] database (hereafter referred to as 'the ECL database') is presented in the following section.

2. Comparison of diagnostics with ECL database

The ECMWF interim reanalysis dataset [9] was used to calculate the diagnostics so as to provide consistent data suitable for systematic analyses. The diagnostics presented are the maximum value at a grid-point within the geographic area bounded by 144°E to 156°E in longitude and 24°S to 36°S in latitude. The exception to this is the Blocking Index which is calculated as a single value based on the average zonal wind speed from 144°E to 156°E. The diagnostics were calculated at 6-hourly intervals from 1989 (marking the start of the ECMWF interim reanalysis dataset) until the end of 2006 (which is currently the end of the ECL database), with a 2-day running mean applied to the resultant time series. As an example of the results obtained, time series of the diagnostics at 500 hPa are shown for the year 2006 in Figure 2. Days on which ECLs occurred (as listed in the ECL database) are shown as vertical red dotted lines, such that events lasting for multiple days appear as wider lines. The 90th percentiles of each diagnostic, calculated for the period 1989-2006, are shown as horizontal blue dotted lines. The green box highlights a time period referred to later in the text.

A summary is presented in Table 1 of the performance of the diagnostics based on the entire period of available data (i.e. 1989-2006) calculated at the 500 hPa pressure level. If a diagnostic is above its threshold (selected here as its 90th percentile for the period 1989-2006) at any time within 1 day of an ECL event occurring, then a 'hit' is counted, otherwise a 'miss' is counted. A 'false alarm' is counted if an ECL event does not occur during the time period that a diagnostic is above its threshold. The number of times that an ECL event did not occur when the diagnostic was lower than its threshold (i.e. 'correct negatives') is not presented since this quantity is difficult to define precisely for variable length events. Consequently, Table 1 represents an events-based perspective, with ECLs that persist for multiple days being counted as single ECL events, as well as continuous periods where a diagnostic is above its threshold being counted as single indicative events. This method allows for the possibility that large-scale drivers of ECL occurrence may occur only at a particular time during a multi-day event, including the likelihood that they are precursors of the ECL occurrence.

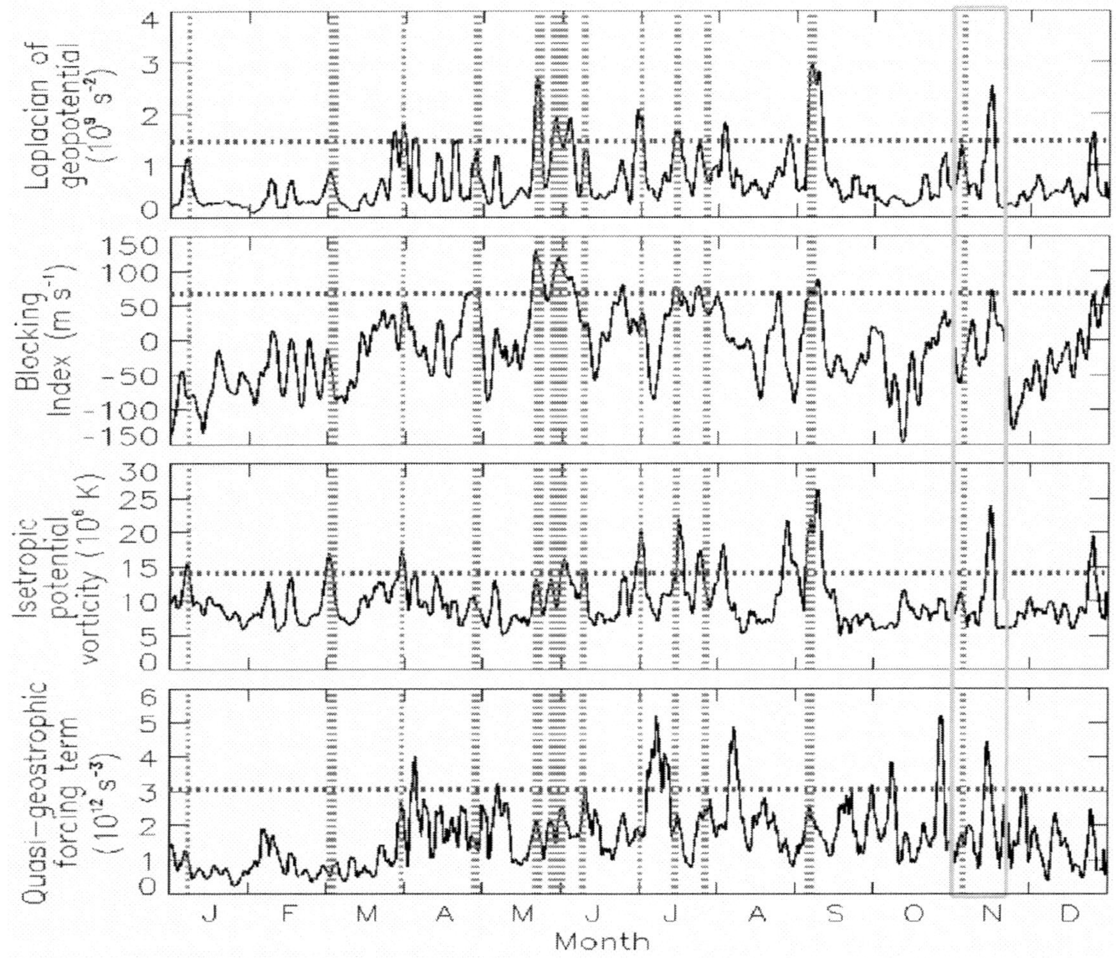

Figure 2. Time series during 2006 at 500 hPa of the Laplacian of geopotential height (upper panel), the Blocking Index (second panel), the magnitude of isentropic potential vorticity (third panel) and the forcing term based on quasi-geostrophic potential vorticity (lower panel). The green box highlights a time period referred to in the text.

Table 1. Example of diagnostic performance for the period 1989-2006 at 500 hPa.

	Hits	Misses	False alarms
Laplacian of geopotential	175	179	104
Blocking Index	135	219	77
IPV	142	212	141
Quasi-geostrophic forcing	103	251	147

A number of statistics commonly used in forecast verification are shown in Table 2. The Critical Success Score measures the fraction of observed and predicted events that were correctly predicted. It is sensitive to hits, while also penalising both misses and false alarms, making it an important statistic to examine. The Laplacian of geopotential height has the highest Critical Success Score, followed by the Blocking Index, then isentropic potential vorticity and then finally the quasi-geostrophic forcing term. The reasons for this order of success can be diagnosed from the other forecast verification measures: the Laplacian of geopotential height has the highest Probability of Detection of the four diagnostics (correctly indicating about half of the total number of ECL events) while also having a low False Alarm Ratio (approximately equal lowest with the Blocking Index).

Table 2. Forecast verification measures based on the results shown in Table 1.

	Probability of Detection $= \frac{hits}{hits + misses}$	*False Alarm Ratio* $= \frac{false\ alarms}{hits + false\ alarms}$	*Critical Success Score* $= \frac{hits}{hits + misses + false\ alarms}$
Laplacian of geopotential	0.49	0.37	0.38
Blocking Index	0.38	0.36	0.31
IPV	0.40	0.50	0.29
Quasi-geostrophic forcing	0.29	0.59	0.21

A closer examination of the hits, misses and false alarms was conducted on a case-by-case basis. It reveals that some of the 'misses' and 'false alarms' may be as much a consequence of the event selection in the ECL database as a deficiency in the diagnostic methodology taken. For example, some 'misses' are relatively weak compared to the events shown in Figure 1: the 'missed' ECL event of 5-6 November 2006 (shown within the green box in Figure 2) produced only moderate rainfall and wind speed impacts (Figure 3) and a lowest recorded mean sea level pressure of 1009 hPa. In contrast, the 'false alarm' indicated by all four diagnostics a week later on 13 November 2006 (also shown within the green box in Figure 2) had stronger maximum wind speeds and only slightly lower rainfall than the 5-6 November 2006 event (Figure 3). This 'false alarm' event was an open trough over the east coast, associated with a closed low further to the south, which is why it was not classed as an ECL even though its lowest MSLP of 1003 hPa along the east coast together with large pressure gradients resulted in severe winds and extensive rainfall along the eastern seaboard.

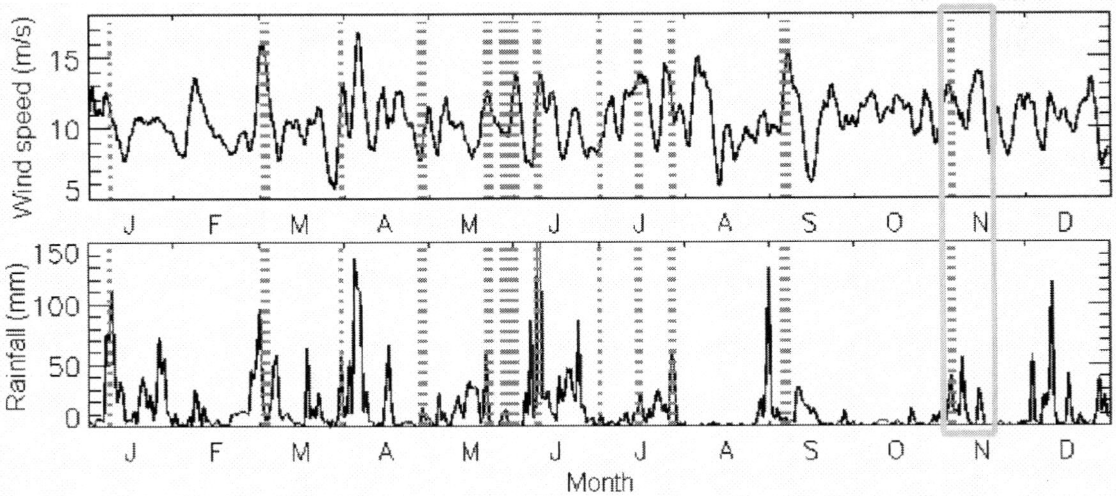

Figure 3. Time series during 2006 of the maximum wind speed in the area 150°E-154.5°E in longitude and 28°S-37.5°S in latitude derived from the ECMWF interim reanalysis dataset (upper panel) and the maximum rainfall within 50 km inland of the NSW coast from a gridded analysis of rainfall observations [10] (lower panel). The green box highlights a time period referred to in the text.

Through consideration of examples such as these, it is apparent that use of a subjective database is problematic as systems with similar impacts (e.g. wind, rain) may or may not be classified as ECLs. It is intended to further examine the diagnostics from an impacts-based perspective. This will allow for the inclusion of significant events that may not necessarily have a strong surface signature, while also avoiding very weak events that may not even be classed as ECLs based on some definitions (e.g. pressure gradient above 4 hPa $(100\ \text{km})^{-1}$ [1]).

3. Conclusions

This paper discusses the methodology and presents initial results of a project examining the feasibility of using large-scale upper-tropospheric diagnostics to indicate ECL occurrence. Early results are encouraging; the next step is to improve the reliability of the diagnostics by examining the role of thresholds, geographic area, time periods and pressure levels used. It is also expected that further improvement in diagnostic performance can be obtained by combining diagnostics and analysing the seasonal dependence of the diagnostic performance. For example, summer ECLs during 2006 (Figure 2) are well represented by isentropic potential vorticity, but not by the other three diagnostics including the best performing diagnostic (i.e. the Laplacian of geopotential height).

Although ECLs are reasonably well represented by Numerical Weather Prediction models [11], the fine spatial and temporal resolutions required to resolve the structure and intensity of these cyclones means that they are poorly represented by current global climate models (GCMs). Existing future climate projections are therefore unlikely to capture well the possible changes to ECLs, resulting in great uncertainty about rainfall projections for the eastern seaboard of Australia where ECLs are such an important feature. It is intended that the ECL diagnostic will be applied to other assimilated data sets with longer records but coarser resolution (e.g. NCEP/NCAR reanalysis [12]) to verify diagnostic performance in relation to the influence of model resolution, and evaluate climate variability of the risk of ECL formation. If found suitable, the diagnostic will then be applied to current and future simulations of the climate by GCMs to infer the influence of climate change on ECL occurrence.

Acknowledgements
This research was undertaken as part of the Australian Climate Change Science Program, with support from the New South Wales Department of Environment, Climate Change and Water. We would also like to acknowledge advice used for this paper provided by Clinton Rakich and Agata Imielska from the NSW Climate Services Centre of the Bureau of Meteorology, Perry Wiles from the National Climate Centre of the Bureau of Meteorology and Milton Speer from the University of NSW.

References

[1] Hopkins LC and Holland GJ 1997 *Journal of Climate* **10** 621-35
[2] Speer MS, Wiles P and Pepler A 2009 *Aust. Met. And Oceanographic Journal* **58** 29-39
[3] Holland GJ, Lynch AH and Leslie LM 1987 *Monthly Weather Review* December 3024-36
[4] Qi L, Leslie LM and Speer MS 2006 *Meteorol. Atmos. Phys.* **91** 201–9
[5] Uppala SM et al. 2005 *Q. J. R. Meteorol. Soc.* **131** 2961–3012
[6] Puri et al. 1998 *Aust. Met. Mag.* **47** 203-23
[7] Pook M and Gibson T 1999 *Aust. Met. Mag.*, Special Edition, 51-60
[8] Bluestein HB1992 *Oxford University Press* Eqn 5.8.15
[9] Uppala et al. 2008 *ECMWF Newsletter* **115** 12–18
[10] Jones DA, Wang W and Fawcett R 2009 *Aust. Met. and Oceanographic Journal* **58** 233-48
[11] Leslie LM and Speer MS 1998 *Weather and Forecasting* 13 822-32
[12] Kalnay et al. 1996 *Bull. Amer. Meteor. Soc.* **77** 437-71

Evaluating a regional climate model's ability to simulate the climate of the South-east coast of Australia

Jason P Evans[1] and Matthew F McCabe[2]

[1]Climate Change Research Centre, UNSW, NSW 2052, Australia.
[2]School of Civil and Environmental Engineering, UNSW, NSW 2052, Australia

E-mail: jason.evans@unsw.edu.au

Abstract. The Weather Research and Forecast (WRF) model was run over South-east Australia from 1985 through 2008. The model used the following physics schemes: WRF Single Moment 5-class microphysics scheme; Rapid Radiative Transfer Model longwave radiation scheme; Dudhia shortwave radiation scheme; Monin-Obukhov surface layer similarity; Noah land-surface scheme; Yonsei University boundary layer scheme and Kain-Fritsch cumulus physics scheme. The model simulation uses boundary conditions from the NCEP/NCAR reanalysis with an outer 50km resolution nest and an inner 10km resolution nest. Both nests used 30 vertical levels. In order to assess WRF's potential for use in investigating the current and future climate, the simulation is evaluated against gridded surface temperature and precipitation observations created as part of the Australian Water Availability Project (AWAP). The WRF simulation is found to reproduce the climate of the south-east Australian coast reasonably well. WRF was able to improve on the climate produced by the NCEP/NCAR reanalysis (NNRP) which provided the boundary conditions for the WRF simulation. Investigation of the time series of precipitation anomalies show that WRF is able to capture the recent drought in the SE Australia. While the overall time series captured the drought well, the spatial patterns associated with the anomalies produced by WRF differed from those found in the AWAP dataset. Further work will investigate the reasons for these spatial differences as well as WRF's performance at shorter time scales.

1. Introduction

The south-east coast of Australia contains most of the nation's population. It contains the largest mountain range and is significantly impacted by events in the nearby ocean. In some areas this coastal region is only a few tens of kilometres wide. Due to these factors, understanding of the current and future climate of the South-east coast of Australia is both important and difficult. A high resolution Regional Climate Model (RCM) simulation has been performed that will be used to address various climate related issues for the region. Initial evaluation of this simulation is presented here.

2. Regional Climate Model

The Weather Research and Forecast (WRF) model is developed and maintained at the National Center for Atmospheric Research in the USA. WRF was run over SE Australia from November 1984 through

2008. The first two months of the simulation are considered spin-up and are not used in the subsequent analysis. The model used the following physics schemes: WRF Single Moment 5-class microphysics scheme; Rapid Radiative Transfer Model longwave radiation scheme; Dudhia shortwave radiation scheme; Monin-Obukhov surface layer similarity; Noah land-surface scheme; Yonsei University boundary layer scheme and Kain-Fritsch cumulus physics scheme. The model simulation uses boundary conditions from the NCEP/NCAR reanalysis project (NNRP) with an outer 50 km resolution nest and an inner 10 km resolution nest that covers SE Australia (see Figure 1). Both nests used 30 vertical levels. Data from the NNRP are on a 2.5° grid.

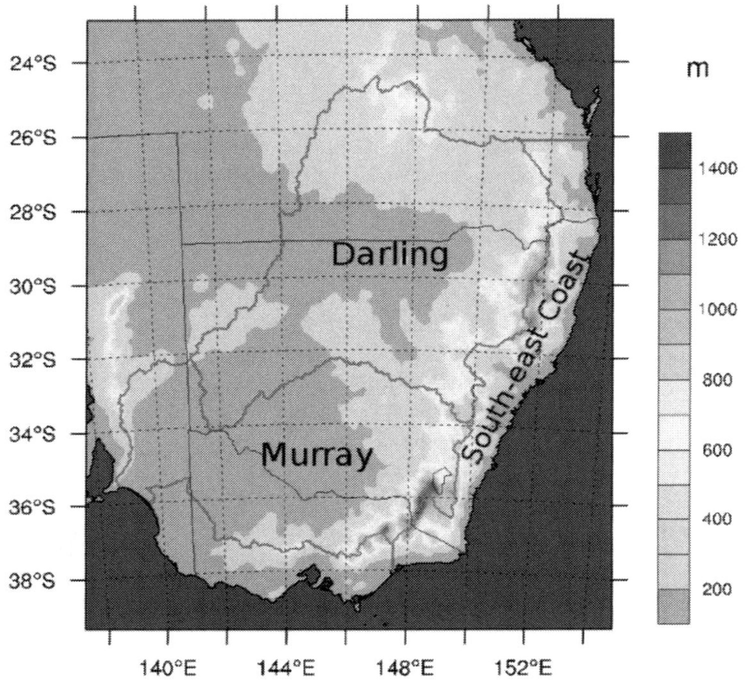

Figure 1. Regional climate model terrain (10km resolution) for south-east Australia. The Murray and Darling basins and the south-east coast are outlined in red.

3. Observations

Observations used for evaluation come from the gridded dataset prepared as part of the Australian Water Availability Project (AWAP). Details of creation of this dataset can be found in [1]. This dataset includes precipitation, maximum and minimum temperature, and vapour pressure surfaces obtained by interpolating surface station measurements onto a 5km grid. Solar radiation is derived from satellite measurements. Various surface hydrology parameters, such as soil moisture and runoff, are obtained using the WaterDyn hydrology model driven by the AWAP meteorological variables. The temperature increases from the high country in the south of the south-east coast region toward the north. The northern end of the coast has a strong precipitation maximum in summer (followed by autumn) while the southern end has a weak precipitation maximum in winter.

4. Statistical measures

Many different statistical measures have been used previously to test the performance of climate models quantitatively. An analyses of the suitability of several of these measures can be found in [2,3].

In this paper the model performance is evaluated against observations using several statistics including the bias

$$Bias = \bar{M} - \bar{O} \tag{1}$$

where $\bar{M}$ is the mean of the modelled values and $\bar{O}$ is the mean of the AWAP observations. The Root Mean Square Error (RMSE) is given by

$$RMSE = \sqrt{\frac{1}{N}\sum_{i=1}^{N}\left(O_i - M_i\right)^2} \tag{2}$$

where N is the number of observed, O, and modelled, M, values being compared. Here N is the number of WRF grid cells in the domain. Climatological values are used in calculating RMSE.

In order to quantitatively evaluate the spatial agreement between model and observations, Walsh and McGegor [4] define the pattern correlation (3), ρ_p, between observed and simulated fields simply as the correlation of a series of data points from the observed field with corresponding values from the modelled field at a fixed time, in this study monthly means are used.

$$\rho_p = \frac{\sum_{i=1}^{N}\left(O_i - \bar{O}\right)\left(M_i - \bar{M}\right)}{\sqrt{\sum_{i=1}^{N}\left(O_i - \bar{O}\right)^2}\sqrt{\sum_{i=1}^{N}\left(M_i - \bar{M}\right)^2}} \tag{3}$$

The anomaly correlation, ρ_a, is similar to the pattern correlation except that fields are replaced by anomalies from climatology. The anomaly correlation provides a more rigorous test of whether the model can capture the spatial pattern of interannual variations. Both the NNRP and AWAP data are interpolated to the WRF grid using inverse distance weighting.

5. Results

5.1. Precipitation and Temperature

Figure 2 shows the difference in seasonal temperature between the simulations and AWAP observations. The reanalysis (NNRP) fails to capture the topography of the Great Dividing Range and hence overestimates temperatures there while WRF is much better at capturing temperatures in that region. Also note that the NNRP generally underestimates temperatures in the low land regions especially in summer. WRF generally performs better at capturing the temperatures in the north though it also contains a summer cold bias. WRF also overestimates the temperatures in the southern coastal region, particularly in winter. This overestimation is inherited from the NNRP boundary conditions but WRF has decreased the warm bias compared to NNRP. These results are reflected in Table 1 which presents averages or sums in time of the spatial statistics. It can be seen that on an annual basis WRF overestimates temperatures in the SE coast by only 0.08K while NNRP overestimates them by 1.02K. WRF produces significant improvements in root mean square error (RMSE), pattern and anomaly correlation over NNRP.

Figure 3 shows the seasonal precipitation differences between the simulations and the AWAP observations. It can be seen that NNRP significantly underestimates the precipitation in the high country in the south in all seasons while WRF produces a much better estimate of precipitation in this area, though containing a small summer overestimate. NNRP also tends to overestimate in the north during Summer and Autumn. WRF demonstrates a better performance in this situation however it tends to underestimate in this area during Autumn. Table 1 shows that WRF produces improvements over NNRP in all statistics in the south-east coast. That is, the precipitation distribution produced by WRF significantly enhances that produced by its driving model (NNRP).

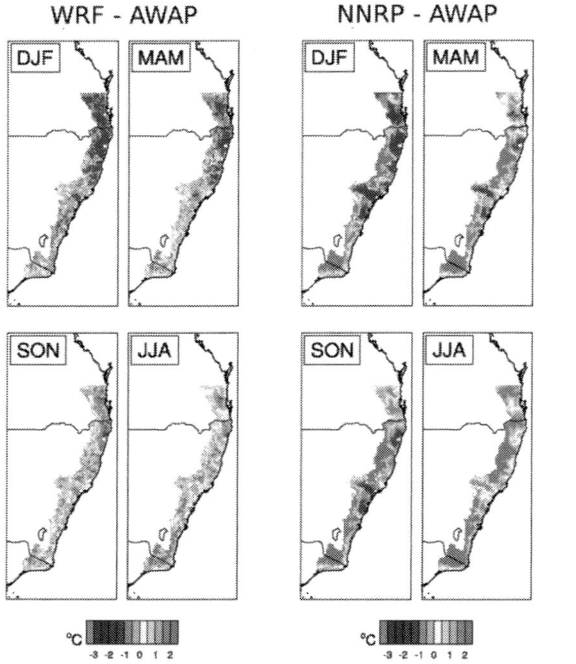

Figure 2. Differences in seasonal 2 m temperature.

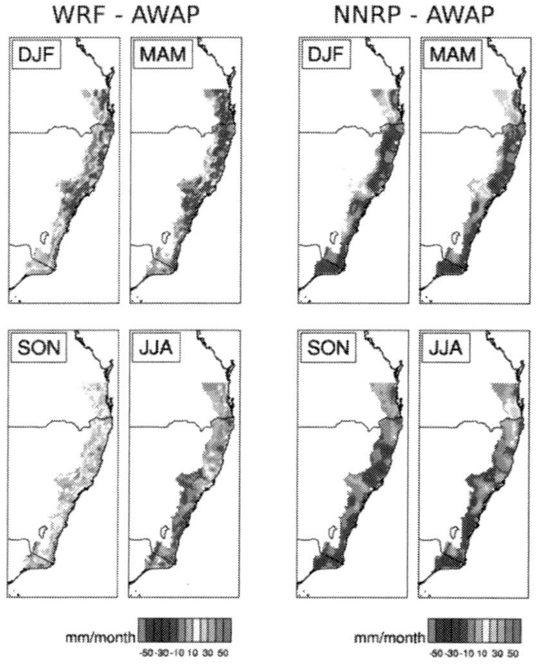

Figure 3. Differences in seasonal precipitation.

Table 1. South-east coast monthly statistics compared to AWAP.

	Temperature (°C)		Precipitation (mm)	
	WRF	NNRP	WRF	NNRP
Bias	0.08	1.02	1.65	-12.5
RMSE	0.83	2.04	13.87	21.56
Pattern Correlation	0.95	0.8	0.68	0.62
Anomaly Correlation	0.27	0.13	0.49	0.36

5.2. ENSO cycles

Much of south-east Australian precipitation is affected by ENSO to some extent. Using the Niño 3.4 index [5], months are defined as warm (El Niño), neutral or cool (La Niña). Warm months have an index greater than 0.5, while cool months have an index less than -0.5. Using this definition warm events are spread evenly through the year (average is 10) while slightly more (less) cool (neutral) events occur in summer compared to winter (average is 7). Figure 4 shows the mean precipitation over all warm months for AWAP, WRF and NNRP. WRF improves on the general trend of a precipitation decrease across the south-east coast, with the largest decreases seen in the northern NSW coast and the NSW-Victoria coastal border and highlands region. Figure 5 Shows the equivalent figure for the cool precipitation anomaly. Again WRF does a good job of reproducing the anomaly associated with cool events with a general increase in precipitation throughout the coast. It also captures the maximum increase along the Queensland-NSW coastal region though not with the same intensity as that seen in the observations.

IOP Conf. Series: Earth and Environmental Science **11** (2010) 012004 doi:10.1088/1755-1315/11/1/012004

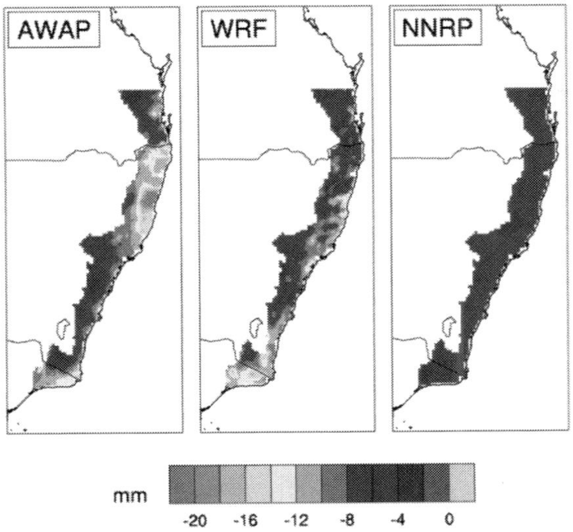

Figure 4. Precipitation anomalies during warm months.

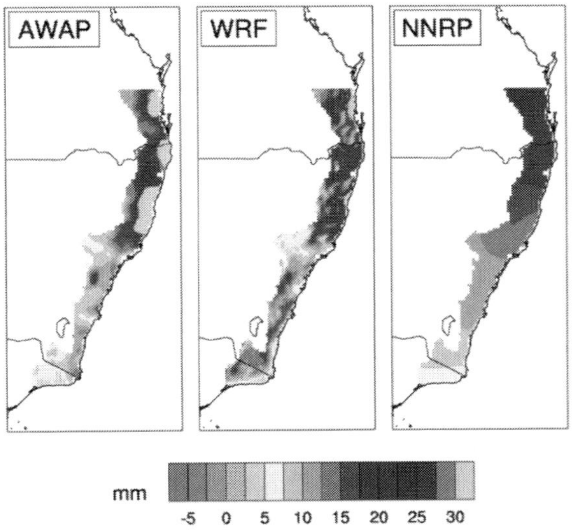

Figure 5. Precipitation anomalies during cool months.

5.3. The recent drought

Monthly precipitation anomaly time series (12 month running average) averaged over the SE Australian coast region can be seen in Figure 6. The recent drought can be clearly seen with an extended period of negative anomalies extending from 2002 through to 2007 in both the AWAP observations and the WRF simulation. Similar low rainfall years occurred in 1991 and 1994 and were simulated well. WRF did not simulate the short dry period in 1997 as well. Overall WRF agrees quite well with the AWAP data in terms of producing below average rainfall. It is worth noting that WRF also does a good job of capturing the variability of the anomaly within the recent drought period with some years being considerably drier than others in agreement with the observations.

The spatial distribution of the precipitation anomalies are shown in Figure 7. There are differences in the distribution of precipitation anomalies though they are of similar size.

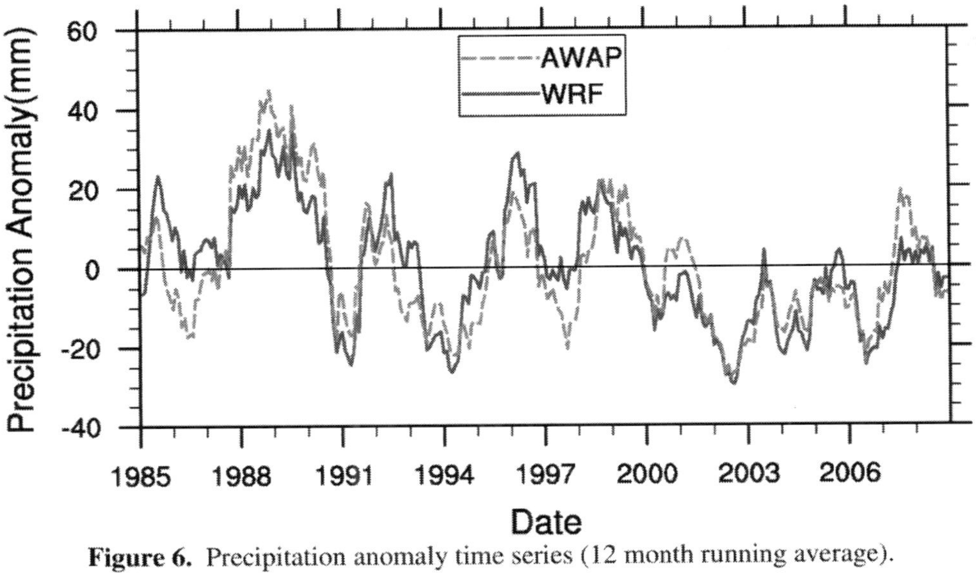

Figure 6. Precipitation anomaly time series (12 month running average).

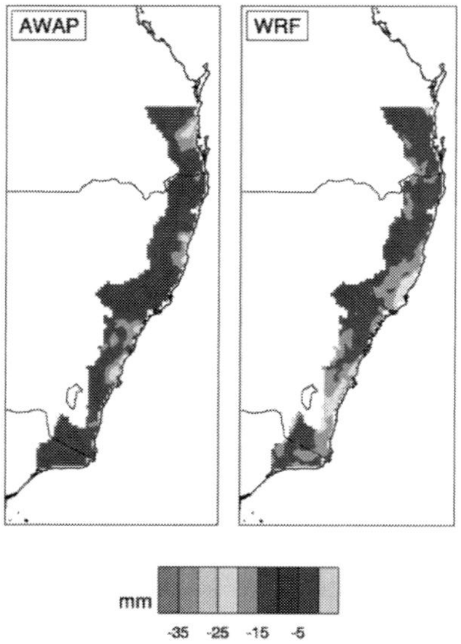

Figure 7. Precipitation anomaly during the recent drought (2002 to 2006 average minus the long term mean).

6. Conclusions

The regional climate model WRF was run for 24 years over SE Australia. The climate simulated by WRF was generally an improvement over that produced by the NNRP despite the fact that WRF does not assimilate any observations while NNRP does. WRF was able to reproduce the inter-annual variability reasonably well. This includes capturing the recent severe drought. While the overall magnitudes and time evolution of precipitation anomalies were captured well, the spatial distribution of these anomalies differed from that observed. WRF provides a significant improvement in capturing the spatial signature of ENSO related precipitation anomalies compared to the reanalysis. Further work will investigate the causes for these spatial differences as well as investigate shorter time-scales from monthly to daily.

Acknowledgements

This work was carried out as part of the ARC Discovery project DP0772665.

References

[1]Jones, D, Wang W, and Fawcett R. (2009), High-quality spatial climate data-sets for Australia, *Aust Meteorol Mag*, 58(4), 233-248.

[2]Legates DR and McCabe GJ. 1999. Evaluating the use of "goodness-of-fit" measures in hydrologic and hydroclimatic model validation. *Water Resour Res* 35: 233-241.

[3]Willmott CJ, Ackleson SG, Davis RE, Feddema JJ, Klink KM, Legates DR, Odonnell J and Rowe CM. 1985. Statistics for the Evaluation and Comparison of Models. *J Geophys Res-Oceans* 90: 8995-9005.

[4]Walsh K and McGregor J. 1997. An assessment of simulations of climate variability over Australia with a limited area model. *International Journal of Climatology* 17: 201-223.

[5]Trenberth, K. 1997. The Definition of El Niño, *B Am Meteorol Soc*, *78*(12), 2771-2777.

IOP Conf. Series: Earth and Environmental Science **11** (2010) 012005 doi:10.1088/1755-1315/11/1/012005

The GOCF/AWAP System – forecasting temperature extremes

Robert Fawcett[1] **and Timothy Hume**[2]

[1] National Climate Centre, Australian Bureau of Meteorology, Docklands, Victoria 3008, Australia.
[2] Centre for Australian Weather and Climate Research, Australian Bureau of Meteorology, Docklands, Victoria 3008, Australia.

E-mail: r.fawcett@bom.gov.au, t.hume@bom.gov.au

Abstract. Gridded hourly temperature forecasts from the Bureau of Meteorology's Gridded Operational Consensus Forecasting (GOCF) system are combined in real time with the Australian Water Availability Project (AWAP) gridded daily temperature analyses to produce gridded daily maximum and minimum temperature forecasts with lead times from one to five days. These forecasts are compared against the historical record of AWAP daily temperature analyses (1911 to present), to identify regions where record or near-record temperatures are predicted to occur. This paper describes the GOCF/AWAP system, showing how the daily maximum and minimum temperature forecasts are prepared from the hourly forecasts, and how they are bias-corrected in real time using the AWAP analyses, against which they are subsequently verified. Using monthly climatologies of long-term daily mean, standard deviation and all-time highest and lowest on record, derived forecast products (for both maximum and minimum temperature) include ordinary and standardised anomalies, "forecast – highest on record" and "forecast – lowest on record". Compensation for the climatological variation across the country is achieved in these last two products, which provide the necessary guidance as to whether or not record-breaking temperatures are expected, by expressing the forecast departure from the previous record in both °C and standard deviations.

1. Introduction

The Bureau of Meteorology's Gridded Operational Consensus Forecast (GOCF) system [1] produces weighted-average forecasts of gridded surface meteorological fields from an ensemble of eight bias-corrected numerical weather prediction (NWP) models. It is run twice daily, initialised at 00 UTC and 12 UTC, by the National Meteorological and Oceanographic Centre (NMOC), but for the purposes of this study, only the 12 UTC (10 pm EST) model runs are used. [UTC is Coordinated Universal Time. Eastern Standard Time (EST) in Australia is UTC+10 hours.] The GOCF system forecasts of temperature are spatially downscaled to 1/24° latitude and longitude resolution across a domain covering Tasmania and the Australian continent. The forecasts are then temporally interpolated to hourly time steps (t = 0, 1, …, 216 hours; t = 0 denoting the analysis). For the 12 UTC model run, eight models contribute to the t = 0, 1, …, 72 hours analysis/forecasts, with six models contributing to the t = 73, 74, …, 120 hours forecasts (see [1] for more detail). The models are bias-corrected with respect to the hourly Mesoscale Surface Analysis System (MSAS) [2], prior to the creation of the GOCF forecasts.

The eight contributing GOCF models [1] are LAPS (Limited Area Prediction System; an Australian regional model), UKMO (United Kingdom Meteorological Office global model), JMA (Japan Meteorological Agency global model), ECMWF-LR (European Centre for Medium Range Weather Forecasts; data on a 1.5° grid), ECMWF-HR (ECMWF data on a 0.5° grid), CMC (Meteorological Service of Canada global model), NCEP GFS (US National Centers for Environmental Prediction Global Forecast System) and GASP (Global Analysis and Prediction System; an Australian global model). [The set of GOCF models changed in December 2009, but we restrict our attention in this paper to the period December 2008 to November 2009.]

The new GOCF/AWAP (henceforth, G/A) system uses the 12 UTC GOCF model run, restricted to $t = 0, 1, \ldots, 119$ hours (i.e., the first five of the nine GOCF days). For compatibility with the Australian Water Availability Project (AWAP) 1/4° daily maximum temperature (Tmax) and minimum temperature (Tmin) analyses [3], the 1/24° resolution GOCF forecast grids are sub-sampled to 1/4° resolution. [The AWAP analyses are also available at 1/20° resolution, and were adopted as the Bureau of Meteorology's operational daily temperature analyses at the start of 2010.] Daily Tmax forecasts are obtained from the hourly forecasts simply as: Day 1 Tmax = $\max(T_0, T_1, \ldots, T_{23})$, Day 2 Tmax = $\max(T_{24}, T_{25}, \ldots, T_{48})$, *etc.*, where T_t denotes the GOCF hourly temperature forecast at hour t. Daily Tmin forecasts are obtained in an analogous fashion. While the 10 pm EST to 9 pm EST forecast "day" obviously does not correspond to the 9 am to 9 am (local time) observation "day", under normal weather conditions the forecasted daily maximum and minimum temperatures will be correctly assigned, in spite of the different time zones across the country.

These "original" GOCF Tmax and Tmin forecasts are subjected to an additional round of bias correction, using the 30 most recently available forecast/analysis pairs, the analyses being taken from the 1/4° AWAP daily temperature analysis grid sets [3] and typically involving a lag of 1 day because of data availability considerations. This second round of bias correction is required for two reasons: (i) the MSAS hourly temperature analysis system [2] and the AWAP daily temperature analysis system [3] treat the elevated parts of the country in different ways, and (ii) the calculation of forecast daily Tmax (*Tmin*) as a simple (uninterpolated) maximum (*minimum*) of 24 hourly values artificially reduces the amplitude of the diurnal temperature range (to some extent).

2. The GOCF/AWAP product set

The G/A system generates a wide range of products, both in "original" and "bias-corrected" forms, where here "bias-corrected" refers to the second (AWAP) round of bias correction, described in the previous section. The first and most obvious product type is the forecast Tmax and Tmin (Figure 1). The "day 1" Tmax forecast, obtained in the morning from the 12 UTC GOCF model run of the night before, isn't of much use under normal circumstances, since it only becomes available a matter of hours before the maximum temperature is expected to be attained (typically in the afternoon), but it still provides guidance to the operational climatologist about that day's workload. The "day 1" Tmin forecast is of even less use, because the forecast minimum typically occurs before the G/A job is actually run (currently mid-morning EST).

Forecast anomalies are obviously useful, and here they are calculated in two ways; (i) based on all the available daily analyses for the relevant calendar month, from 1911 (the start of the AWAP daily analysis grid set) to the present, and (ii) based on a windowed climatology to represent more smoothly the annual cycle on daily temperature (also 1911 to the present). Only results using the calendar month method of calculating the climatology will be presented in this paper. Standardised anomaly forecasts are also useful, because the temporal variance in daily temperature varies markedly across the country. Ordinary (unstandardised) anomalies of a given magnitude are much more "significant" in the tropical north (where the variance is typically smaller) than they are in the mid-latitude south (where the variance is typically larger). Figure 2 shows forecast anomalies and standardised forecast anomalies, using the calendar monthly climatological approach. The standardised anomalies are non-dimensional – a value of +1 means one standard deviation above the climatological mean.

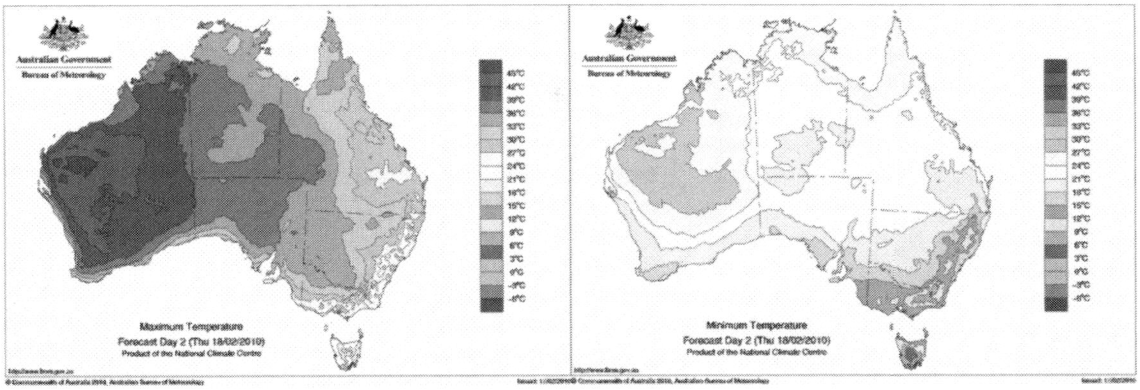

Figure 1. Bias-corrected “day 2” Tmax (left) and Tmin (right) forecasts (in °C) for 18 February 2010. These are derived from the 12 UTC 16 February 2010 GOCF model run.

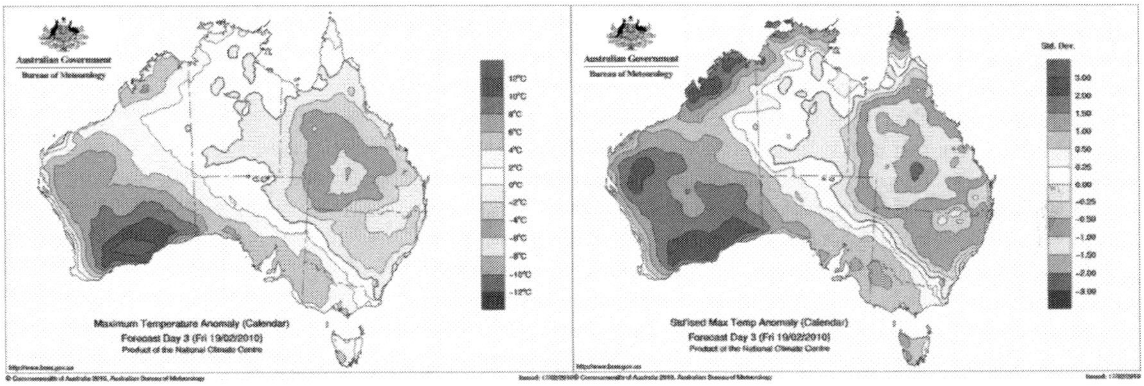

Figure 2. Bias-corrected “day 3” Tmax anomaly forecast for 19 February 2010, ordinary anomalies (in °C) on the left, standardised anomalies on the right. These are derived from the 12 UTC 16 February 2010 GOCF model run.

The G/A system is principally concerned with the forecasting of daily temperature extremes, so the forecast grid is compared against the climatological highest and lowest on record (again, based on a climatology period of 1911 to the present). Figure 3 shows the forecast daily Tmax expressed as a departure from the previous (February daily) highest on record (i.e., {forecast} – {previous highest on record}). Positive values indicate that the record is forecast to be broken, while negative values indicate that the record is forecast to be approached but not broken. [We use the same approach for the other end of the distribution, that is, {forecast} – {previous lowest on record}. Here though, negative values mean the record is forecast to be broken, positive values that the record is forecast to be approached but not broken.] Because of the considerable spatial variation across the country in the variance of daily temperatures, it is useful to also express these departures (from the previous highest or lowest on record) in standard deviations. This helps to guard against the risk of “misinterpretational false alarms” in the tropical north, where the daily variance is small enough that close approaches to the record (in terms of °C) occur quite often.

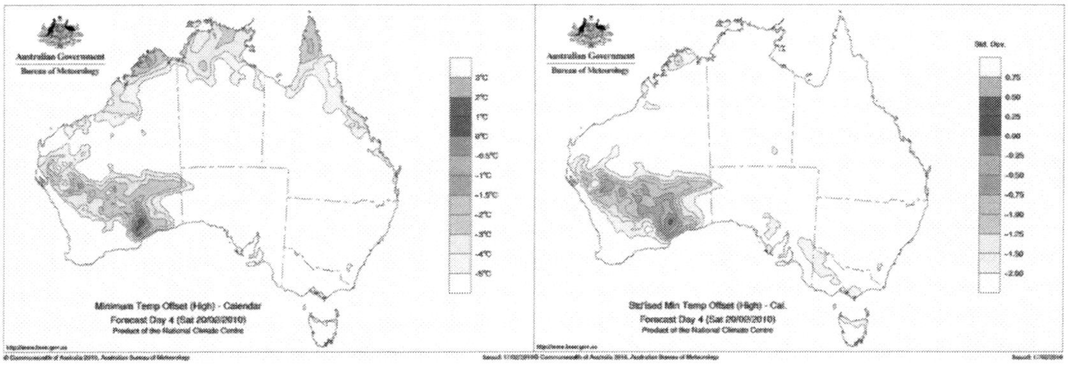

Figure 3. Bias-corrected "day 4" Tmin extremes forecast for 20 February 2010, forecast departures from previous highest on record in °C on the left, and in standard deviations on the right. These are derived from the 12 UTC 16 February 2010 GOCF model run. Positive values mean that the record is forecast to be broken, negative values that the record is forecast to be approached but not broken.

3. Black Saturday – a case study

The example of 7 February 2009 is now presented as a case study. The underlying GOCF system was running quasi-operationally on this day. The AWAP extension was still under construction at that time, so the results presented in this section are a reconstruction of what would have been issued at the time, had the G/A system been available in its current configuration. Figure 4 shows the observed daily maximum temperature for the day, the verifying climatology, and the verifying extremes product. These comprise the daily maximum temperature analysis for 7 February 2010, the previous highest on record (based on all February daily maximum temperatures between 1 February 1911 and 5 February 2009), and the difference between the two of these. The analysis for 6 February 2009 is not included in the calculation of the previous highest on record because it would not have been available at the time the G/A job would have been run on the morning of 7 February. Figure 4 shows that records were broken (ignoring whatever had happened on 6 February) across most of Victoria and adjacent parts of South Australia and New South Wales. Much of northwest Victoria and southwest New South Wales recorded temperatures above 45°C on the day.

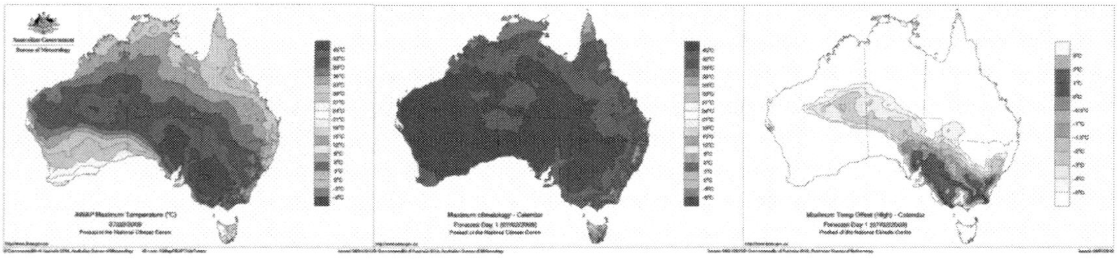

Figure 4. AWAP 1/4° daily maximum temperature analysis for 7 February 2009 (left), climatological highest on record (centre), calculated from all February daily maximum temperature analyses between 1 February 1911 and 5 February 2009, and the difference between the two of these (right), calculated as {analysis} – {previous highest on record}.

Reconstructed G/A Tmax extremes forecasts for 7 February 2009, at lags of 1 day (based on the 12 UTC 6 February 2009 model run) to 5 days (based on the 12 UTC 2 February 2009 model run) are shown in Figure 5, together with the "day 1" verification from Figure 4. The occurrence of record breaking temperatures is already predicted in the "day 5" forecast, and useful guidance about the extent of the temperature pattern is available in the "day 2" and "day 1" forecasts, although the actual outcome exceeded the prediction in areal extent, particularly along the Victorian coast where the record-breaking temperatures extended all the way to the coast.

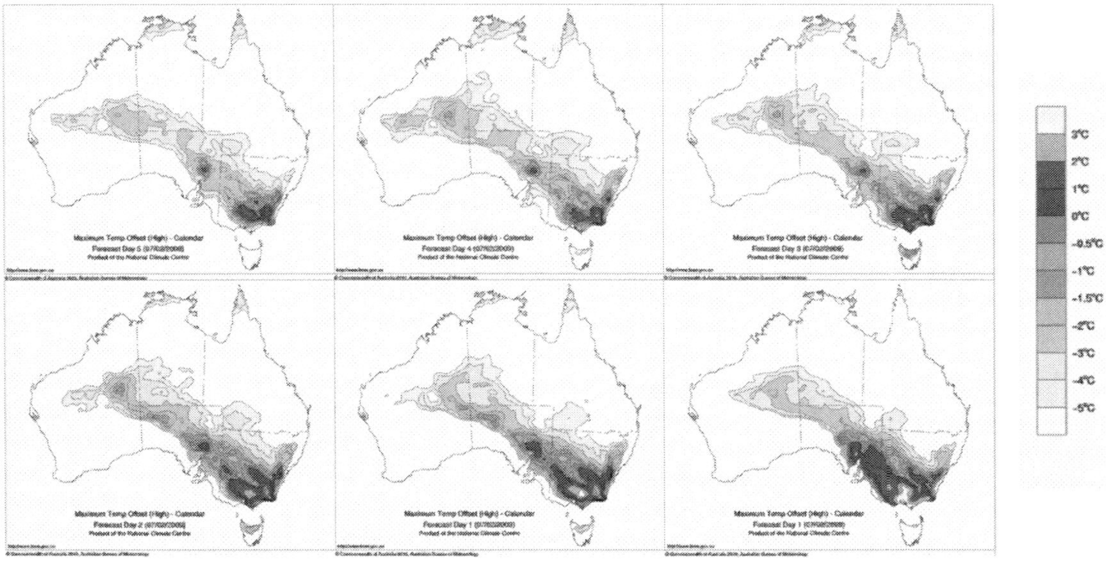

Figure 5. Reconstructed Tmax extremes forecasts for 7 February 2009, at lags of 5 days (top left), 4 days (top centre), 3 days (top right), 2 days (bottom left) and 1 day (bottom centre), together with the "day 1" outcome (bottom right).

4. Verification results

Verification results presented in this section for the bias-corrected forecasts have in each case been compiled on a monthly basis for each calendar month from December 2008 to November 2009, with the monthly statistics being averaged to produce annualised results. Results averaged across all Australian grid points are shown in Table 1. Average mean errors or biases are quite small (less than 0.05°C across a lot of the country), indicating that the 30-day bias correction procedure as implemented in real time is functioning adequately – an important point because the "original" forecasts obtained from the GOCF system have substantial biases with respect to the AWAP analyses. Mean errors in the corrected forecasts are largest in magnitude in regions of sparse temperature observation.

Table 1. Annualised verification scores (ME = mean error or bias, MAE = mean absolute error, RMSE = root mean square error, R = correlation coefficient), averaged across all Australian grid points, for the bias-corrected forecasts. Scores are calculated in the usual way for each month (separately) between December 2008 and November 2009, then averaged across the 12 months. "Error" is defined as {forecast} – {analysis} in the calculation of ME, MAE and RMSE.

Score	Variable	Day 1	Day 2	Day 3	Day 4	Day 5
ME	Tmax	0.00°C	0.00°C	0.02°C	0.03°C	0.00°C
	Tmin	–0.01°C	0.00°C	0.00°C	0.01°C	0.03°C
MAE	Tmax	0.96°C	1.08°C	1.26°C	1.40°C	1.83°C
	Tmin	1.28°C	1.37°C	1.48°C	1.57°C	2.04°C
RMSE	Tmax	1.26°C	1.40°C	1.62°C	1.81°C	2.32°C
	Tmin	1.58°C	1.69°C	1.83°C	1.95°C	2.49°C
R	Tmax	0.89	0.87	0.83	0.80	0.66
	Tmin	0.83	0.80	0.77	0.74	0.60

Mean absolute errors (MAEs) for Tmax at "day 2" (Figure 6) are less than 1°C across much of the tropical north, and less than 1.5°C across most of the rest of the country. A small region at the top of the Great Australian Bight has MAEs around 2°C. Averaged across the country, the "Day 2" MAE is 1.08°C. "Day 2" MAEs for Tmin are typically somewhat larger than those for Tmax, being between 1° and 2°C across most of the country. This may reflect the greater dependence of Tmin on local topography. "Day 2" root-mean-square errors (RMSEs) are typically larger than the MAEs for both Tmax and Tmin, but those for Tmax are still between 1° and 2°C across most of the country (the national average is 1.40°C). Across the five-day forecast period, forecast skill not surprisingly degrades with increased lead time in both variables.

As previously indicated, the "Day 1" minimum temperatures typically occur before the G/A job has been run, while the "Day 1" maximum temperatures typically occur some hours after the job run. Therefore the "Day 2" results are operationally of most importance.

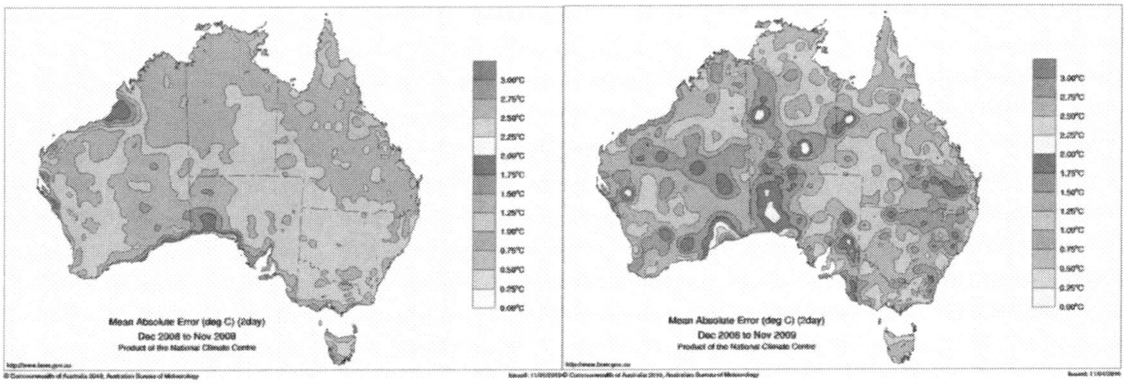

Figure 6. Mean absolute errors (in °C) for the bias-corrected Tmax (left) and Tmin (right) forecasts (December 2008 to November 2009).

5. Concluding remarks

The G/A system described in this paper has been running experimentally in real time for nearly a year. Together with reconstructed forecasts using archived GOCF forecasts (where the GOCF system was running operationally in real time, before the G/A system was constructed), there are now about 16 months of assessable results. Verification statistics, comprising forecasts aggregated into calendar months, are routinely generated. Based on experience so far, the new system provides useful guidance for when daily temperature records are predicted to be approached or broken.

One anticipated future development is tying the forecast "day" more closely to the observational "day" (9 am to 9 am local time in the various States/Territories). Such a change should result in some improvement in forecast skill, particularly in the south of the country. Other possible extensions include extremes forecast products relating to temperature levels such as the 1st and 99th percentiles, in addition to those currently relating to the 0th (lowest on record) and 100th (highest on record) percentiles. Another development is a closer inspection of the performance of the new forecast system under extreme conditions.

References

[1] Bureau of Meteorology 2008 *Analysis and Prediction Operations Bulletin No. 74: Operational Implementation of the Gridded OCF System* (Melbourne, Australia: Australian Bureau of Meteorology)

[2] Bureau of Meteorology 2007 *Operational Implementation of the Mesoscale Surface Analysis System* (Melbourne, Australia: Australian Bureau of Meteorology)

[3] Jones D A, Wang W and Fawcett R 2009 High-quality spatial climate data-sets for Australia *Aust. Meteor. Ocean. J.* **58** 233-248

IOP Conf. Series: Earth and Environmental Science 11 (2010) 012006 doi:10.1088/1755-1315/11/1/012006

Network-derived inhomogeneity in monthly rainfall analyses over western Tasmania

Robert Fawcett[1], Blair Trewin[1] and Ian Barnes-Keoghan[2]

[1] National Climate Centre, Australian Bureau of Meteorology, Docklands, Victoria 3008, Australia.
[2] Tasmanian Regional Office, Australian Bureau of Meteorology, Hobart, Tasmania 3000, Australia.

E-mail: r.fawcett@bom.gov.au, b.trewin@bom.gov.au, i.barnes-keoghan@bom.gov.au

Abstract. Monthly rainfall in the wetter western half of Tasmania was relatively poorly observed in the early to middle parts of the 20th century, and this causes a marked inhomogeneity in the operational gridded monthly rainfall analyses generated by the Australian Bureau of Meteorology up until the end of 2009. These monthly rainfall analyses were generated for the period 1900 to 2009 in two forms; a national analysis at 0.25° latitude-longitude resolution, and a southeastern Australia regional analysis at 0.1° resolution. For any given month, they used all the monthly data from the standard Bureau rainfall gauge network available in the Australian Data Archive for Meteorology. Since this network has changed markedly since Federation (1901), there is obvious scope for network-derived inhomogeneities in the analyses. In this study, we show that the topography-resolving techniques of the new Australian Water Availability Project analyses, adopted as the official operational analyses from the start of 2010, substantially diminish those inhomogeneities, while using largely the same observation network. One result is an improved characterisation of recent rainfall declines across Tasmania. The new analyses are available at two resolutions, 0.25° and 0.05°.

1. Introduction

The Bureau of Meteorology started generating operational national gridded monthly rainfall analyses using the work of Jones and Weymouth [1] in 1996. These were at 0.25° latitude-longitude resolution, and employed the two-dimensional Barnes successive correction technique [2]. An additional set of regional analyses at 0.1° resolution, covering southeastern Australia (Tasmania, Victoria and adjacent parts of South Australia and New South Wales), was also constructed to take account of the greater station density in that part of the country. Historical analyses were generated back to 1900, and for any given month, they used all the monthly data from the standard Bureau rainfall gauge network available in the Australian Data Archive for Meteorology (ADAM) at the time of the analysis. This network has changed markedly, with numbers of available stations rising from around 3300 per month in 1900 to a peak of around 7700 per month in 1973, thereafter declining to around 6000 per month in 2008. Because of these changes, there is obvious scope for network-derived inhomogeneities (i.e., changes arising purely from changes in the observation network, and not from changes in the actual rainfall) in the analyses, although how those inhomogeneities might manifest themselves could depend on the particular analysis technique employed. The Barnes analyses remained the Bureau's operational gridded monthly rainfall analyses until the end of 2009. At the start of 2010, they were superseded by a new set of three-dimensional topography-resolving analyses [3], created as part of the Australian Water Availability Project (AWAP). The AWAP rainfall analyses are produced nationally at two resolutions, 0.25° and 0.05°. They also use all the monthly data in the standard Bureau rainfall gauge network in ADAM, subject to the caveat that site elevations above sea level (ASL) must be either available or satisfactorily estimated.

2. Methodology

The Barnes analysis methodology contains no climatological information and depends on the network to resolve rainfall gradients, such as that between the climatologically wet western Tasmania and dry eastern Tasmania. A sparse gauge network may mean that those gradients are inadequately represented in the analyses. The absence of stations in southwestern Tasmania (Figure 1) for much of the 20th century means that the Barnes analyses are artificially dry when such sites are absent. Unlike the Barnes analyses, the AWAP analyses have the climatological rainfall gradients "built in", and are therefore less dependent on the network available in any given month. The technique involves a two-dimensional Barnes analysis of the fraction of the climatological monthly mean, coupled with a three-dimensional topography-resolving climatological analysis – see [3] for more detail on how this is done.

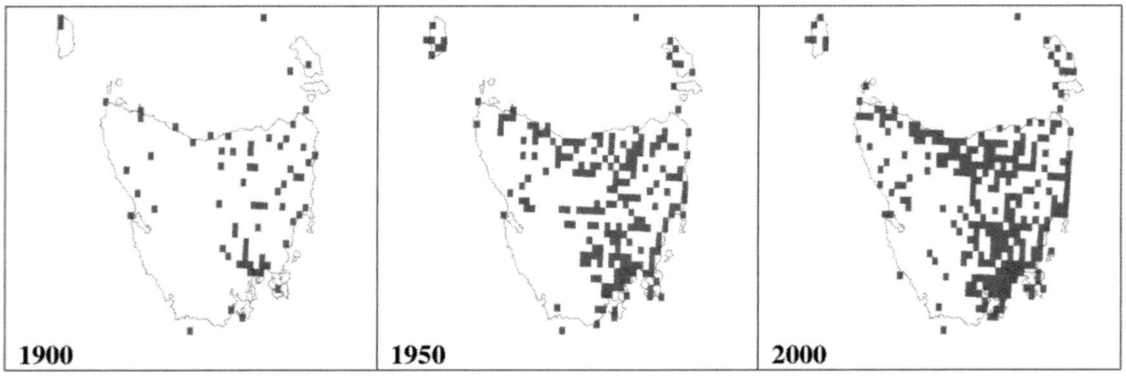

Figure 1. State of the July monthly rainfall gauge network across Tasmania in 1900 (left), 1950 (centre) and 2000 (right). Each blue square indicates at least one reporting station within the 0.1° latitude-longitude grid cell in the indicated month.

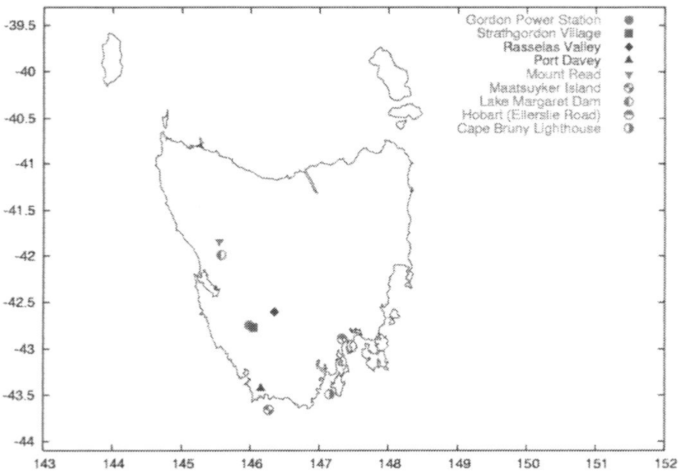

Figure 2. Locations of rainfall stations mentioned in this study.

In this study, we show by means of some illustrative examples that the network-driven inhomogeneities present in the Barnes analyses for southwestern Tasmania are very much reduced in the newer AWAP analyses. We do this by a comparison of observed station data and grid interpolations at the observation locations. In doing this, however, we restrict our attention to network-driven inhomogeneities. It is likely that rainfall analyses in southwestern Tasmania are also affected by inhomogeneities at individual stations. In particular, some potential inhomogeneities have been identified in the rainfall record at Maatsuyker Island, off the far southwest coast of Tasmania [4]. The locations of Maatsuyker Island and the other rainfall stations mentioned in this study are shown in Figure 2.

3. Results

3.1. Strathgordon Village

A clear example of network-based inhomogeneity in the Barnes analyses can be seen by interpolating the analyses to Strathgordon Village (BoM station number 097053; 42.768°S, 146.046°E, 322 m ASL). We use the method of bi-cubic polynomial interpolation (see [5] for example) for this purpose. Interpolations from four grid sets (the two superseded Barnes sets and the two operational AWAP sets) are shown in Figure 3, together with annual data from Strathgordon Village which opened in June 1968. [A near neighbour, Strathgordon (Gordon Power Station) (BoM #097061; 42.743°S, 145.981°E, 310 m ASL), opened a few years later in 1971.] The two Barnes analyses show a substantial rise in their typical values at precisely the time the station opened (Table 1), with the 0.1° analyses showing an additional inhomogeneity in the late 1930s and 1940s. The two AWAP analyses do not show this major inhomogeneity, and in addition they show much more realistic values than the 0.25° Barnes analyses in those years for which station data exist.

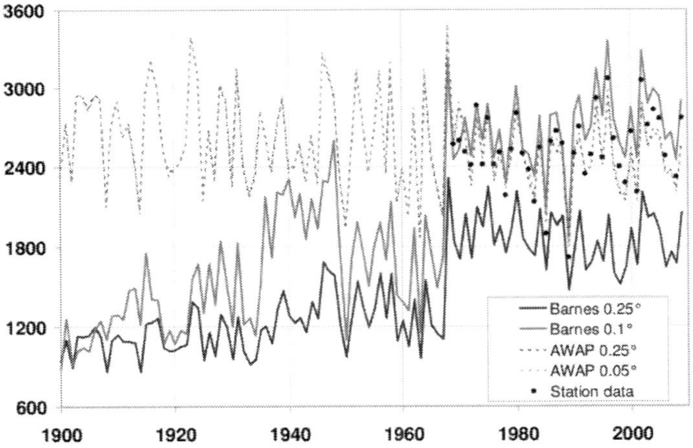

Figure 3. Annual rainfall (1900-2009, in mm) from the four gridded analysis sets (lines) interpolated to Strathgordon Village, and the station data (dots) from 1969. Barnes analyses are shown in thick solid lines, AWAP analyses in thin dashed lines.

Table 1. Average annual rainfall (in mm) at Strathgordon Village for two periods. The ratio of the second period to the first period is also shown. Values interpolated from the four gridded analysis sets.

Period	Barnes 0.25°	Barnes 0.1°	AWAP 0.25°	AWAP 0.05°
1900-1967	1192	1569	2607	2614
1969-2009	1860	2683	2479	2482
Ratio	1.56	1.71	0.95	0.95

The additional inhomogeneity in the late 1930s and 1940s appears to be associated with the Rasselas Valley site (BoM #097009; 42.6°S, 146.35°E, approximately 480 m ASL), whose first monthly rainfall observation was for May 1936 and last observation for July 1949. Table 2 shows the results of interpolating the four analysis grid sets to Rasselas Valley for three periods; 1900-1935, 1937-1948 and 1950-1967. The last of these three periods is truncated at 1967 because of the opening of nearby Strathgordon Village in 1968. In the 0.25° Barnes analyses, the second period is 32% wetter than the first, while in the 0.1° Barnes analyses it is 60% wetter. The two AWAP analyses show no such change in the local rainfall climatology.

Table 2. Average annual rainfall (in mm) at Rasselas Valley for three periods, and the ratios of the second to first, and third to second periods. Values interpolated from the four gridded analysis sets.

Period	Barnes 0.25°	Barnes 0.1°	AWAP 0.25°	AWAP 0.05°
1900-1935	992	1015	1840	1724
1937-1948	1309	1627	1798	1724
1950-1967	1188	1310	1658	1604
Ratio (2nd to 1st)	1.32	1.60	0.98	1.00
Ratio (3rd to 2nd)	0.91	0.80	0.92	0.93

3.2. Port Davey

Another example occurs with the opening of Port Davey (BoM #097007; 43.43°S, 146.16°E, 6 m ASL) in 1946 in the far southwest of the State, as can be seen in Figure 4 which shows the ratio of rainfall during the ten years after the opening of the site to that of the ten years before the opening. There is a significant local inhomogeneity in both Barnes analyses which is not present in the AWAP analyses, and consistent with the Strathgordon Village example the inhomogeneity is stronger in the higher resolution Barnes analysis.

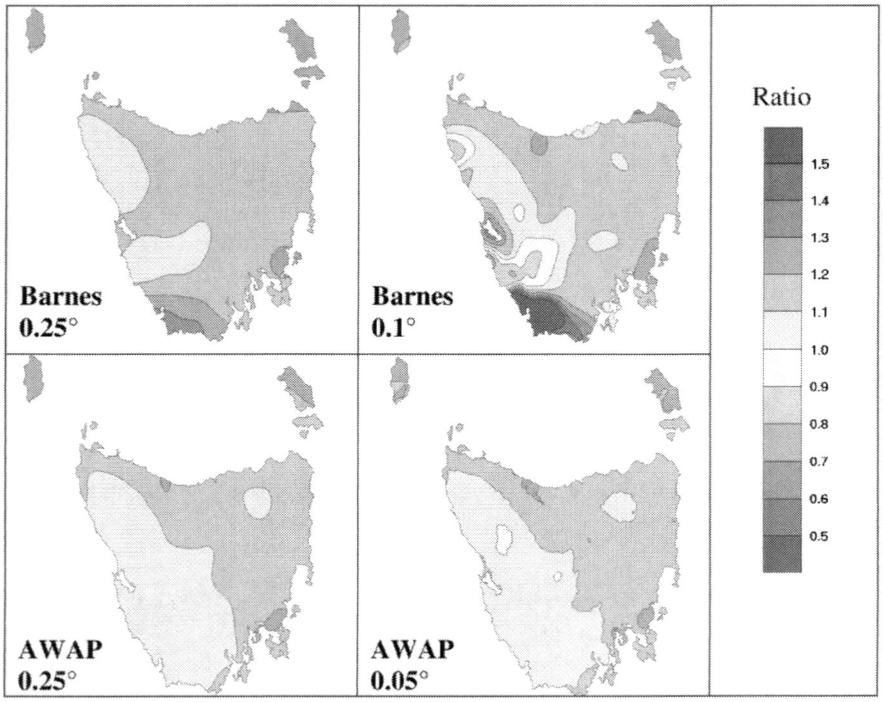

Figure 4. Change in analysed average rainfall in the decades before and after the inclusion of Port Davey observations (as expressed by the ratio of 1947-56 to 1936-45 rainfall).

Additional large inhomogeneities (Figure 5) are also evident in the 0.1° Barnes analyses, and to a lesser extent in the 0.25° Barnes analyses, coincident with the cessation of Port Davey's observations in the second half of 1970 (the Barnes analyses becoming drier) and their intermittent reappearances in 1986-1988, and more reliably in late 1991 (the 0.1° Barnes analyses becoming wetter).

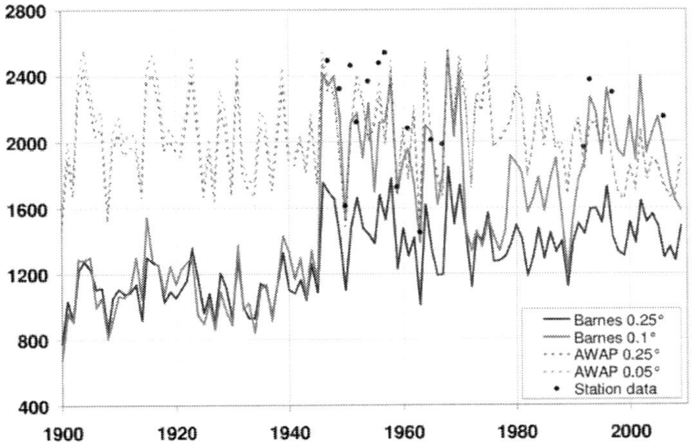

Figure 5. Annual rainfall (1900-2009, in mm) from the four gridded analysis sets (lines) interpolated to Port Davey, and the available station data (dots). A large number of missing monthly totals at the site results in annual station totals being available for relatively few years.

3.3. Mount Read

Mount Read (BoM #097085; 41.844°S, 145.542°E, 1119 m ASL) is a high-elevation site which opened in 1996. The site record is not of the highest quality – it has many missing monthly totals in its record since it opened and has complete annual data for just two years – but on a monthly basis it is frequently the wettest site in Tasmania. Interpolating the analyses to its location and comparing the annual averages over the periods 1970-1995 and 1997-2009 provides little evidence of inhomogeneity (Figure 6). The ratios of the second to first periods are 1.02 (0.25° Barnes), 1.04 (0.1° Barnes), 0.99 (0.25° AWAP) and 1.01 (0.05° AWAP). The 0.05° AWAP analyses are rather wetter than the other three grid sets, although this is not surprising because gridded analyses in effect represent a spatial average, with the finer resolution analyses representing spatial averages over smaller areas. Observations from other sites in the area, including Lake Margaret Dam (BoM #097006; 41.993°S, 145.573°E, 665 m ASL), another "wet" site with more or less continuous observations from 1912, would appear to have provided sufficient information to the AWAP analysis system to estimate rainfall at Mount Read.

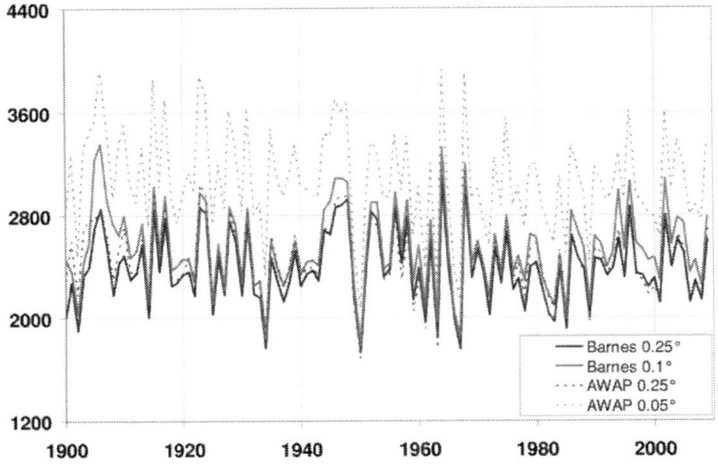

Figure 6. Annual rainfall (1900-2009, in mm) from the four gridded analysis sets (lines) interpolated to Mount Read. A large number of missing monthly totals at the site results in annual station totals being available for only two years.

4. Discussion

The examples of Strathgordon Village and Port Davey illustrate the inability of the Barnes analyses, and in particular the 0.1° resolution set, to cope with the changing characteristics of the rainfall gauge network across Tasmania, particularly across the wetter parts of the southwest. The newer AWAP analyses perform rather better. This leads to a much improved characterization of the recent rainfall declines over much of Tasmania (Figure 7). The artificial wetting of the Barnes analyses across the 20th century in the southwest has masked the recent drying trend evident in the AWAP analyses and individual site data.

A direct comparison of Figure 7 with individual site data is of limited value, however, as there are only three Tasmanian sites which have complete monthly rainfall data in ADAM for the period 1900-2009, and two of these (Maatsuyker Island, as discussed earlier, and Cape Bruny) are highly exposed locations with potentially major homogeneity issues as a result [4]. Their rainfall percentiles for the 1997-2009 period are (a) 43.3% at Cape Bruny Lighthouse (BoM #094010), (b) 4.12% at Hobart (Ellerslie Road, BoM #094029), and (c) 0.0% at Maatsuyker Island Lighthouse (BoM #094041). Only the Hobart data, which are consistent with the AWAP assessment in Figure 7, can be considered sufficiently homogeneous to provide a useful comparison with the long-term rainfall analyses.

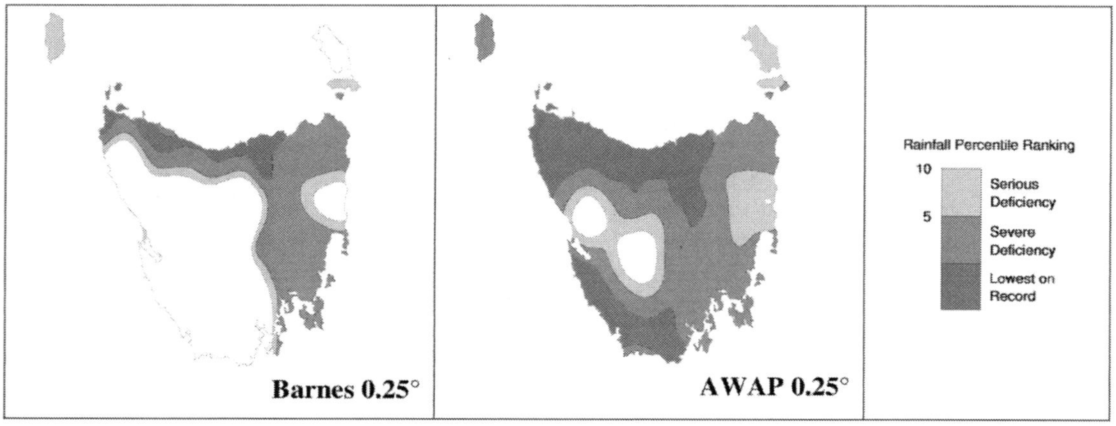

Figure 7. Rainfall percentiles across Tasmania for the 13-year period 1997-2009, calculated with respect to the 1900-2009 period, 0.25° Barnes analyses (left) and 0.25° AWAP analyses (right). White regions denote rainfalls above the 10th percentile. Red shades denote rainfall below the 10th (serious deficiency) and 5th percentile (severe deficiency), with the strongest shade denoting areas of lowest on record. Agreement between the two assessments is closer in the more densely observed parts of the State.

References

[1] Jones D A and Weymouth G 1997 *An Australian monthly rainfall data set* (Melbourne, Australia: Australian Bureau of Meteorology)

[2] Koch S E, desJardins M and Kocin P J 1983 An interactive Barnes objective map analysis scheme for use with satellite and conventional data *J. Clim. Appl. Meteor.* **22** 1487-1503

[3] Jones D A, Wang W and Fawcett R 2009 High-quality spatial climate data-sets for Australia *Aust. Meteor. Ocean. J.* **58** 233-248

[4] Trewin B C 2010 Site location effects on measured precipitation in highly exposed coastal and alpine environments *Conference abstracts 17th National Conference of the Australian Meteorological and Oceanographic Society, Canberra, 27-29 January 2010*

[5] Fawcett R J B, Jones D A and Shitaye B 2006 *Climatological rainfall analyses for southeast South Australia* (Adelaide, Australia: South Australian Department of Water, Land and Biodiversity Conservation)

A methodology of tracking transitioning Cyclones

L A Garde[1], A B Pezza[1], I Simmonds[1] and N E Davidson[2]
[1] The Department of Earth Sciences, The University of Melbourne, Victoria 3010, Australia
[2] Centre for Australian Weather and Climate Research, Bureau of Meteorology, Victoria, 3000, Australia

E-mail: l.garde@pgrad.unimelb.edu.au

Abstract. "Cyclone transition" is a process in which either a tropical cyclone or an extratropical cyclone undergoes a fundamental dynamic and thermodynamic transformation. A "tropical transition" is a process in which a cold-core extratropical cyclone transforms to a warm-core tropical cyclone, while "extratropical transition" refers to the opposite mechanism. The transition process is gradual, with no sharp defining threshold representing its commencement or completion. This study develops and explores a methodology of identifying tropical transition and extratropical transition events. Using the 1.5° ERA-Interim reanalysis dataset over the period from 1989 to present, cyclone tracks are derived using The Melbourne University Tracking Scheme. The average temperature difference between the cyclone core and an environmental ring, set at a distance corresponding to the cyclones size, is then calculated throughout the troposphere during the life cycle of the cyclone. This difference represents the index or phase of the cyclone relative to its immediate environment. The lack of a complete cyclone transition climatology and the uncertainty behind the transition process is the motivation for this research, with the overarching aim of placing these systems in climatic perspective. This methodology seeks to create the foundations for further cyclone research.

1. Introduction

The term "cyclone transition" refers to a gradual dynamic and thermodynamic process, which occurs as a cyclone's core temperature and its frontal characteristics transform. Over the last century, numerous studies have analysed tropical cyclones (TCs) and extratropical cyclones (ECs) as two mutually exclusive groups [1]. As improvements were made to satellite and observational networks, attention shifted to the "gray" areas in between these groups [1].

Research into understanding the dynamics involved with cyclone transition is very important, as history has shown that the most intense of these systems pose a hazard to infrastructure and the general population of which there are many recent examples. An extratropical transition (ET) is the process in which a well developed warm-core TC evolves to a cold-core EC. The converse is true for a tropical transition (TT). The transition concept appreciates that there are no physical constraints in the free atmosphere to prevent one given type of cyclone from transforming into a different type or from having hybrid characteristics. It is important to stress that cyclone transition is a dynamic and thermodynamic transformation and not simply a geographical shift.

Work conducted many decades ago and research undertaken towards the beginning of the twenty-first century turned to objectively quantifying the gray areas and the transitional processes between TCs, ECs and Hybrid cyclones (HCs) [1, 2, 3, 4, 5, 6, 7, 8, 9, 10].

Beven [3] proposed an expanded classification system which utilises core temperature and frontal characteristics. TCs are warm and non-frontal, whilst ECs are frontal but can possess either warm- or cold-cores. In addition, Beven's scheme would see HCs categorised separately as "strange hybrids", emphasizing their structural variety and uncertainty, due to their ability to draw energy by multiple processes. This characteristic makes these systems extremely difficult to categorise. Although not commonly adopted in the Southern Hemisphere (SH), from the Northern Hemisphere's perspective, subtropical cyclones are given special consideration as cold, non-frontal systems occurring over the subtropics. Examples of such systems include the Hawaiian Kona Low [11]. Beven's scheme poses no sharp boundaries between classification groups, rather a continuous spectrum.

Hart [1] developed the concept of "Cyclone Phase Space". This methodology proposed a concise means of objectively quantifying several parameters relating to the structure of a cyclone, allowing for a more flexible approach toward cyclone classification [1]. Utilising phase diagrams calculated throughout the cyclone life cycle, this method in principle discriminates between cyclones by assessing the thermal symmetry or frontal nature (lower-tropospheric thickness) and tropospheric thermal wind (cold – versus warm core structure). Hart makes a distinction between warm-cold cores based upon the examination of two tropospheric layers of equal mass: 900–600 and 600–300 hPa [1]. Hart's phase diagrams illustrate the transition between warm- and cold-core systems, effectively highlighting the onset and completion of ET, TT, and the development of HCs (Fig. 1).

The following discusses a methodology which builds on previous work by Garde et al. [12] and [1, 3], by identifying and mapping cyclone transition events for the SH.

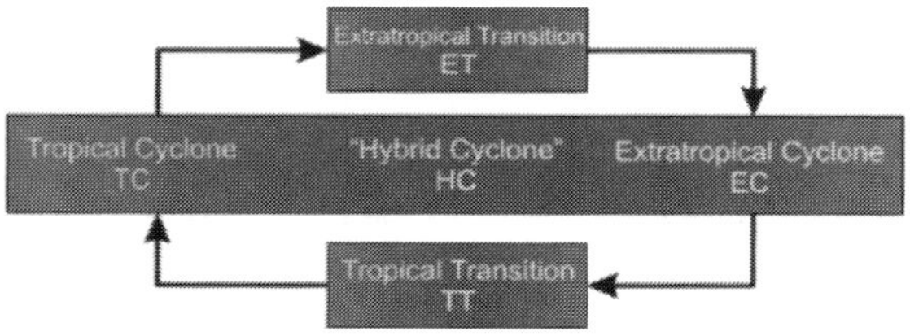

Figure 1. Cyclone Transition spectrum.

2. Data and Methodology

2.1. Data

Research concerning cyclone thermodynamics requires the need to resolve mesoscale features in order to obtain a better understanding of internal structures. However, at this stage the only datasets available of sufficient temporal duration to conduct this climatological study are the reanalysis products. Covering the data-rich period since 1989, the European Centres Medium-Range Weather Forecasts' ERA-Interim dataset was chosen as it provides a reasonable compromise between horizontal resolution (1.5°) and temporal resolution. This dataset is used in both the cyclone tracking and thermodynamic analysis. For details regarding the ERA-interim dataset see [13].

2.2. Methodology

2.2.1. Cyclone Detection

The identification and tracking of cyclone trajectories are derived using The Melbourne University Tracking Scheme (MUTS). MUTS is a proven method of locating and tracking cyclone (anticyclone) centres on a sphere [14, 15, 16, 17, 18] based on mean sea level pressure (MSLP) data. From this algorithm, many cyclone statistics are calculated which include cyclone intensity, cyclone depth and cyclone radius (*RD*). MUTS objectively defines *RD* as a weighted mean distance from the cyclone center to the points at which the Laplacian of the pressure ($\nabla^2 P$) is zero around the edge of the cyclone, thus providing an objective location of the cyclone's immediate environment.

2.2.2. Cyclone Transition Index

To objectively develop the transition climatology, an extension module of MUTS is needed which utilises analysis similar to that done by [1] and [3]. Initially all cyclone locations (x_{core}) are paired with the ERA-Interim temperature grids. Once this has been achieved, thermodynamic analysis of the cyclone core and the environment can begin.

Referring to Figure 2a, the extension module evaluates the temperature of the cyclone core (T_{core}) by performing a bilinear interpolation with the nearest 4 grid points. This step is then repeated on all environmental points ($x_{env\text{-}n}$), where n represents the total number of environmental sample points set at a distance *RD* from x_{core}. The ring of environmental points is averaged, representing the average environmental temperature ($\overline{T_{env}}$). Repeating the above steps, T_{core} and $\overline{T_{env}}$ are calculated at multiple levels (Fig. 2b). Care is taken when evaluating these values outside the 900 to 300 hPa range as to avoid extrapolation below ground or within the boundary layer, and to prevent inclusion of the stratospheric phase, which is often the opposite of the tropospheric phase [1].

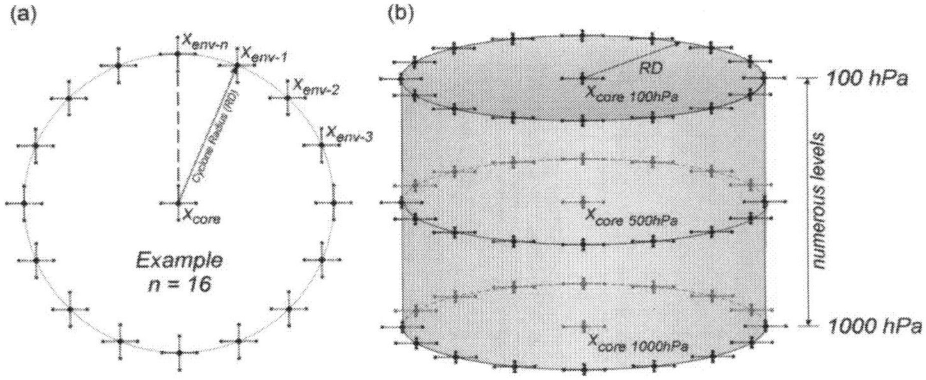

Figure 2. (a) Schematic of the cyclone core (x_{core}) and environmental ring ($x_{env\text{-}n}$) sample points set at a distance objectively determined by *RD*. (b) As in (a) computed over several levels from 1000 hPa to 100 hPa. Image not to scale.

Following equation (1), the cyclone transition index is then established by evaluating the temperature difference between T_{core} and $\overline{T_{env}}$. The two possible outcomes represent whether the cyclone is in a warm-core ($index > 0$) or a cold-core ($index < 0$) phase relative to the cyclone's immediate environment. By establishing a further index between the 900-600 and 600-300 hPa levels, the depth of the warm- or cold-core can be evaluated. For example, a deep warm-core would result from higher core temperatures in the upper troposphere between 600-300 hPa.

The final stage of this module is to map the cyclone tracks. This climatological analysis will reveal the location of "transition hotspots", and highlight potential cyclone transition trends for the SH. Once completed this module will form the foundations of other insightful analysis which will investigate the role large scale mechanisms and air-sea interactions have in the transition process.

$$T_{core} - \overline{T_{env}} = \begin{cases} > 0 & warm\ core \\ < 0 & cold\ core \end{cases} \tag{1}$$

2.2.3. Cyclone Radius and Detection Speed analysis

Figure 3 shows a histogram analysis which was conducted to gain insight into MUTS ability to detect cyclones of various sizes. 1025 cyclones were tracked from four selected months (March 2001, March

2004, July 2007 and July 2008), creating a database of 14021 cyclone radii. Systems with a life-cycle of less than 1 day were not considered. It can be seen that the average cyclone radius is 573.12 km with minimum and maximum cyclone radii of 199.31 and 1478.07 km respectively. It is clear that MUTS is adept at tracking a range of cyclone sizes.

Whilst testing the above methodology, an analysis into the computation speed was undertaken. Figure 4 shows an exponential trend as the number of environmental sample points (x_{env-n}) is increased. A further configuration used a dynamic *RD* sampling setting, where the number of x_{env-n} points was determined based on the *RD* (i.e. larger cyclones require more environmental sampling). Results showed that there was no performance improvement as these computation times reflected the fixed 64 point configuration (averaged 101.95 minutes/month). However, these computation times will decrease as improvements are made to the computer hardware.

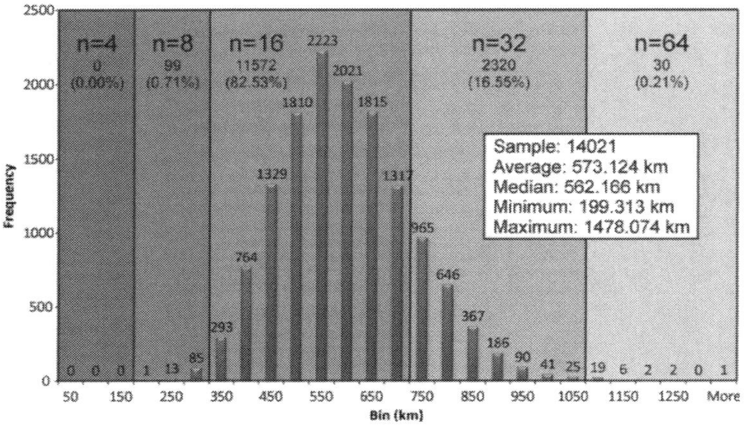

Figure 3. Cyclone radius histogram analyzing 1025 cyclone tracks corresponding to 14021 cyclone radii. The 5 regions shaded in gray represent the total number of environmental sample points (x_{env-n}) used for various cyclone radii. These categories are used when dynamic sampling is enabled.

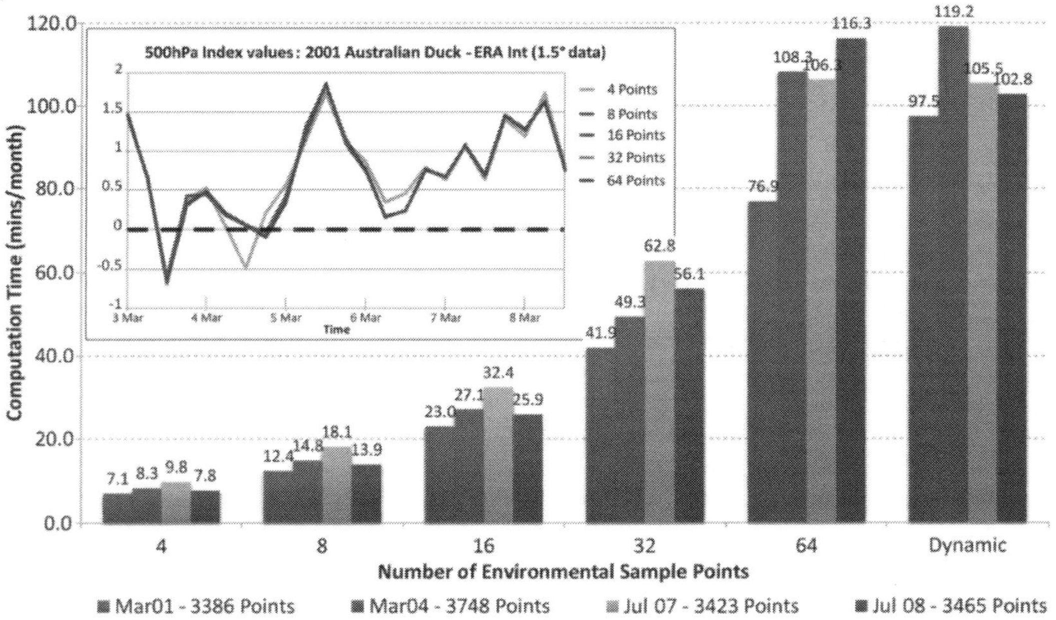

Figure 4. Computation Speed tests (minutes/month) analyzing 14021 cyclone radii (1025 cyclone tracks). Insert chart highlights the performance for each environmental point sample test for the 500hPa level (2001 Australian Duck case study).

IOP Conf. Series: Earth and Environmental Science 11 (2010) 012007 doi:10.1088/1755-1315/11/1/012007

The insert in Figure 4 illustrates a time series of the 500 hPa index values for the 2001 Australian Duck cyclone (see [19] for a detailed analysis regarding this cyclone). This time series highlights that similar index values are calculated when using 16, 32 or 64 environmental sample points, which is in contrast to 4 and 8 points. Trials concerning other cyclones also reflect these results. Moreover, Figure 3 highlights that of the 14021 cyclone points tracked during the 4 month study, 11572 (82.53%) required 16 point sampling, whilst only 30 (0.21%) cyclones required 64 point sampling.

3. Transition Index Results

Numerous cyclones were analysed to test the validity of this methodology. Figure 5 illustrates four transition index time series for the following cyclones: Typhoon David (1997), the First South Atlantic hurricane Catarina (2004), Severe TC Nicholas (2008) and a random subpolar low (2008). These systems are selected and presented here as they illustrate transition signatures for (a) ET, (b) TT, (c) a TC, and (d) an EC. Note that the horizontal dashed line about zero represents the boundary between the warm- and cold-core index. It can be seen that the amplitude of the transition signal varies from cyclone to cyclone.

In cases (a) and (b), key periods can be identified as the cyclone undergoes transition. The ET of Typhoon David is evident around September 19th. An abrupt index shift in all levels is detected as the mid to upper troposphere cyclone core becomes cooler (+13 to -6°C) than the environment. Furthermore, the TT of Catarina (b) is also evident as the cyclone's upper levels (300 and 200 hPa) become warm-core (+2.5°C) around March 22nd-23rd. The transition periods identified by this methodology for both David and Catarina are in agreement with other independent research [4, 20].

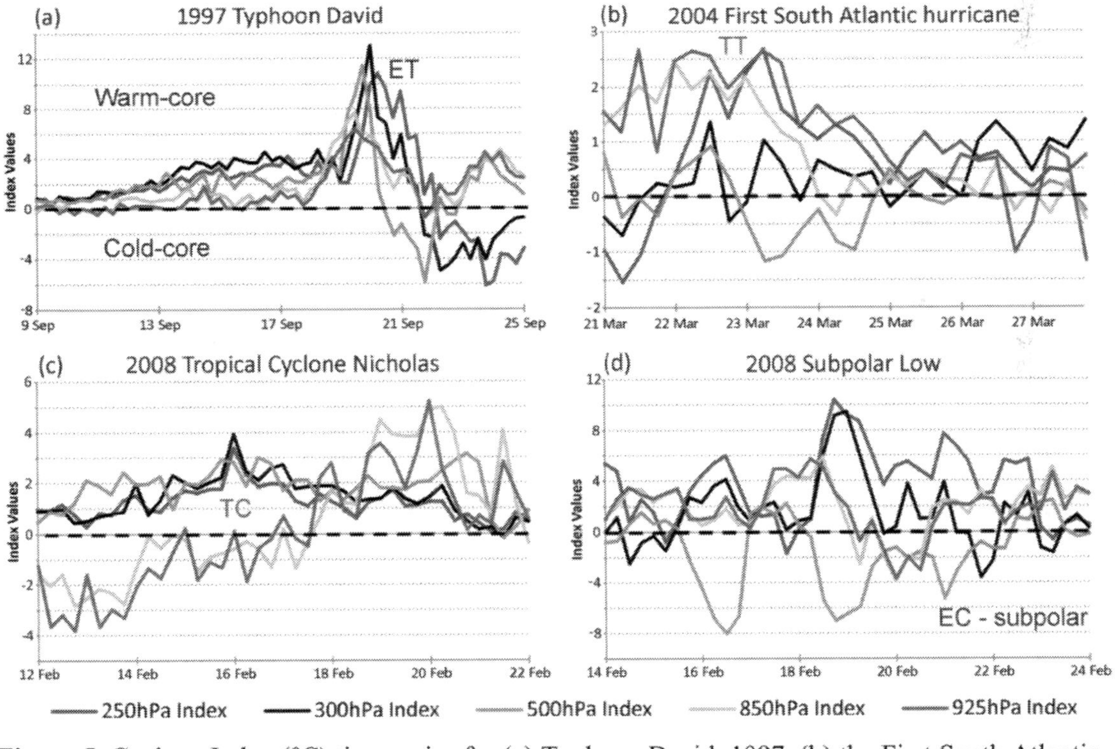

Figure 5. Cyclone Index (°C) time series for (a) Typhoon David, 1997, (b) the First South Atlantic hurricane Catarina, 2004, (c) Severe TC Nicholas, 2008 and a (d) subpolar low, 2008. Horizontal dashed line about zero represents the boundary between the warm and cold core index. 1.5° ERA-interim data used.

In addition, two "non-transitioning" systems are presented. In the case of Severe TC Nicholas (c), the upper level warm-core begins to weaken around February 18th, coinciding with station MSLP observations as the cyclone weakened and made landfall along the Western Australian coastline. The time series of the subpolar low (d) is more difficult to examine. This extremely high latitude system shows warming in the low and upper levels, and an oscillation in the mid 500 hPa level. Further analysis of the EC species is needed.

4. Conclusions and Further work

A method of objectively diagnosing cyclone transition has been presented here. Current research is focussed on developing a set of recognition criteria based on this transition index methodology to automatically classify cyclones into "non-transitioning" and "transitioning" groups. This will enable an improved assessment of the entire cyclone transition spectrum. Attention will also be given to analysing the development of HCs, which can undergo explosive development, posing a significant risk to the Australia-New Zealand region.

This research aims to place the development of transitioning cyclone in climatic perspective as understanding these systems is extremely important, particularly for the SH where many aspects of their formation are still unknown.

References

[1] Hart R E 2003 *Mon. Weather Rev.* **131** 585–616
[2] DiMego G J and Bosart L F 1982 *Mon. Weather Rev.* **110** 385–411
[3] Beven J L 1997 *Preprints, 22nd Conf. on Hurricanes and Tropical Meteorology* 645–46
[4] Klein P M, Harr P A and Elsberry R L 2000 *Weather and Forecasting* **15** 373–95
[5] Hart R E and Evans J L 2001 *J. Climate* **14** 546–64
[6] Sinclair M R 2002 *Mon. Weather Rev.* **130** 590–609
[7] Davis C A and Bosart L F 2003 *Mon. Weather Rev.* **131** 2730–47
[8] Jones S C, *et al* 2003 *Weather and Forecasting* **18** 1052–92
[9] Davis C A and Bosart L F 2004 *Bull. Amer. Meteorol. Soc.* **85** 1657–62
[10] Guishard M P, Nelson E A, Evans J L, Hart R E and O'Connell D G 2007 *Meteorol. and Atmos. Phys.* **97** 239–53
[11] Simpson R H 1952 *J. Meteorol.* **9** 24–35
[12] Garde L A, Pezza A B, Simmonds I and Davidson N E 2009 *Ninth Int. Conf. on Southern Hemisphere Meteorology and Oceanography,* http://www.bom.gov.au/events/9icshmo/ last checked in June 2010
[13] Berrisford P, Dee D, Fielding K, Fuentes M, Kallberg P, Kobayshi S and Uppala S 2009 *The ERA-Interim archive – Version 1.0 ERA report series,* http://www.ecmwf.int/publications/library/do/references/list/782009 last checked in June 2010
[14] Murray R J and Simmonds I 1991 *Aust. Meteorol. Mag.* **39** 155–66
[15] Murray R J and Simmonds I 1991 *Aust. Meteorol. Mag.* **39** 167–80
[16] Jones D A and Simmonds I 1993 *Clim. Dyn.* **9** 131–45
[17] Simmonds I and Keay K 2000 *J. Climate* **13** 550–61
[18] Lim E P and Simmonds I 2007 *J. Climate* **20** 2675–90
[19] Garde L A, Pezza A B and Bye J A T 2010 *Mon. Weather Rev.* **138** 2038–57
[20] Pezza A B and Simmonds I 2005 *Geo. Res. Lett.* **32** L15712, DOI:15710.11029/12005GL023390.

IOP Conf. Series: Earth and Environmental Science 11 (2010) 012008 doi:10.1088/1755-1315/11/1/012008

Onset, active and break periods of the Australian monsoon

Hakeem A Shaik[1] **and Samuel J Cleland**[2]

[1]Northern Territory Regional Office,
Australian Bureau of Meteorology,
PO Box 40050, NT Regional Office, Bureau of Meteorology, Casuarina
NT 0811, Darwin, Australia

[2] Bureau of Meteorology, Cape Grim BAPS, Smithton
Tasmania TAS 7300, Australia

E-mail: h.shaik@bom.gov.au

Abstract. Four operational techniques of monsoon monitoring the Australian monsoon at Darwin have been developed in the Darwin Regional Specialised Meteorological Centre. Two techniques used the rainfall only criteria and look into the onset of wet season rainfall/monsoon rainfall. The other two techniques are based purely on Darwin wind data. The data used for the study ranges from 14 to 21 years. The main purpose of the study is to develop near-real time monitoring tools for the Australian monsoon at Darwin.

The average date of onset of the monsoon ranges from 19 December to 30 December. The average date of monsoon onset is 28 December. In eleven out of twenty-one years the onset date remained within three days range between the two rainfall techniques, whereas it is eleven out of fourteen years between the wind techniques. The median number of active monsoon spells in a wet season is 3 for the rainfall techniques and 6 for the wind techniques. The average length of each active monsoon spell is around 4 days for all of the techniques.

The date of onset of the monsoon has shown negative correlation with the Southern Oscillation Index (SOI) that is late onset is found to occur in El Niño years while early onset is more likely in La Niña years.

1. Introduction

Tropical Australia has a monsoon climate with seasons of wet and dry weather. The Australian monsoon is often weaker than the Indian/Southeast Asian monsoon and the timing with respect to the seasonal cycle of rain and wind is less reliable. The weather over northern parts of Australia has a build-up period prior to widespread monsoonal rainfall.

Four techniques have been developed in the Darwin Regional Specialised Meteorological Centre to monitor the north Australian monsoon progress at Darwin. The techniques are based on the previous works such as Nicholls [1], Shaik and Bate [2], Holland [3] and Drosdowsky [4].

2. Data

Rainfall from seven locations in the Northern Territory (Australia) for the period 1987 to 2008 was used in the current study. Wind data was obtained from the Darwin Airport observing site for the period 1990 to 2008.

3. Monsoon onset monitoring techniques

3.1. Accumulated rainfall – Technique 1 (Accum rain)

The monsoon onset date is the date on which the accumulated rainfall amount starting from 1 September reached 15 percent of the mean annual station rainfall for all the seven stations used in the study [1].

3.2. Widespread rainfall – Technique 2 (Australian Monsoon Rainfall Index (AMRI)

The AMRI for a station is equal to 1 if the 5-day running mean daily rainfall exceeds 150% of the long-term daily mean. If not, the AMRI will be zero. The date of onset or active spell of the monsoon in a region will be the day when the sum of AMRI from the stations within the area is 5 or more. In the present study that will be at least five out of 7 stations report AMRI equal to one [2].

3.3. 850 and 200 hPa level winds at Darwin – Technique 3 (U wind)

The date of the onset or active spell of monsoon will be the first occurrence of $\geq$ 5 knots westerly winds at 850 hPa level, $\geq$ 5 knots easterlies at 200 hPa level and Darwin 3-day running mean daily rainfall is $\geq$ 0.2mm. A 3-day running mean was applied to 850-hPa level and 200 hPa level wind data from Darwin Airport starting from 1 November each year.

3.4. Deep-layer mean zonal wind – Technique 4 (DLM wind)

The date of the onset or active spell was defined as the first day of the first spell of westerly burst in the deep westerly stream that is greater than 8 knots and that lasts two or more consecutive days, overlain by upper level easterlies and Darwin 3-day running mean daily rainfall is $\geq$ 0.2mm. Pressure-weighted deep layer mean wind was calculated as per [4].

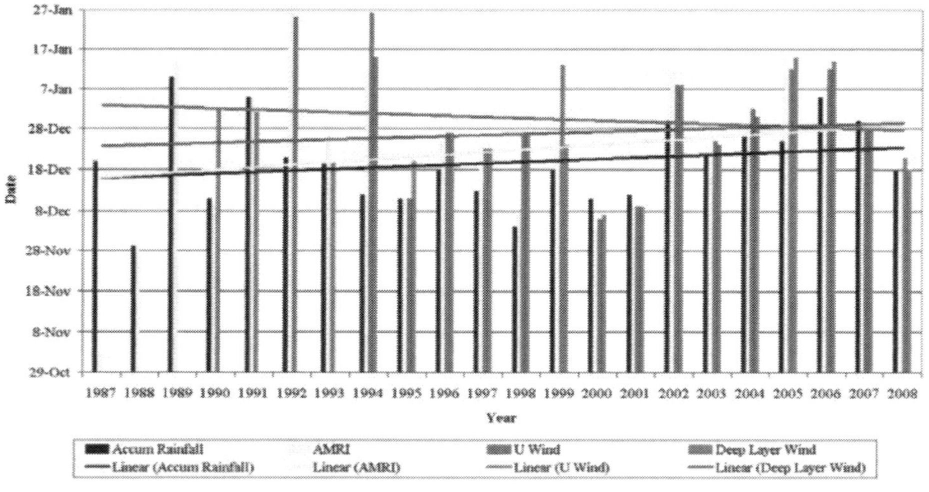

Figure 1. Dates of monsoon onset in different years from the four techniques.

4. Date of Onset

Figures 1 shows the onset dates for each year and for each technique. The onset date remained same for both wind techniques in 6 years and the onset date remained within three days range in eleven out of fourteen years. In rainfall techniques, out of twenty-one years the onset date was same in 3 years while in 10 years it was within 3 days range.

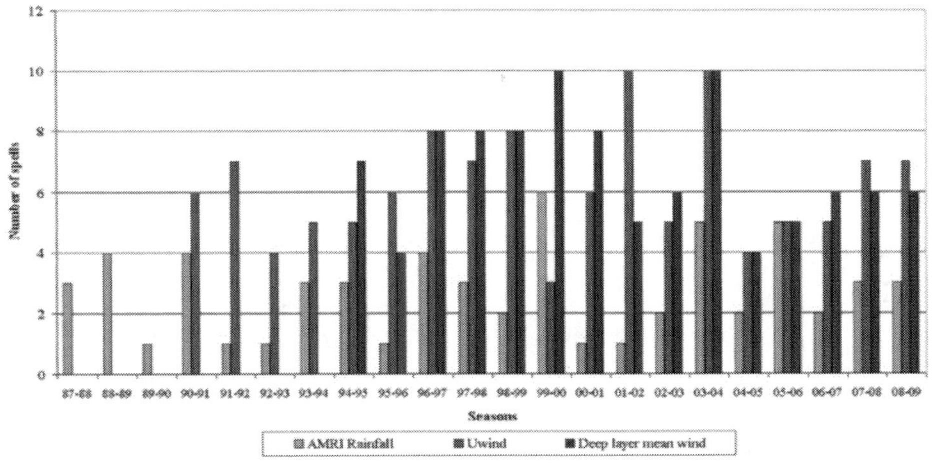

Figure 2. Number of active monsoon spells in each season.

The average and median date of onset with technique 1 was 19 December and 20th December. The average and median date of onset for technique 2 was 25 December. The dates of the onset analysed by technique 3 has the mean as well as the median date of onset was 30 December. The average and the median date of onset remained as 28 and 27 December respectively with the DLM wind technique. The difference in monsoon onset dates between the techniques is mostly due to the synoptic patterns of individual years.

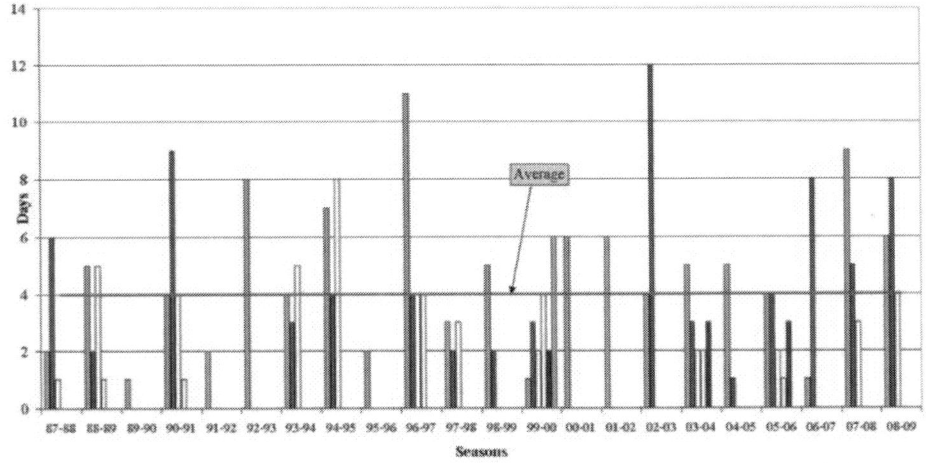

Figure 3. Length of each monsoon spell in each season for AMRI technique.

5. Active spells of monsoon

Figure 2 shows the number of active monsoon spells in each season for the three techniques. The number of active monsoon spells in a season ranged from 1 to 6 in AMRI (rainfall) technique whereas the wind techniques had 3 to 10 active spells each year. In some years the rainfall had only one active spell but the wind technique had 10 spells (2001-02 season). These years are dominated by dry monsoon periods where the wind pattern persists without much rainfall. Some of these years are prominent with short period wind bursts or not enough deep layer monsoon wind structure. The median number of active spells was 3 with the AMRI technique and for the wind techniques it was 6 spells.

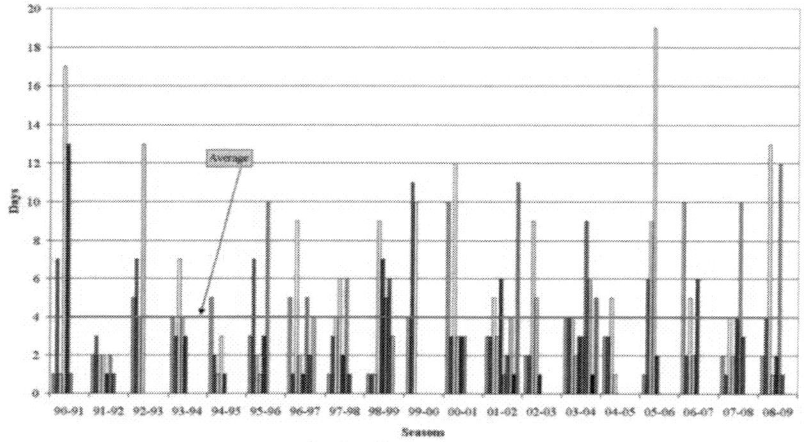

Figure 4. Length of each monsoon spell in each season for Uwind technique.

6. Length of active monsoon spell

Figures 3 to 5 show the length of each wet spell (in days) using different techniques. With AMRI the wet spell length ranged from one day to 12 days and the average spell length was 4 days. With the U wind technique the length of the maximum monsoon spell was 19 days though the average length of the spell was 4 same as of AMRI technique. With the DLM wind technique the average spell length 3.3 days was highly modulated by several one-day spells. Several short duration spells in the DLM wind technique indicate that the deep layer monsoon structure was not prevailing for a longer duration or the strength of the deep layer was not enough to sustain for longer periods.

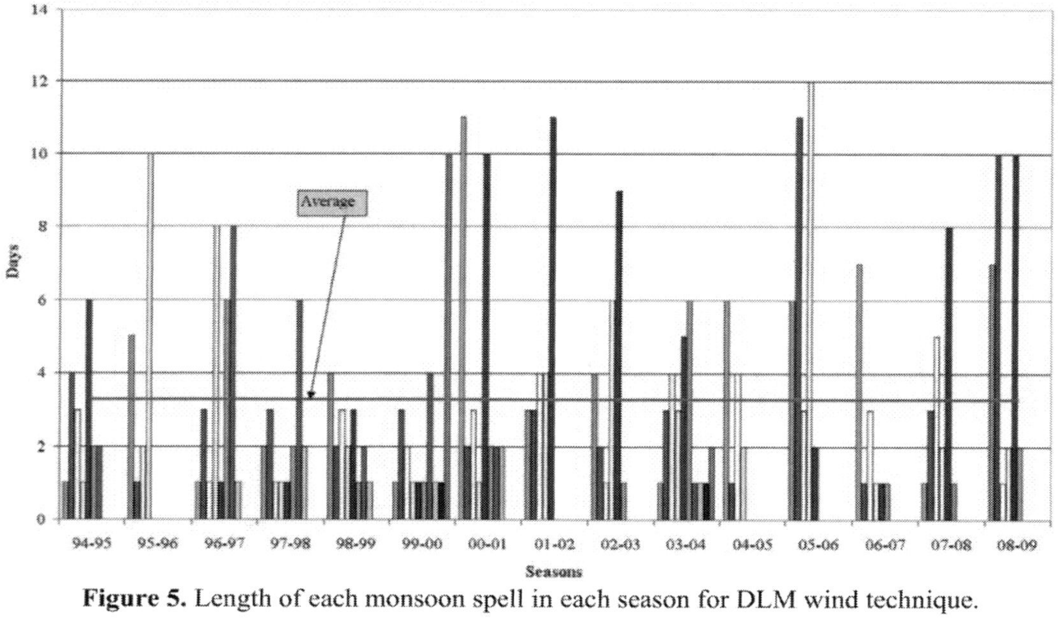

Figure 5. Length of each monsoon spell in each season for DLM wind technique.

7. Date of Monsoon onset and Southern Oscillation Index (SOI)

Previous studies indicate a strong relationship between the Australian monsoon and the ENSO (El Niño - Southern Oscillation (Meehl and Arblaster [5]). The relationship between monsoon onset and various types of SOI values is studied. The types of SOI include October SOI, November SOI, mean of September to November SOI, past month SOI, past 3 months mean and past 6 months mean to the onset month. For example, past 3 months SOI for an onset date in

January will be the mean of the October, November and December SOI. Figure 6 presents the relationship between the November SOI and the date of onset. Figure 7 presents the relationship between the September-November (3 month) mean SOI and the date of onset. The lines on the graph indicate the linear best fit (BF) for each technique. The correlation coefficients obtained are presented in figure 8.

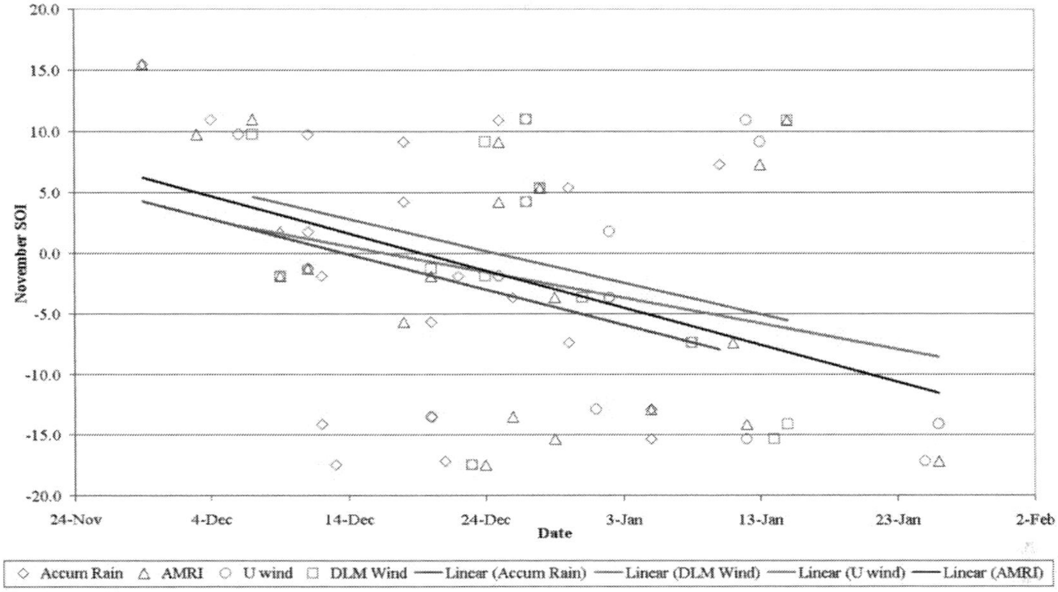

Figure 6. Relationship between November SOI and the monsoon onset.

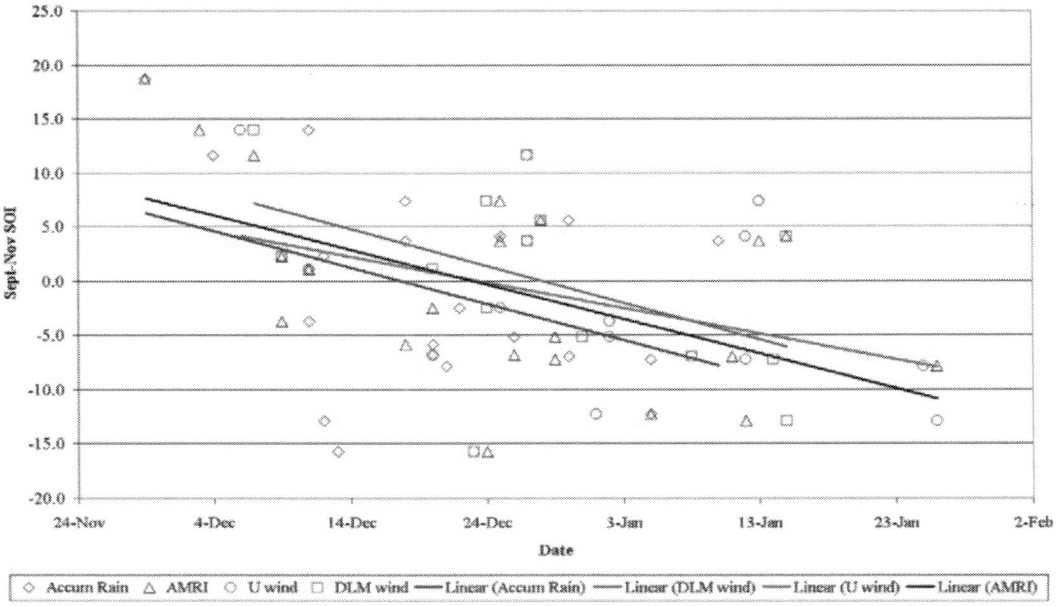

Figure 7. Relationship between September to November (3 month mean) SOI and the monsoon onset.

All of the three techniques have shown a general trend of late onset in El Niño years and early onset during La Niña events. The AMRI technique has shown a better correlation with the past three months SOI and also with the September to November SOI.

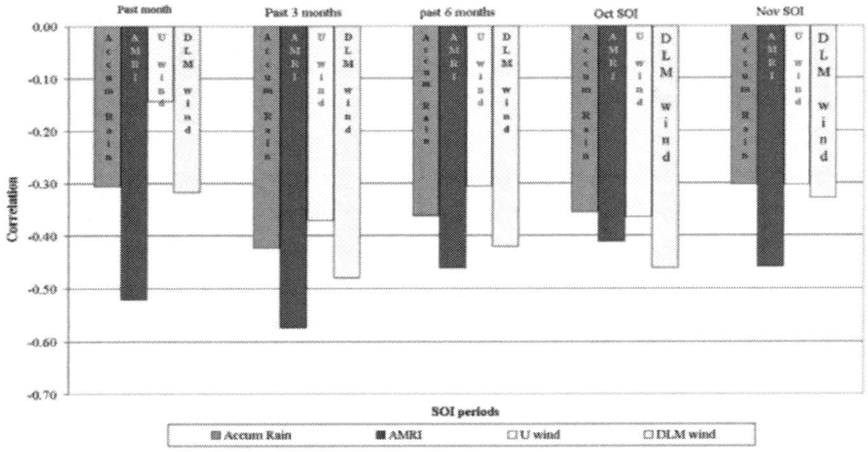

Figure 8. Correlation between SOI and onset dates.

8. Conclusions

The average date of onset of the monsoon ranges from 19 December to 30 December. The average onset date using only AMRI, U wind and DLM wind techniques is 28 December. In eleven out of twenty-one years the onset date remained within three days range between the two rainfall techniques. Eleven out of fourteen years the onset date remained within three days range in the wind techniques.

Out of the four techniques, the DLM wind and the AMRI techniques represent the monsoon onset better than the other techniques.

The median number of active monsoon spells in a wet season is 3 for the rainfall techniques and 6 for the wind techniques. The average length of each active monsoon spell is around 4 days for all of the techniques.

The date of onset of the monsoon has shown negative correlation with the Southern Oscillation Index (SOI), that is late onset is found to occur in El Niño years while early onset is more likely in La Niña years.

9. References:

[1] Nicholls N 1984 A system for Predicting the onset of the north Australian wet season *J. Climatol.* **4** 425-35

[2] Shaik A H and Bate P 2000 On the estimation of the date of the monsoon onset and withdrawal over Northern Australia *Australia New Zealand Climate forum, Hobart April 2000*

[3] Holland G J 1986 Interannual Variability of the Australian Summer Monsoon at Darwin:1952-8 *Mon.Wea.Rev.* **114** 594-604

[4] Drosdowsky W 1996 Variability of the Australian Summer Monsoon at Darwin: 1957–1992 *J. Climate* **9** 85-96

[5] Meehl G A and Arblaster J M 1998 The Asian-Australian monsoon and El Niño-Southern Oscillation in the NCAR Climate System Model *J. Climate* **11** 1356-85

IOP Conf. Series: Earth and Environmental Science **11** (2010) 012009 doi:10.1088/1755-1315/11/1/012009

Improving Weather Forecasts for Wind Energy Applications

Merlinde Kay[1] and Iain MacGill[2]

[1]School of Photovoltaic and Renewable Energy Engineering and Centre for Energy and Environmental Markets, University of New South Wales, Sydney, NSW 2052, Australia
[2]School of Electrical Engineering and Telecommunications, and Centre for Energy and Environmental Markets, University of New South Wales, Sydney, NSW 2052, Australia

E-mail: m.kay@unsw.edu.au

Abstract. Weather forecasts play an important role in the energy industry particularly because of the impact of temperature on electrical demand. Power system operation requires that this variable and somewhat unpredictable demand be precisely met at all times and locations from available generation. As wind generation makes up a growing component of electricity supply around the world, it has become increasingly important to be able to provide useful forecasting for this highly variable and uncertain energy resource. Of particular interest are forecasts of weather events that rapidly change wind energy production from one or more wind farms. In this paper we describe work underway to improve the wind forecasts currently available from standard Numerical Weather Prediction (NWP) through a bias correction methodology. Our study has used the Australian Bureau of Meteorology MesoLAPS 5 km limited domain model over the Victoria/Tasmania region, providing forecasts for the Woolnorth wind farm, situated in Tasmania, Australia. The accuracy of these forecasts has been investigated, concentrating on the key wind speed ranges 5 – 15 ms^{-1} and around 25 ms^{-1}. A bias correction methodology was applied to the NWP hourly forecasts to help account for systematic issues such as the NWP grid point not being at the exact location of the wind farm. An additional correction was applied for timing issues by using meteorological data from the wind farm. Results to date show a reduction in spread of forecast error for hour ahead forecasts by as much as half using this double correction methodology – a combination of both bias correction and timing correction.

1. Introduction

The last ten years has seen substantial growth in wind energy, to the extent that it has become an important component to commercial electricity generation around the globe [1]. Australia is blessed with extensive wind resources [2] and therefore well placed to pursue the energy security and climate change benefits of major wind deployment. For this to become a reality, however, there is a need to provide accurate, near-real-time, wind-focused forecasts for wind-farm sites. For many within the population, weather forecasts provide nothing more than the information needed to choose appropriate attire for the day. However, in the energy industry, forecasts play a vital role – accurate weather information and forecasts, updated on an ongoing basis, is central to managing supply demand balance within the electricity industry [3]. For example, on a hot summer afternoon that is followed by an evening cool change, electricity demand on the system changes dramatically as large numbers of air-conditioning units go from full capacity operation to an off-state. The industry must ensure that supply precisely meets demand at all times at all locations within the network, and key generation technologies such as coal and gas-fired plant have significant constraints on their operation – start up times and expense, minimum operating levels and ramp rates. Accurate forecasts of demand are vital contributors to industry decision making on the operation of these plants.

Demand side variability and unpredictability has been the main focus for forecasting within the electricity sector. However, as highly variable and somewhat unpredictable renewable generation sources such as wind achieve higher penetrations within power systems, it becomes increasingly important to provide useful forecasts of this generation. For wind farms, numerous weather phenomena have the ability to rapidly change wind conditions and hence affect wind-farm operation. Extreme wind events can cause power output for wind farms to suddenly drop from full rated output to zero power in a matter of minutes or even less. Even wind speed and direction variability within relatively normal weather conditions can drive considerable power generation variability, which provides challenges for maintaining secure and efficient power system operation. This paper investigates the wind speed regimes that are of importance to wind energy forecasts and concentrates on improving the hourly predictions of an extreme wind event by applying a form of bias correction methodology to a time series of 24 hour NWP forecasts. In Section 2 we present the case study for this work. A description of the methodology is provided in Section 3, and Section 4 investigates the results from applying the bias methodology.

2. Case study region and forecast model

The case study for this work has focused on the Woolnorth wind farm, situated on the north-western tip of Tasmania. It was developed in three stages across two nearby sites known as Bluff point and Studland Bay. The Bluff point site has a generation capacity of 64.75 MW and our study focuses on the Bluff point site alone. More details on the layout and specifics of the wind farm monitoring arrangements can be found in [4]. Measurements of wind speed and direction for our study are taken as the average wind speed from each of the 37 wind turbines (measured at the nacelles residing approximately 60m a.g.l.). All data is available at 10 minute intervals. The wind speed has been averaged over the preceding hour to be consistent with the NWP forecast hours. The NWP used for this analysis is the limited domain Vic/Tas 5 km model from the Australian Bureau of Meteorology (BoM) [5]. Two NWP grid points lie in the location of the wind farm as seen in Figure 1. All forecasts are made at 12UTC and are a 24 hour time series. We have concentrated on forecasts made at a height of 45m within the NWP because this is the grid height closest to the hub height of the wind turbine.

2.1 Important wind speed regimes for this site

The basic relationship between the wind's speed and its power, P, is non-linear (see Figure 2) and is a function of density, ρ, swept area of the turbine blades, A, and the wind speed, v, cubed, and is given by:

$$P = \frac{1}{2}\rho A v^3 \qquad (1)$$

Figure 2 is referred to as a power curve, and describes the electrical power you might expect from a turbine (or multiple turbines) at a particular wind speed over the wind farm site. Note that the curve for the wind farm represents the aggregated output of the 37 turbines on site. The actual wind farm power curve can be impacted by turbine availability and each wind turbine will see a slightly different wind speed from the others due to terrain, turbulence and interactions such as wake effects.

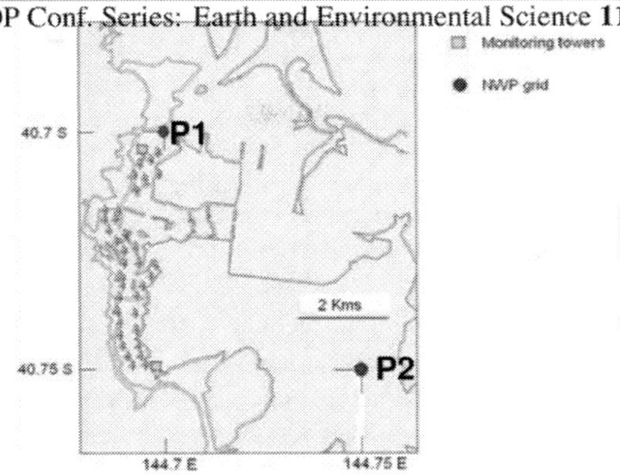

Figure 1. Layout of the 37 wind turbines at Bluff Point in North Western Tasmania and their physical relation to the two NWP grid points used in this study. The green squares represent the original north and south meteorological towers.

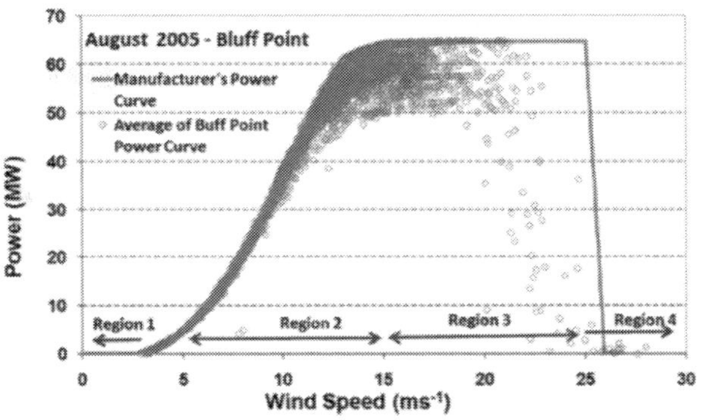

Figure 2. Power curve for August 2005 for the Woolnorth wind farm. The red line displays the manufacturers power curve for a single turbine (scaled to represent the 37 turbines from Figure 1).

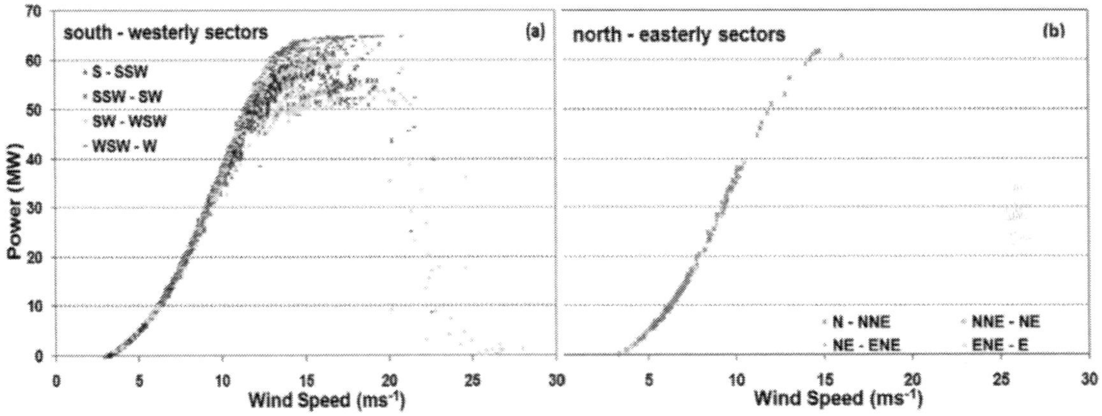

Figure 3. (a) Directional power curve for the Woolnorth wind farm, August 2005 for the south to westerly sector, (b) for the east to northerly sector.

The key weather-related issues identified from Figure 2 for wind power generation are: Region 1 – wind speeds less than 3 ms^{-1} (between 3 – 5 ms^{-1} is termed the cut-in wind speed); Region 2 – wind speeds between cut-in to rated (typically 5 – 15 ms^{-1}); Region 3 – is where generator output does not vary from rated output (typically 15 – 25 ms^{-1}) and Region 4 – wind speeds greater than 25 ms^{-1} where power output is rapidly curtailed as these wind speeds are sufficiently high that the turbine needs to be shut down to protect itself from possible damage. We can immediately identify that generation variability and predictability challenges can arise in region 2 due to the highly non-linear relationship between wind speed and power output, and for the boundary between region 3 and 4 (note the manufacturers power curve) where the units may go from full output to zero with only a small change in wind speed. Notably, even small changes in wind speed within this region can lead to large changes in wind power output. As Figure 2 represents the average of the 37 turbines the sudden drop in power output occurs more in region 3 than the boundary of region 3 and 4. Figure 3(a) and (b) display the directional power curve for the south to westerly sectors and north to easterly sectors respectively for Bluff Point and illustrate how direction dependant the wind speeds are. Strong winds and much of the longer-term average energy output for the wind farm comes when the wind is from the south to the westerly sectors, whereas winds from the north to easterly sectors are less common and typically have lower speeds. For our results we have therefore chosen to concentrate on wind speeds that fall between cut-in to rated (region 2) and the boundary of region 3 and 4.

3. Methodology to improve key wind speed regimes

One of the main errors found in NWP forecasts is bias [6]. Bias causes the mean of the historical distribution to deviate from the true value (i.e., produce readings that systematically over-predict or under-predict). Bias correction utilises the knowledge of the bias from past forecasts by correcting for the under-prediction or over-prediction seen in a previous forecast. This is achieved by adding or removing the bias factor to the next forecast. In developing a bias methodology one must consider issues including choice of bias method, the historical data sample size (bias window), and what parameter will best represent the centre of the historical distribution. The bias parameter chosen for our study is the trimean or Best Easy Systematic (BES) estimator [7, 8].

$$BES = (Q1 + 2Q2 + Q3)/4 \quad (2)$$

where Q1, Q2 and Q3 are the 25%, 50% and 75% quartiles respectively.

The quartiles statistically divide the distribution into the lowest 25% of the data, the median and the highest 25%. The BES is a measure of the centre of a distribution. It is calculated as a weighted average of the distribution's quartiles and median, weighting the median more heavily. An advantage of using the BES is that it is more resistant to fluctuations within a historical data sample [8].

Once the bias parameter has been chosen, it is necessary to determine the optimum bias window. The historical sample or bias window can affect the accuracy or improvement of the new forecast [6]. The BES is then added or subtracted from the next hour's original forecast to produce a bias Corrected Prediction (CP).

NWP models have not generally been developed with a view to making the detailed wind predictions required for wind energy forecasting purposes, and the resolution of the models is often too coarse to predict the fine scale winds that a wind farm may experience [2]. The NWP also typically produces forecasts hourly, which often can miss the fine scale changes that occur when wind speed and direction change extremely quickly. Another problem that occurs in using standard NWP models is that the forecast grid point is not at the exact location of the wind farm. These factors can lead to less than ideal forecasts, and therefore after the initial bias correction, an extra correction (referred to in the text below as a Double Corrected Prediction – DCP) was also made to the bias CP to improve not just the accuracy but also the timing and location errors associated with the forecast. The correction simply makes use of a weighted observation from the wind farm taken 10 minutes before the hour and combines that with a weighted CP forecast to implement a smoothing process to take into account any errors that may be associated with the forecast timing and sudden wind speed changes. This smoothing function relies on the smoothing factor α to control the closeness of the most recent observation and the CP forecast:

$$DCP = \alpha O_t + (1 - \alpha)CP \quad (3)$$

where $0<\alpha<1$. DCP is the new prediction, and the observations from the wind farm are represented by O_t. Various smoothing factors were investigated and it was found that setting $\alpha = 0.5$, evenly weighting the observation and CP, provided results that best smoothed out timing errors and large errors in the forecast, but also worked well when the forecast errors were small.

4. Results

Sudden and extreme wind events can rapidly change the power production of a wind farm from operating at full capacity to no power in a matter of minutes. Such an event occurred at the Woolnorth wind farm during the 30-31 August 2005. Figure 4, shows the wind speed observations – blue line, NWP prediction from a 12UTC forecast – red line and corresponding bias corrected predictions, CP and DCP for both 48 hour and 7 day bias windows for the Bluff point wind-farm for the extreme wind event that occurred. A range of bias windows were investigated with results indicating that increasing the bias window past 48 hours made a minimal difference to the results. All predictions and measurements have been hourly averaged for consistency with the NWP. The important features that the hourly data illustrates are the prolonged high wind speeds which exceeded 25 ms^{-1}, when the wind turbines shut-down to prevent them from being damaged. This in turn caused a sudden and complete drop in power output around 00:00 UTC on the 31 August. The meteorological features causing this extreme event were a deepening low pressure system, accompanied by a strong front and trough following closely behind.

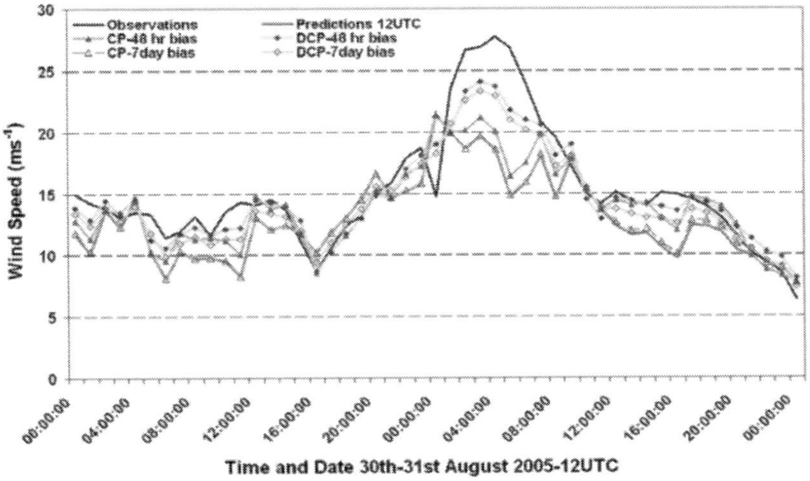

Figure 4. Observations for the Woolnorth wind farm, predictions and bias corrected forecasts (made with a 48 hour and 7 day bias window) at 12 UTC for the 30 - 31 August 2005.

The NWP model mis-timed the increase in wind speed, and also the duration of the high wind event along with under-predicting the wind speed for a majority of the time over the 48 hour period. The 48 hour bias window performs better overall, and making use of the wind farm observations for the DCP predictions has smoothed some of the timing problems that occurred. The major dip in wind speed before the sudden increase at 0:00 UTC on the 31 August has eliminated the 'out of phase' error that originally occurred with the NWP predictions, however, the smoothed correction does not fully capture the wind speed dip. There are however noticeable improvements in the magnitude, and timing of other wind speed changes. On average the DCP predictions improved the original predictions by 2 ms^{-1} during the 48 hour period, with maximum improvements of up to 7 ms^{-1} during the peak of the high wind speed event.

To quantify how the bias forecasts have performed, the error statistics for the original predictions and DCP are shown in Figure 5 (a) and (b) respectively. Results quantifying the bias forecast performance have concentrated on the wind speed ranges 5 – 15 ms^{-1}. This wind speed range is important because small changes in wind speed within this range can cause large changes in power output due to the highly non-linear relationship between wind speed and power noted earlier. The error statistics shown are for a more extensive time series, covering Summer (DJF) 2005/2006. A spread of values outside the wind speed range on the x-axis indicates an incorrect prediction, and the y-axis indicates whether it was a positive error (under- prediction) or negative error (over-prediction) that occurred. By applying the bias correction we see a reduction in the spread of errors from the original prediction and this is especially evident for the DCP prediction. We have successfully managed to reduce the spread of forecast error in the y-axis by half when implementing the DCP correction. This is particularly important for periods where forecasts over predicted wind speeds greater than 15ms^{-1} as this would erroneously indicate to the wind farm operators that the wind farm should be at full power when it actually is not.

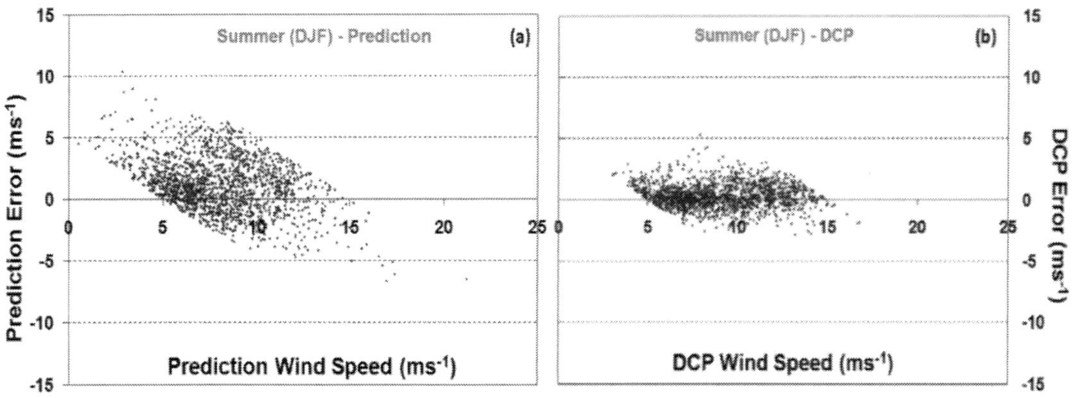

Figure 5. Error statistics for (a) the NWP predictions, and (b) DCP Summer 2005/2006 in the wind speed ranges 5 – 15 ms^{-1}. All predictions made at 12 UTC.

5. Conclusion

The ability to accurately forecast the power output of a wind-farm can greatly aid power system operators and market participants in ensuring a reliable and secure electricity supply to the consumer. This work has investigated the application of a correction to standard NWP model forecasts, making use of actual wind farm measurements to improve these wind speed forecasts in terms of their accuracy and, to some extent, timing. Often these forecast biases can be quite location dependent, and in terms of high wind speed ranges, there is often a systematic under prediction of wind speed events. Initial results from applying the bias correction methodology to an extreme wind event have proved promising for hourly forecasts. Forecasts for the wind energy sector are typically made for the short to medium term, usually up to 72 hours ahead. Other timescales may also be required depending on the power system requirements. A continuation of this work will aim to apply the methodology for longer forecast periods and include a correction for wind direction. As the bias will vary seasonally, longer time scales and seasons will also need to be validated.

References

[1] Sanchez I 2006 Short-term prediction of wind energy production *Int. J of Forecasting* **22** 43-65

[2] Archer C L and Jacobson M Z 2005 Evaluation of global wind power *J. Geophys. Res* **110** *D1211, doi:10.1029/2004JD005462*

[3] Feinberg E A and Genethliou D 2005 Load Forecasting *Applied Mathematics for Restructured Electric Power Systems: Optimization, Control, and Computational Intelligence* ed J H Chow, F F Wu and J J Momoh (Berlin: Springer) pp 269-285

[4] Kay M J, Cutler N, Micolich A, MacGill I and Outhred H 2009 Emerging Challenges in Wind Energy Forecasting for Australia *Aust. Meteor. and Ocean. Journal,* **58(2)** 99-106

[5] Puri K, Dietachmayer G, Mills G, Davidson N, Bowen R and Logan L 1998 The new BMRC Limited Area Prediction System, LAPS *Aust. Meteor. Mag.* **47(3)** 203-223

[6] Woodcock F and Engel C 2005 Operational Consensus Forecasts *Wea. and Forecasting* **20** 101-111

[7] Wonnacott T H and Wonnacott R J 1977 *Introductory Statistics* (New York:Wiley) p 650

[8] Tukey J W 1977 *Exploratory Data Analysis* (Reading MA : Addison-Wesley) p 688

IOP Conf. Series: Earth and Environmental Science **11** (2010) 012010 doi:10.1088/1755-1315/11/1/012010

Extreme inflow events and synoptic forcing in Sydney catchments

Acacia S Pepler and Clinton S Rakich

NSW Climate Services Section, Bureau of Meteorology
P.O. Box 413, Darlinghurst, NSW 1300

Email: a.pepler@bom.gov.au

Abstract. The Sydney catchment region encompasses over 16,000km^2, supplying water to over 4 million inhabitants. However, few studies have investigated the synoptic and climatic influences on inflow in this region, which are crucial for understanding the vulnerability of water supply in a changing climate. This study identifies extremely high and low inflow events between 1960 and 2008 based on catchment averages. The focus of the study is an analysis of the synoptic cause/s of each extreme inflow event. The events are evaluated to identify any trends and also to determine the concurrent significant climatic influences on rainfall over the catchments. Relationships between catchment inflow, rainfall, tropical SST indices, and other influencing factors such as observed wind and temperatures are investigated. Our results show that East Coast Lows and anomalously easterly flow are the drivers of high inflow events, with low inflow events dominated by westerly wind patterns and the El Niño-Southern Oscillation.

1. Introduction

The Sydney catchment region covers 16248km^2, encompassing the major Shoalhaven (5820km^2) and Warragamba (9850km^2) catchments and five minor catchments in the Hawkesbury-Nepean and Woronora systems. It extends from near Cooma to Lithgow, and is capable of holding 2,608,730ML and supplying water to more than 4 million people. Consequently, understanding the nature and trends in rainfall in this area is crucial to understanding the vulnerability of future water supply for the Sydney Metropolitan region. The region experiences on average 1100mm of rainfall per year, with average inflows of 2740GL; however, this varies significantly between years. Although rainfall is highest in summer, extreme rain events and overall inflow is concentrated in winter months (Table 1).

Table 1. Seasonal rainfall and inflow averages in Sydney catchment region

Season	Summer	Autumn	Winter	Spring
Average rainfall (mm)	314	268	239	287
Average inflow (GL)	595	688	895	560
Days >= 10mm	9.6	7.3	6.0	8.7
Days >= 50mm	0.4	0.5	0.6	0.3

Very few studies have investigated the synoptic and climatic influences on inflow in the Sydney region, most focusing on the state as a whole. On a synoptic level, the most pertinent studies are by Abbs et al. [1,2], which examined the synoptic typing of extreme rainfall events in the Central Coast and Upper Parramatta River areas. These studies identified East Coast Lows (ECLs), intense cyclonic systems off the coast of NSW, to be responsible for 78% of high rainfall events in Winter and 29% in Summer, with the remainder caused by synoptic patterns bringing onshore easterly windflow. These results are validated by numerous studies, with ECLs found responsible for ~7 significant and ~2 heavy rain events on the NSW coast p.a. [3] and approximately 7% of Australian disasters [4]. Onshore easterly flow is also known to have a strong impact on NSW rainfall, with a strong relationship found between anomalous easterly flow and coastal rainfall [5,6]. Statistically significant correlations also exist between Sydney area rainfall and anticyclonicity south of Tasmania in all seasons [7], a common source of such wind patterns.

On a seasonal basis, the El Niño-Southern Oscillation (ENSO) is the strongest influence on rainfall and runoff in NSW, particularly during winter and spring. A 50-100% increase in rainfall and over 200% increase in streamflow is found for many locations during La Niña years [8], including in the Hunter Valley, just north of Sydney [9,10]. However, this influence may be weaker along the east coast, with some studies finding lower or absent correlations [11]. This is particularly true for warm events, as the relationship with La Niña is more consistent [10,12]. Other indices have also been found to impact NSW rainfall, including the Indian Ocean Dipole (IOD) and the Southern Annular Mode (SAM); however, studies also show the effects of these reduced in coastal regions [13,14]. Due to the poor relationships of climatic indices, therefore, and the capacity for isolated events to cause extremely high rainfall and inflow, it becomes necessary to look at the synoptic systems responsible for inflow and also the lack of inflow.

2. Data

24-hour (9am to 9am) inflow and catchment averaged rainfall data were supplied by the Sydney Catchment Authority for the period January 1960 – April 2008, with the total inflow and area-weighted average determined for all catchments. For Section 3, daily total inflow data was sorted by magnitude, with inflow above 100,000ML identified as 'high' inflow. 119 days in the 48 year period exceeded this amount. These high inflow days were grouped when necessary to identify 41 contiguous high-inflow events, of lengths between 1 and 8 days. For Section 5, total inflow and rainfall were analysed on a seasonal basis to determine the fifteen lowest seasonal inflow totals, ranging between 19 and 80GL.

Various climatic influences were examined, including the El Niño-Southern Oscillation (ENSO), Southern Annular Mode (SAM), Indian Ocean Dipole (IOD) and the Gayndah-Deniliquin index [5], (GDI). Daily and seasonal mean sea level pressure (MSLP) anomalies, 500hPa geopotential height, seasonal precipitable water anomalies, and vector wind data were retrieved from the NCEP reanalysis dataset [15,16]. Charts of daily, seasonal and annual rainfall totals, anomalies and deciles were retrieved from the Australian Water Availability Project [17]. These were supplemented by synoptic information available from the Australian Bureau of Meteorology's (BoM's) Maritime Low Database [3] and BoM Monthly Weather Reviews.

3. High Inflow Events

The majority of high inflow events occurred during Autumn and Winter, comprising 29% and 53% of high inflow days respectively, although events occurred in every month except December. These events showed strong interdecadal patterns, with clusters of many events interspersed with periods of none (Figure 1). However, in the 16 years 1993-2008 there have been only three high inflow events in two months, well below the average of 8.5 events per decade. Interestingly, ten of the fourteen events since 1980, and six of the seven events since 1990, have occurred during winter, but this is true for only nine of the twenty seven events prior to 1980, which were more evenly distributed between seasons. This suggests that the decrease in frequency of events may be largely due to a decrease in autumn and spring events.

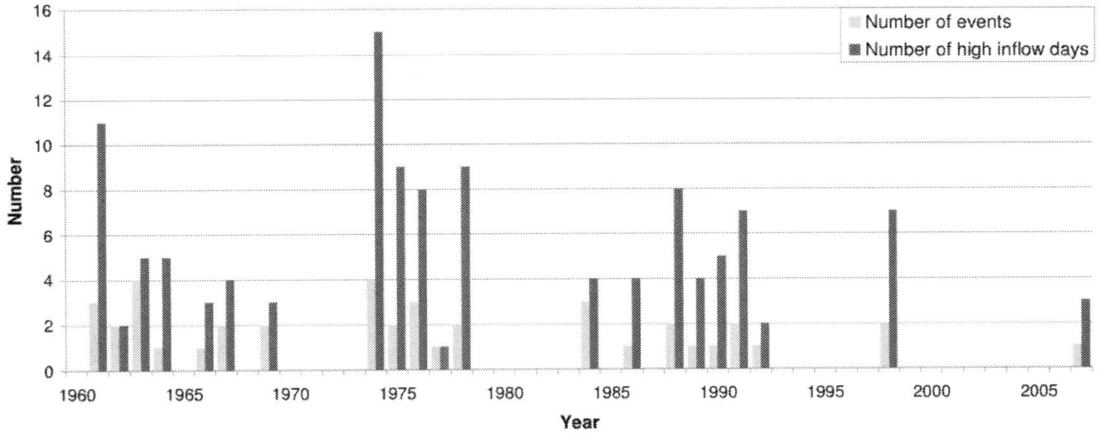

Figure 1. Annual frequency of days >100,000ML and contiguous low inflow events.

No clear relationships were seen between heavy inflow events and SAM or IOD, both of which were evenly distributed between positive and negative values. Between October and March, however, there seems to be a strong relationship with the GDI, as only three of fourteen events had slightly negative anomalies. Of the 41 events, 27 also occurred during months of positive SOI, indicating an increased potential for heavy inflow events when SOI is high. There are distinct exceptions, however, including one event occurring with an SOI of -20 and two events during El Niño year 1991. No relationship is of sufficient strength to enable any predictive skill, with synoptic typing proving more informative. The events are therefore subclassified into the two major synoptic types associated with them, either ECLs or onshore flow, confirming the studies by Abbs et al. ([1,2]).

East Coast Lows were found to be responsible for 61% of events and 66% of high inflow days, and occurred predominantly between May and August, where they formed 19 of the 22 events. They were also disproportionately responsible for the most extreme events, causing the four highest daily totals, seven of the nine events of five or more days, and thirteen of the seventeen events with maximum inflow exceeding 200,000ML. The majority of these fell into the types Easterly Trough and Inland Trough [3], with the remaining three consisting of two secondary lows and one decaying tropical cyclone. Interestingly, East Coast Lows formed 10 of the 14 events since 1980 and 6 of the 7 since 1990, compared to only 15 of the 27 events between 1960 and 1979. This suggests that much of the decline in events is due to a decrease in onshore events [3]. ECLs are also responsible for the two major dam filling events since 1992, in August 1998 and June 2007. These recharged Sydney dam levels by 40% and 20% respectively, making such events critically important for water supply.

Onshore flow events formed the remainder of the database. These events occurred predominantly during September-April and were associated with easterly flow and positive monthly GDI anomalies. These were divided into three subcategories by type, with an additional event associated with windflow from a decaying tropical cyclone in SE QLD:

1. Six Inland Low events, with a surface low over central NSW and high pressure to the southeast, usually associated with an upper cold pool. This category included all the significant non-ECL events, including two events with peak inflows above 400,000ML and duration of at least four days.
2. Four Inland Trough events, featuring high pressure to the southeast, a trough over central to eastern NSW and an upper cold pool directing easterly to north-easterly flow, which

failed to develop closed circulation. These formed only small events, with maximum inflows below 200,000ML.

3. Five Tasman High events, featuring a strong anticyclonic circulation in the Bass Strait or Tasman Sea, directing north-easterly to south-easterly winds. As with category 2 events, these only caused smaller events.

4. Low Inflow Events

In comparison, the majority of low inflow events occurred during Spring and Summer, with only two recorded in Autumn and three in Winter. Seasonal inflow also showed similar decadal patterns to the frequency of high inflow events (Figure 2), with much more significant inflow decrease for the equivalent trend in rainfall. This decrease is increasingly evident in recent years, with twenty of the lowest fifty events occurring since 1992, and four of the fifteen lowest events occurring since 2002. Only two seasons have occurred since 1992 with total inflows above 1000GL.

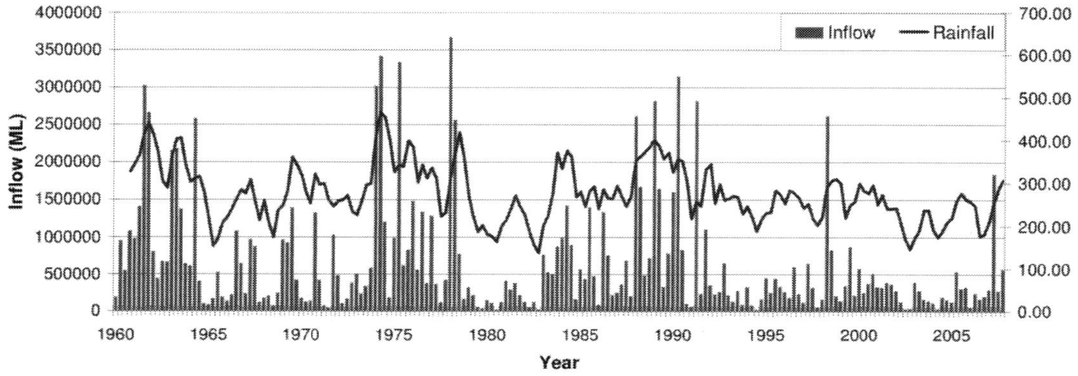

Figure 2. Seasonal inflow and 4-period moving average of seasonal rainfall in Sydney Catchments

As expected, there is a strong relationship between rainfall and inflow. Only three events did not have 3-month or 12-month rainfall in the lowest 10% of all seasons, two of which followed a preceding low inflow event. However, this relationship is nonlinear, with only two of the six seasons with rainfall below 100mm also among the lowest inflow events. Low inflow could also result from only moderate rain decreases, with rainfall as high as 244mm in Summer 2002/3, only 70mm below average. The link to rainfall was strongest in winter and autumn, where all events had three-monthly rainfall in the lowest decile.

Relationships with climatic indices are variable, but are stronger than for high inflow events. The majority of events (12/15) occurred with negative SOI, indicating a strong ENSO influence; however, only 8 occurred during El Niño years. No clear signal is seen with SAM over all events since 1979. However, four of the five spring events are negative, expected with decreased rainfall. Nine of the fifteen events also have negative monthly GDI, including eight of the ten events in spring/summer. Interestingly, eight of the fifteen events occurred under positive IOD, which would be expected to enhance rainfall. This is consistent seasonally, and indicates this index has little validity in coastal NSW. However, climatic indices are clearly insufficient to explain all variability, as all indices were conducive to high rainfall in Winter 1971, but only 105mm of rain was recorded.

Performing synoptic typing for extended dry periods is problematic, as the underlying pressure patterns vary significantly between seasons. In particular, no consistent trend can be seen for synoptic features in autumn and winter due to the small dataset for such seasons, with rainfall showing a stronger influence than pressure features. However, it is possible to broadly define the features of low inflow seasons in Summer and Spring. The majority of summer events occurred in El Niño years, with high pressure anomalies over the continent and anomalously hot, dry westerly winds. In comparison, spring events, and the remaining summer events, tend to have strong negative pressure anomalies to the south of the continent. These patterns also bring westerly wind anomalies, leading to a suppression of the easterly winds that bring the majority of the rain in these seasons (Figure 3).

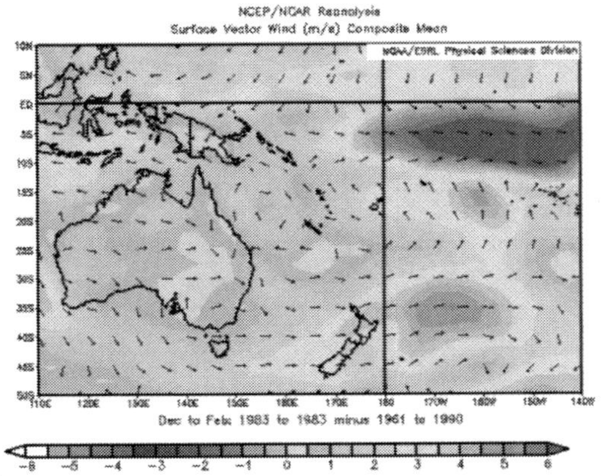

a)

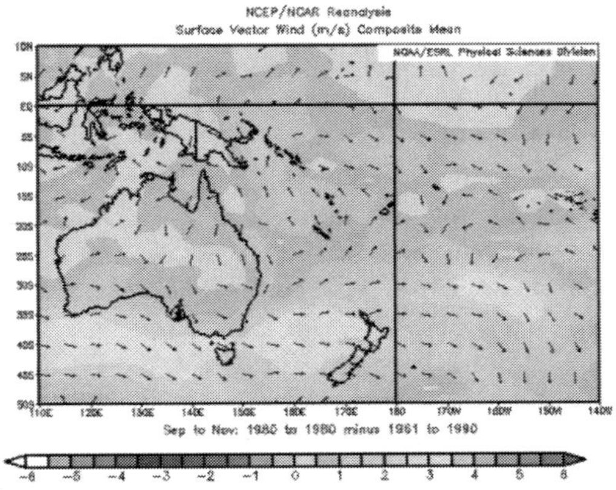

b)

Figure 3. Surface wind anomalies (m/s) in low inflow seasons: a) Summer 1982/83, b) Spring 1980

In a surprising outcome, 13 of the events had maximum temperatures over the catchment in the highest 30% of years, and 8 in the highest 10% of years, as strong a relationship as seen for rainfall. This was particularly evident in the warmer months, with the high temperatures potentially enhancing the impact of only moderately decreased rainfalls via impacts on evapotranspiration. Interestingly, the one event with negative temperature anomalies, Winter 1971, has both one of the lowest rainfalls and the highest inflow of the low events. Comparatively, Winter 2004, with similar rain but some of the highest temperatures on record, has the fifth lowest inflow on record. However, as high temperatures may also be a consequence of the clear synoptic patterns leading to reduced rain, more investigation of this relationship is crucial, with potentially significant implications for future water availability in a warming climate.

5. Conclusions

An investigation of high and low inflow events has confirmed the results of many previous studies [8,9], finding a decrease in inflow in Sydney Catchments since 1992 far larger than the equivalent decreases in rainfall. A large portion of this decline appears to be related to decreases in onshore flow events in spring and autumn. Results from studies by Abbs et al. [1,2] and others have been confirmed, with East Coast Lows found to be the most critical weather system for water supply in Sydney. This is particularly evident over recent decades. Indices such as the SOI and SAM may have some impact on inflows, although no influence of IOD is noted. This climatological influence is more significant for low inflow periods than isolated high inflow events. However, the largest impact appears to be from increased westerly wind flow and reduced winds, represented by negative GDI anomalies. This appears to be enhanced by high temperatures, which increase evapotranspiration and exacerbate the impact of dry seasons.

Acknowledgements

This work was funded by the NSW Department of Water and Energy

References

[1] Abbs D, Aryal S, Campbell E, McGregor J, Nguyen K., Palmer M, Rafter A,Watterson I and Bates B 2006 *Projections of Extreme Rainfall and Cyclones: Final Report to the Australian Greenhouse Office* (Aspendale, Victoria: CSIRO Marine and Atmospheric Research.)

[2] Abbs D and Rafter T 2008 *Impact of Climate Variability and Climate Change on Rainfall Extremes in Western Sydney and Surrounding Areas: Component 4 - Dynamical Downscaling: Report to the Sydney Metro Catchment Management Authority and Partners* (Aspendale, Victoria: CSIRO Marine and Atmospheric Research.)

[3] Speer M S, Wiles P and Pepler A 2009 *Aust. Met. Mag.* **58** 29-39

[4] Hopkins L C and Holland G J 1997 *J. Clim.* **10** 621-35

[5] Rakich C S, Holbrook N J and Timbal B 2008 *Geophys. Res. Lett.* **35** L08713

[6] Speer M S 2008 *Atmos. Sci. Lett.* **9** 160-70

[7] Nowak H and Leighton R M 1997 *Aust. Met. Mag.* **46** 267-76

[8] Verdon D C, Wyatt A M, Kiem A S and Franks S W 2004 *Water Res. Res.* **40** W10201

[9] Wooldridge S A, Franks S W and Kalma J D 2001 *Hydrol. Sci.* **46** 73-88

[10] Kiem A and Franks S 2001 *Hydrol. Sci.* **46** 715-27

[11] McBride J L, Nicholls N 1983 *Mon. Wea. Rev.* **111** 1998-2004

[12] Chiew F H S, Piechota T C, Dracup J A and McMahon T A 1998 *J. Hydrol.* **204** 138-49

[13] Meyers G, McIntosh P, Pigot L and Pook M 2007 *J. Clim.* **20** 2872-80

[14] Meneghini B, Simmonds I and Smith I N 2007 *Int. J. Clim.* **27** 109-21

[15] NOAA/ESRL Physical Sciences Division, Boulder Colorado (2010) *The NCEP Reanalysis Dataset.* Available from http://www.esrl.noaa.gov/psd/

[16] Kalnay, E *et al.* 1996 *Bull. Amer. Meteor. Soc.* **77** 437-71

[17] Australian Bureau of Meteorology (2010) *Australian Water Avaialability Project.* Available from http://www.bom.gov.au/jsp/awap/rain/index.jsp

17th National Conference of the Australian Meteorological and Oceanographic Society IOP Publishing
IOP Conf. Series: Earth and Environmental Science **11** (2010) 012011 doi:10.1088/1755-1315/11/1/012011

East coast lows, atmospheric blocking and rainfall: a Tasmanian perspective

Michael Pook, James Risbey and Peter McIntosh

The Centre for Australian Weather and Climate Research (A partnership between CSIRO and the Bureau of Meteorology), Castray Esplanade, Hobart, Tasmania 7000.

E-mail: Mike.Pook@csiro.au

Abstract. Although the term 'east coast low' is normally associated with intense cyclones near the east coast of mainland Australia, cutoff lows of similar type also affect Tasmania. This paper demonstrates that the cutoff low is a major source of rainfall for the agricultural districts and water catchments of eastern Tasmania. In particular, an analysis of synoptic systems and daily rainfall reveals that cutoff lows are responsible for almost 50% of April to October rainfall in parts of the northeast and a slightly lower proportion in the southeast. The other large contribution to rainfall is from frontal systems but the relative effects of the various synoptic types vary widely across the state as a result of the complex topography. Cutoff lows commonly form the cyclonic portion of a blocking dipole which can have opposing influences on Tasmanian rainfall. The high latitude anticyclone suppresses rainfall in western and southwestern Tasmania, while the cutting off of a relatively small cyclonic component equatorwards of the high frequently results in enhanced rainfall over eastern Tasmania. Results from two climate models indicate that the accurate simulation of blocking and cutoff lows remains difficult to achieve and this has implications for projections of Tasmanian rainfall on seasonal and longer time scales.

1. Introduction

The term 'east coast low' is commonly applied to intense cyclones experienced along the east coast of mainland Australia where they frequently produce flooding rains, gales and high seas [1]. However, it can be readily demonstrated that similar lows are also important contributors to heavy rainfall events and severe gales in eastern Tasmania [2]. The disastrous storm during the 1998 Sydney to Hobart Yacht race is one such example [1, 3]. Others include the major flooding events in southern Tasmania in April, 1960 and in the northeast of the state in November, 1975 [2].

These extreme east coast lows belong to a broader class of synoptic systems known as cutoff lows which have been shown to have a major influence on growing season (April to October) rainfall in some of the cropping regions of southeast Australia [4, 5]. In this paper we demonstrate that the cutoff low is also a major source of rainfall for the agricultural districts and water catchments of central and eastern Tasmania but emphasise that the relative contributions from these systems vary substantially from district to district.

Cutoff lows commonly form the cyclonic portion of a blocking dipole which can have opposing influences on Tasmanian rainfall. On the one hand, the high latitude anticyclonic portion tends to suppress rainfall in western and southwestern Tasmania where the major hydroelectric catchments are located, by interrupting the prevailing westerly airstreams and frontal systems [6]. On the other hand, the cutting off or shearing of a relatively small cyclonic component equatorwards of the high frequently results in the formation of an east coast low which brings rainfall to eastern Tasmania. The accurate simulation of blocking and cutoff lows is a critical requirement of climate models if realistic projections of rainfall are to be achieved on seasonal and longer time scales. A preliminary assessment of the ability of selected climate models to simulate these processes in the Tasmanian region is presented in this paper.

2. Data and Method

2.1. *Data Sources*

Four rainfall stations were selected to represent three important agricultural regions of Tasmania. Cressy and Launceston Airport were chosen for the Central North, Fingal (Legge Street) for the Fingal Valley and Richmond (Brookbanks) in the Southeast. Daily rainfall totals from these stations were obtained from the Patched Point Dataset (PPD) supplied by the Queensland Department of Natural Resources and Mines [7] to provide a consistent technique for inserting estimates of missing data. The PPD uses original Bureau of Meteorology (BoM) measurements for a particular meteorological station but interpolated data are inserted to fill any gaps in the observation record. Monthly rainfall data for Tasmania have been extracted from the data set constructed at the Queensland Department of Natural Resources [7] and displayed on a 0.5° grid. We will refer to the gridded rainfall as SILO data. The locations of the four stations and the gross features of Tasmanian topography are shown in Figure 1.

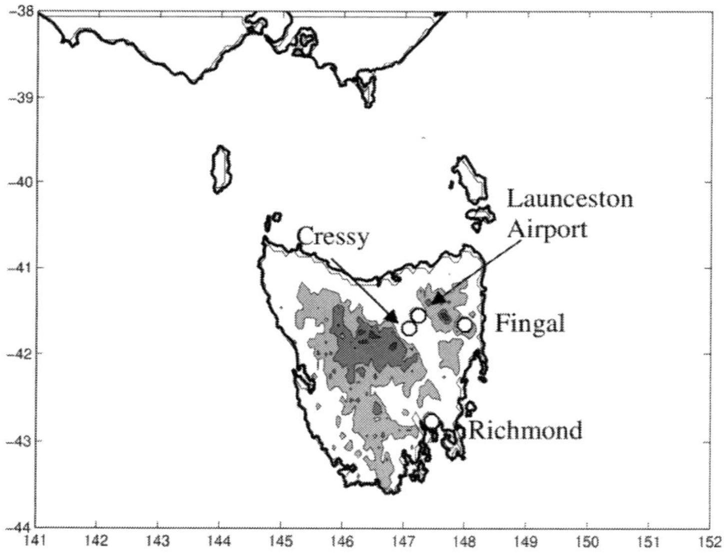

Figure 1. Map of Tasmania in relation to the southeast mainland of Australia showing the four rainfall stations referred to in the text. Shades of grey indicate elevations at 400 m intervals.

The synoptic analysis was conducted using the National Centers for Environmental Prediction (NCEP) - National Center for Atmospheric Research (NCAR) climate reanalysis data set (Reanalysis 1) [10, 11]. The data consist of 4 analyses per day (at six-hourly intervals from 0000 UTC) at a resolution of 2.5° latitude by 2.5° longitude for the standard atmospheric levels from the surface to the lower stratosphere. Fields selected for examination were mean sea level pressure (MSLP), the 500 hPa pressure surface, the (computed) 1000-500 hPa atmospheric thickness and the 1000-500 hPa thickness anomaly relative to the long-term climatology for a particular month. Figure 2 shows one slice in time

IOP Conf. Series: Earth and Environmental Science **11** (2010) 012011 doi:10.1088/1755-1315/11/1/012011

for the 1998 storm (1800 UTC 26 December) and illustrates the analysis box designed to cover the Tasmanian region.

The NCEP data set was supplemented with daily weather maps at 0000 UTC published in the BoM 'Monthly Weather Review' series [9]. These charts include frontal analysis which has been performed manually by expert analysts employing interpretation of satellite imagery in addition to standard analysis of synoptic data [see 4, 5]. We have confidence that these charts provide the most reliable method of determining the occurrence of frontal systems which are important rain-producing systems in southern Australia. The synoptic analysis presented here covers the months of April to October for each year from 1991 to 2008.

2.2. Method

For each day on which rainfall was recorded at any of the four rainfall stations, an assessment was made of the nature of the synoptic system responsible for the precipitation event. Having regard to typical velocities of synoptic systems the analysis region was defined by a fixed box with limits, 35°S, 135°E; 35°S, 155°E; 50°S, 155°E and 50°S, 135°E as shown in Figure 2. Synoptic systems were classified according to the classification scheme developed and applied in two previous studies [4, 5]. A rain-producing system was required to have been located within the box during the 24 hour period to which the rainfall applied. The categories found to be significant in this analysis are cold frontal systems of all types, cold-cored lows which have become cut-off from the westerly airstream (cutoff lows), northwest airstreams, southwest airstreams and troughs of low pressure which were found to be associated with thermal ridges. Low pressure systems forming within the latter are excluded from the cutoff low category because they have a warm-cored structure with height and never satisfy the negative thermal anomaly condition specified in the cutoff low definition.

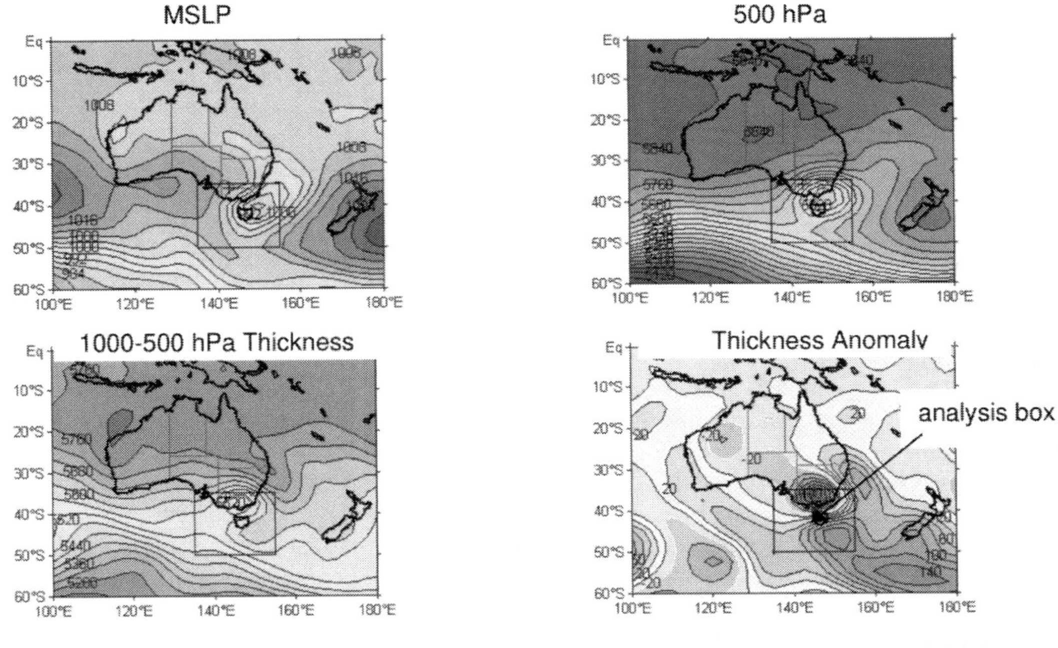

Figure 2. NCEP/NCAR fields of MSLP, 500 hPa geopotential, 1000-500 hPa thickness and thickness anomaly at 1800 UTC 26 December 1998 showing the intense cutoff low pressure system which had a severe impact on the 1998 Sydney to Hobart Yacht Race. The co-ordinates of the 'analysis box' are given in the text.

3. Results

3.1. Synoptic analysis

The synoptic analysis reveals that the relative contributions made to daily rainfall by the selected synoptic types vary substantially across the cropping regions (Figure 3). In particular, cutoff lows are responsible for approximately 50% of April to October rainfall in parts of the northeast (Fingal), a lower proportion (39%) in the southeast (Richmond) while the proportion of rainfall attributed to cutoff lows declines to about 35% in the central north. The other large contribution to rainfall is from frontal systems which account for approximately 50% of growing season precipitation in the central north but less than 30% a short distance to the east at Fingal. The relative effects of the two main synoptic types vary widely across the state, mainly as a result of the complex topography. Clearly, the effect of topography is exemplified by the contrast between rain associated with northwest and southwest airstreams. In the central north, northwest airstreams contribute about 8% of rainfall but only 3% in the Fingal Valley and about 4% in the southeast. By way of contrast, the southwest airstream is associated with 7% of rainfall in the southeast but only around 1% in the sheltered north and northeast.

At the northeastern site of Fingal, total April to October rainfall is highly correlated with cutoff low rainfall for the same period (r=0.9, n=18) and displays a high degree of interannual variability. Figure 4 shows the time series of total April to October rainfall at Fingal, and as separate plots, cutoff rainfall and frontal rainfall for the period 1991 to 2008.

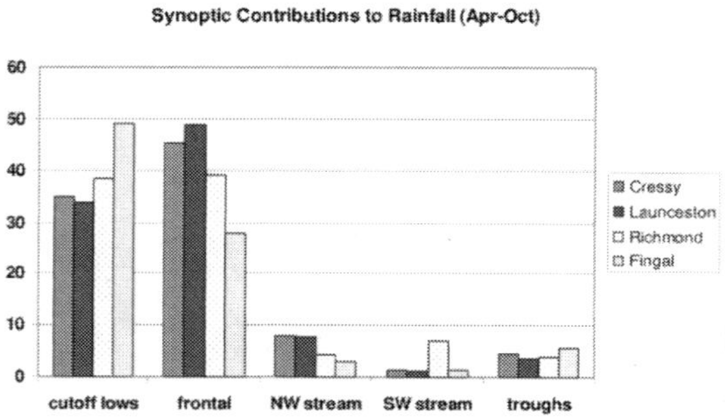

Figure 3. The relative contributions to April to October rainfall by synoptic systems at the four rainfall stations.

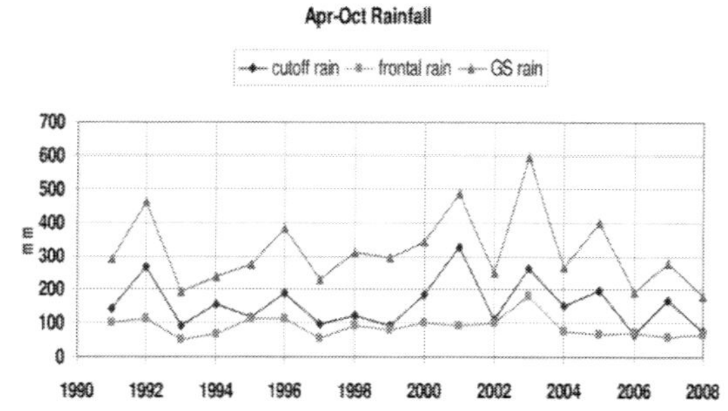

Figure 4. Time series of April to October rainfall at Fingal (top line), cutoff rainfall (middle line) and frontal rainfall (bottom line).

3.2. Cutoff lows and Blocking

Atmospheric blocking has been identified as an important climate driver for Australia's agricultural districts and water catchments [8]. Individual occurrences of blocking are influenced by the configuration of the atmospheric long waves around the Southern Hemisphere at any time but its onset and maintenance in the Australasian region is not well understood and is not treated adequately in climate models. Atmospheric blocking is associated with a splitting of the upper tropospheric westerly airstream into two distinct branches and the degree of this splitting of the westerlies can be represented by a simple blocking index (BI) which has been employed in previous studies [4, 5]. Blocking in the Australian region is regularly accompanied by the development of a cyclone or low pressure trough equatorward of the anticyclone and the majority of these low pressure systems could be classified as cutoff lows [4].

Figure 5 demonstrates that there is a significant positive correlation between blocking which occurs near Tasmania (as represented by the BI at 140°E) and the growing season (April to October) rainfall in the northeast of the state. Additionally, there is a weak negative correlation with rainfall in southwestern Tasmania. This pattern can be interpreted as a response to the circulation change induced by the blocking dipole structure which leads to increased stability and anomalous easterly flow over western Tasmania and the formation of cutoff lows near the northeast.

3.3. Cutoff lows in atmospheric numerical models

In an attempt to gauge the ability of numerical climate models to simulate faithfully the mix of synoptic systems we have carried out tests of two well-known models for their ability to generate cutoff lows. The first model is the CSIRO Mark 3 coupled ocean/atmosphere model, integrated over the 20th century with realistic greenhouse and aerosol forcing. This is a climate change model that, once started, is allowed to run without re-initialization or flux correction [12]. We analyse the last 60 years of the run. The second model is an atmosphere-only version of the NCAR CCSM3 model (identified as CAM3) using a monthly sea surface temperature (SST) climatology and run for 30 years [see 13]. Note that these are climatological years, and are not related to particular historical years.

In this case, the models are compared to two different reanalysis data sets, NCEP-NCAR [10, 11] and ERA-40 [14] in addition to the manual analysis. An automatic recognition system for cutoff lows has been developed in a previous study [15] and applied here in the Tasmanian region box defined in Section 2. This system has been calibrated against a 50 year manual analysis in SE Australia.

Figure 6 reveals that the numbers of cutoff lows in the Tasmanian analysis box which were identified by the automatic recognition system applied to the two climate models are markedly lower than the results of the manual analysis. When the automatic recognition system is applied to the two reanalysis products the results agree well with the manual analysis. For this study region the CSIRO Mark 3 model gives distinctly better results than CAM 3.

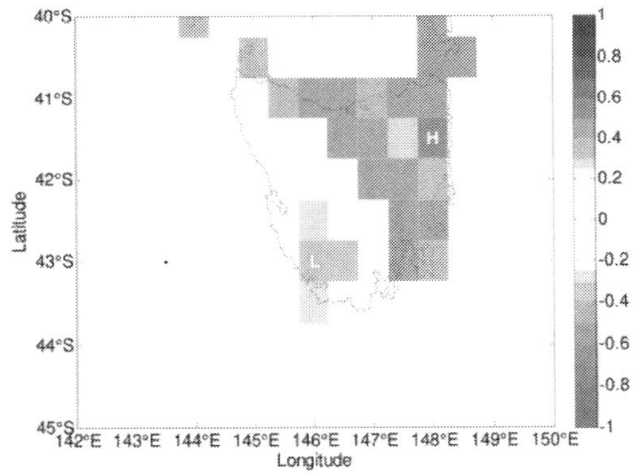

Figure 5. Map of Tasmania showing the correlation between the mean blocking index at 140°E and SILO rainfall for the months of April to October. The 'H' ('L') represents the maximum positive (negative) correlation. Shading indicates a confidence level of at least 95%.

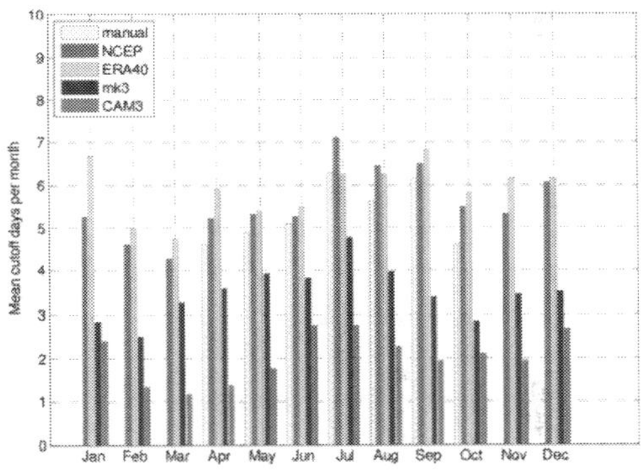

Figure 6. Monthly mean number of days on which cutoff lows are identified in the analysis box for the manual analysis, the reanalysis models and two climate models. Note that the manual analysis is for April to October only.

4. Discussion and conclusion

In addition to their well-known effects on the coast of New South Wales and southeastern Queensland we have shown that east coast lows are significant causes of severe weather in Bass Strait and eastern Tasmania. However, in a more general sense we have also shown that these severe weather systems form a subset of the cutoff low, a synoptic system that has a major influence on rainfall over the extratropical regions of Australia. In Tasmania, the situation is further complicated by the complex topography which suggests that there will be large spatial differences in the influences exerted by the synoptic mix of rain-producing systems. This is confirmed by our 18 year synoptic analysis of rainfall in representative cropping areas. A key result is that cutoff lows produce a higher proportion of April to October rainfall than do fronts in the northeast but frontal rain dominates in the nearby Central North.

There is a close relationship between cutoff lows and blocking although not every cutoff low is found with an associated block. The blocking Index at 140E (Apr-Oct) is positively correlated with rainfall in the northeast of Tasmania and negatively correlated in the southwest, where significant hydroelectric storages exist. We interpret this as resulting from the circulation change associated with the blocking dipole and interactions with topography and speculatively, oceanic influences off the northeast coast.

Models must be able to reproduce realistic synoptic systems in order for their rainfall forecasts to be credible. It is not sufficient for a model to get the right amount of rainfall via the wrong physical mechanism. The challenge for climate models is to simulate faithfully the observed mix of synoptic systems if forecasts and projections of regional rainfall are to have acceptable accuracy.

Acknowledgements

This work was partly funded by grants from the Managing Climate Variability Program of the Australian Government and the Climate Futures for Tasmania Project and carried out within the CSIRO Climate Adaptation Research Flagship.

References

[1] Wright W (ed) 2004 *Drought, dust and deluge* (Melbourne, Vic: Bureau of Meteorology) 78pp
[2] Bureau of Meteorology 1993 Climate of Tasmania 30pp
[3] Mills G A 2000 BMRC Research Report No.76, Bureau of Meteorology Research Centre
[4] Pook M J, McIntosh P C and Meyers G A 2006 *J. Appl. Meteorol. Clim.* **45** 1156–70
[5] Pook M J, Lisson S, Risbey J, Ummenhofer C C, McIntosh P and Rebbeck M 2009 *Int. J. Climatol.* **29** 2012-26
[6] Pook M J and Gibson T T 1999 *Aust. Meteorol. Mag.* Special Edition 51-60
[7] Jeffrey S J, Carter J O, Moodie K M and Beswick A R 2001 *Environ. Modell. Softw.* **16** 309-30
[8] Risbey J S, Pook M J, McIntosh P C, Wheeler M C and Hendon H H 2009 *Mon. Weather. Rev.* **137** 3233-53
[9] Simmonds I and Richter T 2000 *Theor. Appl. Climatol.* **67** 19-32
[10] Kalnay E *et al* 1996 *B. Am. Meteorol. Soc.* **77** 437-71
[11] Kistler R *et al* 2001 *B. Am. Meteorol. Soc.* **82** 247-67
[12] Gordon H B *et al* 2002 CSIRO Atmospheric Research Technical Paper No. 60 130pp
[13] Hurrell J W, Hack J J, Phillips A S, Caron J and Yin J 2006 *J. Climate* **19** 2162–83
[14] Kallberg P, Simmons A, Uppala S and Fuentest M 2004 ERA-40 Project Report Series No. 17, ECMWF
[15] McIntosh P, Pook M, Risbey J, Lisson S, Stephens D and Rebbeck M 2007 Oceans to Grains: A New Approach to Targeted Seasonal Forecasts. Final Report to Land and Water, Australia, Australian Government

IOP Conf. Series: Earth and Environmental Science **11** (2010) 012012 doi:10.1088/1755-1315/11/1/012012

Determination of visual range during fog and mist using digital camera images

John R Taylor and Jamie C Moogan

School of Physical, Environmental and Mathematical Sciences, UNSW@ADFA, Canberra ACT, 2600, Australia

E-mail: j.taylor@adfa.edu.au

Abstract. During the winter of 2008, daily time series of images of five "unit-cell chequerboard" targets were acquired using a digital camera. The camera and targets were located in the Majura Valley approximately 3 km from Canberra airport. We show how the contrast between the black and white sections of the targets is related to the meteorological range (or standard visual range), and compare estimates of this quantity derived from images acquired during fog and mist conditions with those from the Vaisala FD-12 visibility meter operated by the Bureau of Meteorology at Canberra Airport. The two sets of ranges are consistent but show the variability of visibility in the patchy fog conditions that often prevail in the Majura Valley. Significant spatial variations of the light extinction coefficient were found to occur over the longest 570 m optical path sampled by the imaging system. Visual ranges could be estimated out to ten times the distance to the furthest target, or approximately 6 km, in these experiments. Image saturation of the white sections of the targets was the major limitation on the quantitative interpretation of the images. In the future, the camera images will be processed in real time so that the camera exposure can be adjusted to avoid saturation.

1. Introduction

Daytime visibility is the distance at which a dark object can just be recognised against the sky at the horizon [1]. Traditionally, visibility was assessed by a human observer's ability to distinguish large dark objects at known distances from their observing site. However, with the trend to automated observations, the human based approach has often been replaced, or supplemented, by an instrumental determination of the extinction co-efficient due to scattering, generally over a small (of order 1 m) optical path. The meteorological range or standard visual range, V, is then calculated from

$$V = -\frac{1}{k}\ln\varepsilon. \tag{1}$$

Here k is the extinction coefficient and ε is the contrast threshold, which is generally taken as 0.02 [1]. These automated values are essentially point measurements, so there has been on going interest in automatic techniques which mimic the human observation process using digital imaging [2]. In many of these applications, visibilities in clear sky conditions are of interest (pollution and visual amenity studies), while in others, particularly those involving air or road transport, measurements during restricted visibility conditions, like precipitation, mist and fog, are required. Our interest was in measurements during fog and mist. In particular, we wished to use the visual range to infer the liquid water content of fogs [3] which was a parameter required

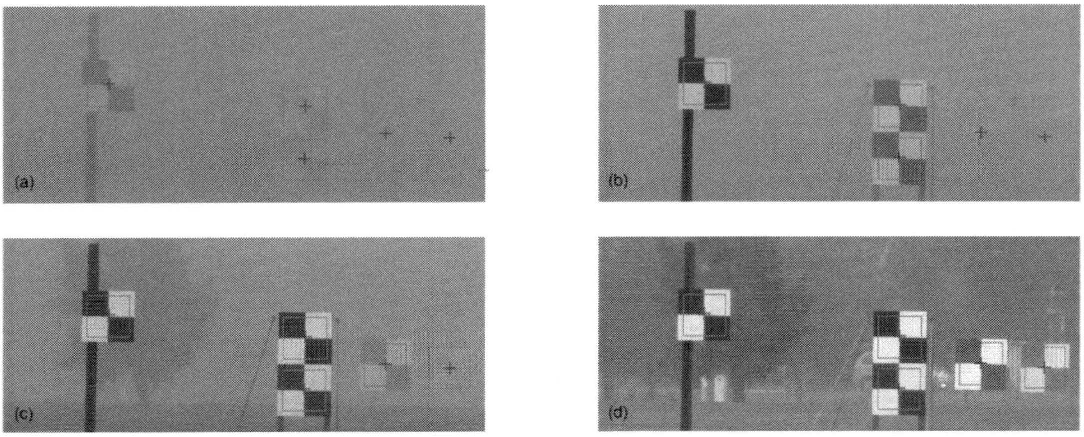

Figure 1. Images acquired on 26th May 2008, (a) 8:18:08 am LT, (b) 8:30:08 am LT, (c) 8:36:08 am LT, (d) 8:54:07 am LT. See text for the definition of LT. Blue crosses show the starting position for locating the target centre, red crosses the position that maximises the correlation between the target and a target template. Red rectangles show the extent of the target used in the histogram analysis (118 by 116 pixels). If no rectangle is shown (as in a and b) the correlation analysis failed to locate the target. In these situations the red cross is located at one extreme of the search area for the correlation.

to implement a one-dimensional fog clearance model [4]. Although the imaging approach used here provides a greater degree of spatial averaging than that from a single point measurement of extinction, the degree of averaging is limited by the small visual range (of order one hundred to several hundred of metres) in fog.

2. Experiment

In operation the instrumentation is quite simple. A consumer grade, digital still camera (Canon Power Shot SX100IS) was set up to acquire images at times controlled by a computer. Four targets located 50, 100, 355 and 571 m to the south-west of the camera were included in each image. The targets (figure 1) were divided into four panels, upper left/lower right painted black and lower left/upper right painted white, hence forming a "unit cell" of a chequerboard pattern. The target dimensions were set so that each target occupied a similar number of pixels in the images independent of their distance from the camera. Under computer control, the camera acquired images at 6 minute intervals from 6 am LT (Local Time, LT, is Australian Eastern Standard Time, AEST=UTC+10) to noon LT and the images were archived in the camera's default JPEG format for subsequent processing. Apart from interruptions due to power failures, the camera operated from 24th April to 31st August 2008. Figure 1 shows a sequence of images taken during fog on 26th May.

3. Image interpretation

The luminance of an object, L, at distance d from an observer is given by the Koschmeider equation [5]

$$L = L_0 \exp(-kd) + L_f\,(1 - \exp(-kd))\,, \tag{2}$$

where L_0 is the intrinsic luminance of the object, L_f is the luminance of the sky, and k is the extinction coefficient of the atmosphere (defined previously). The first term, $L_0 \exp(-kd)$, represents the decrease of luminance due to light from the target being scattered away from the observer along the path between them. The second term, $L_f(1-\exp(-kd))$, represents an increase in luminance due to light from the sky being scattered into the path between the object and observer; this is commonly called the "airlight" [6]. For a target composed of black and white segments the luminance of the white and black segments (L_w and L_b respectively) measured at d is

$$\begin{aligned} L_w &= R_w L_i \exp(-kd) + L_f(1-\exp(-kd)), \\ L_b &= R_b L_i \exp(-kd) + L_f(1-\exp(-kd)), \end{aligned}$$

where L_i is the luminance incident on the target and R_w and R_b are the reflectances of the white and black segments respectively. With these definitions of the segment luminances, the Michelson contrast, C, (appropriate for patterns where the areas of the light and dark are similar [7]), is

$$C = \frac{L_w - L_b}{L_w + L_b} = \frac{(R_w - R_b)L_i \exp(-kd)}{(R_w + R_b)L_i \exp(-kd) + 2L_f(1-\exp(-kd))}. \tag{3}$$

This expression can be simplified since $L_i = L_f$ or, if the light field is uniform over the scene, the luminance incident on the target is equal to the sky luminance at the target. The black and white segments will no longer be discernable when C falls to the threshold value, ε. We rearrange the expression for C above to find an expression for the distance d_ε at which this occurs,

$$d_\varepsilon = -\frac{1}{k}\ln\left[\frac{\varepsilon}{C_0\overline{R} - \varepsilon(\overline{R}-1)}\right]. \tag{4}$$

Here we have introduced the intrinsic contrast of the target, $C_0 = (R_w - R_b)/(R_w + R_b)$, and the average reflectance, which for a target with the dark and light segments having equal areas, is $\overline{R} = (R_w + R_b)/2$. For a perfect target, $C_0 = 1$ and $\overline{R} = 0.5$, equation 4 for d_ε reduces to

$$d_\varepsilon = -\frac{1}{k}\ln\left[\frac{2\varepsilon}{1+\varepsilon}\right]. \tag{5}$$

A comparison with the definition for V (equation 1) shows that the segments of a perfect black and white target will not be discernable at a closer distance than would a perfectly black target viewed against the horizon sky.

In practice, we have measurements of C (derived from the images) at a given distance d which can be used to calculate k given the target properties ($\overline{R}$ and C_0) using

$$k = \frac{1}{d}\ln\left[\frac{C_0\overline{R} - C(\overline{R}-1)}{C}\right]. \tag{6}$$

This equation follows from equation 3 using a similar approach to that which lead to equation 4. Using the experimentally derived values of the extinction coefficient, k, the meteorological range, V, is then calculated using equation 1.

4. Image Processing

Consumer digital cameras are designed to produce good quality colour images, not quantitatively measure contrast or luminance, and the digital value of the pixels, representing the intensity in an image, is not linearly related to the luminance of an object. To show this, the camera used for this study was set up to view an object of (approximately) uniform luminance, an LCD monitor screen set to a uniform grey level. A series of images was then taken with a

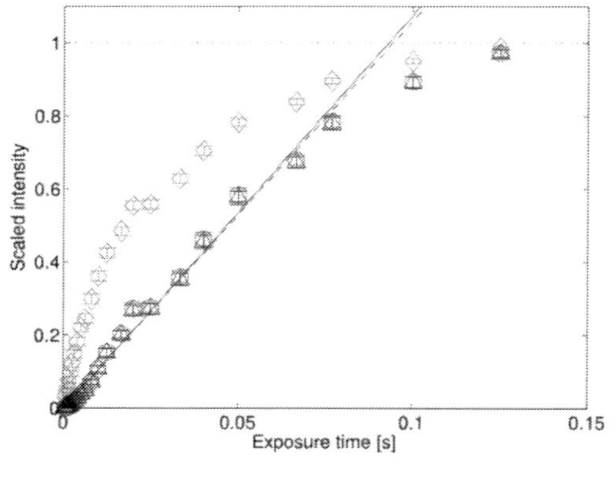

Figure 2. Plot of average pixel value for a uniform screen at a range of exposure times. Pixel value is normalised to the maximum value. Symbols: Image averages, ◇, inverse sRGB transform, △, inverse gamma transform, ○. Error bars (smaller than the plot symbols) were derived from the standard deviation of the pixel values. The two straight lines show least square fits to the transformed values. Note how the raw pixel values stretch the intensity range at low intensities, and the start of the saturation of the image for exposure times between 0.08 and 0.1 s.

fixed lens aperture over the camera's full range of exposure times. Figure 2 shows a plot of the average intensity in the central 25 % of the area of the image. The relationship between image luminance (directly proportional to the exposure time) and pixel value is not linear. However, we can see that, if the average pixel values are transformed with the expected sRGB transform [8] or, as a close approximation, an inverse gamma transform with gamma of 2.2, an approximately linear intensity scale results until the image starts to saturate at approximately 0.8 of full scale. Saturation of the images of the white section of the targets was the major limitation on the results obtained using the present technique. Under clear sky conditions, the automatic exposure function of the camera overexposed these because they were much brighter than the average of the scene but only occupied a small fraction of the area of the image.

Processing of the camera images was carried out in MATLAB and started with conversion of the colour images to greyscale (using the MATLAB function rgb2gray) and subsampling to the image area shown in figure 1. The subsampled image was then median filtered to reduce noise that became apparent when the greyscale images were displayed in MATLAB. Subsequent processing proceeded on a target by target basis. The inverse sRGB transform was used to linearise the intensity scale and the target centre was located by finding the maximum correlation with an ideal target image. Next the central region of the target (red squares in figure 1) was extracted and the intensity histogram calculated. Screen contrast was calculated based on the mean intensities in the peaks of the histogram corresponding to the black and white sections of the target. Quality checks on the calculated contrast included: dark and light peaks in the intensity histogram not resolved; histogram peaks too wide (generally because of some part of the target was obscured in the image); and white section of the image saturated or in the nonlinear portion of the intensity scale. Following the analysis, we have a coarse estimate that the visibility should lie between the distance of the furthest target detected (by the correlation) and the first target not detected. A finer scale calculation was then made for each screen using the calculated contrast and the method described at the end of section 3 using equation 6.

5. Results

Figure 3 shows time series of meteorological range from the camera and from the Australian Bureau of Meteorology's (BOM) Vaisala FD-12 visibility meter for May 26th and May 29th, 2008. The Vaisala instrument is located at the BOM observing site at Canberra airport, about 3 km from the location of our camera and visibility screens. Note that the 3 km separation of the instruments means that the details of the time variation of the meteorological range could be quite different at the two locations. For example, on the 29th (figure 3b) visual range at the airport dropped rapidly before 7:00 am; a similar drop occurred approximately 30 minutes

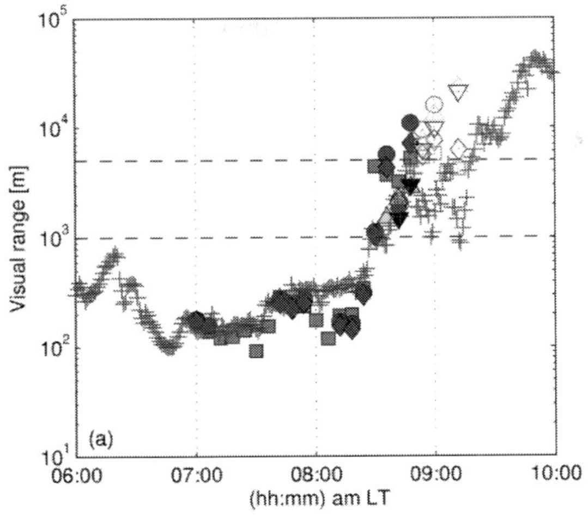

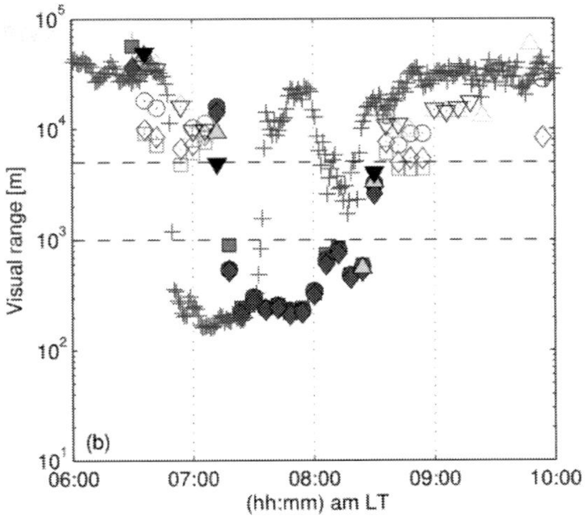

Figure 3. Comparisons of meteorological range from the BOM Vaisala FD-12 instrument at Canberra Airport and the target contrast analysis for (a) 26th May and (b) 29th May 2008. Symbol types are from targets at distance: 50 m, □, 100 m, ○ and ◇, 355 m, △ and 571 m, ▽. Unfilled symbols had an intensity in the light section greater than 0.8, but were not fully saturated. BOM meteorological range values, +. Lower dashed line shows a meteorological range of 1 km, the accepted boundary between fog and mist conditions.

later at the camera location. Visual range at the airport then increased rapidly to greater than 5 km; a similar increase did not occur until an hour later at the camera location. In contrast, on the 26th May (figure 3a) the fog clearance time (marked by the increase of meteorological range to > 1 km) at the two locations was similar, although thicker fog persisted longer at the camera location. This persistence is shown by the smaller visual ranges between 8 am LT and fog clearance.

While the detailed differences in the time variation of visual range detailed above, and similar differences on other foggy days for which data is available, preclude a direct comparison between the visual ranges derived from the two systems, the values of visual range during fog events (in particular, the minimum values of just under 100 m) are similar for both systems. Also, the rapid onset of reductions in visibility and clearance are captured by both systems. Because of saturation effects, meteorological ranges greater than 5 to 10 km could not be reliably determined from the camera images. This is demonstrated by the fact that visual ranges from the camera system when the images are close to saturation (unfilled sysmbols on figure 3b) are generally low compared to the values from the Bureau's instrument for the same time. On the 29th, the values of visual range from the five targets are very similar and give an idea of the reliability of the technique (see the spread of estimates from the last two sets of filled symbols on figure 3b at approximately 8:30 am LT). On figure 3a the results for the targets at different ranges are more inconsistent as the fog clears around 8:30 am LT, with the closer targets giving significantly greater visibilities. The consistency of the results for the different targets on the 29th May leads us to conclude that the variation on the 26th May is caused by the spatial variations in the extinction coefficient, with the more distant targets remaining in conditions of lower visibility after fog clearance at the closer targets. This would be consistent with the general light south-easterly winds that are observed under strongly stable conditions at this location [9] advecting fog towards the more distant targets.

6. Conclusions / Further work

Our results, and previous work, have shown that quantitative estimates of meteorological range can be derived from digital camera imagery, particularly under conditions of restricted visibility such as fog and mist. Further, the uniformity of meteorological range estimates from targets at different ranges can give information about the spatial variability of fog. In the future we will improve our measurements in three ways. Firstly, using a camera with the capacity to capture raw, rather than compressed images, will reduce the uncertainty resulting from the need to linearise the camera's response. Secondly, implementation of real time image analysis will overcome the problem of pixel saturation. With these two changes, it should be possible to extend the meteorological ranges that can be reliably determined with the present targets out to tens of km, values similar to those that the Vaisala meter returns under clear sky conditions. Finally, by setting up a self-contained light source at some of the targets, we will be able to derive night time estimates of meteorological range.

Acknowledgments

The Bureau of Meteorology kindly supplied the Canberra Airport meteorological range data plotted in figure 3. Dr Stephen James and two anonymous referees made helpful coments on earlier versions of this paper.

References

[1] American Meteorological Society 2000 *Glossary of Meteorology* (Boston, MA: Allen Press) http://amsglossary.allenpress.com last checked in February 2010

[2] Xie L, Chiu A and Newsam S 2008 Estimating visibility using general purpose cameras *Proc. Adv. in Visual Computing 4th Int. Symp. (Las Vegas NV USA, December 1-3 2008)* part III ed G Bebis *et al* (Berlin: Springer) pp 356-367

[3] Kunkel B A 1984 Parameterization of droplet terminal velocity and extinction coefficient in fog models *J. Clim. Appl. Meteorol.* **23** 34–41

[4] Reudenbach C H, Bendix J 1998 Experiments with a straightforward model for the spatial forecast of fog/low stratus clearance based on multisource data *Meteorol. Appl.* **5** 205–216

[5] Hautiere N, Tarel J, Lavenant J and Auert D 2006 Automatic fog detection and estimation of visibility distance through use of an onboard camera *Mach. Vision Appl.* **17** 8–20

[6] Bohren C F 1987 *Clouds in a Glass of Beer, Simple Experiments in Atmospheric Physics* (New York: Wiley)

[7] Peli E 1990 Contrast in complex images *J. Opt. Soc. Am.* **7** 2032–2040

[8] Stokes M, Chandrasekar S and Motta R 1996 *A Standard Default Color Space for the Internet - sRGB* http://www.w3.org/Graphics/Color/sRGB last checked in February 2010

[9] Taylor J R, Hirsch A and Burns B A 2009 Modelling low-level boundary layer structure in complex terrain: verification of TAPM meteorological prediction in the Canberra Region, submitted to *Aust. Meteorol. Oceanogr. J.*

IOP Conf. Series: Earth and Environmental Science 11 (2010) 012013 doi:10.1088/1755-1315/11/1/012013

The climate of the Eastern Seaboard of Australia: a challenging entity now and for future projections

Bertrand Timbal

The Centre for Australian Weather and Climate Research - a partnership between CSIRO and the Bureau of Meteorology, Box 1289 Melbourne VIC 3001 Australia.

E-mail: B.Timbal@bom.gov.au

Abstract. The Eastern SeaBoard (ESB) of Australia has long been recognised as a separate climate entity. Using the latest gridded observations from the Bureau of Meteorology, a definition of the spatial extent of the ESB is proposed. It appears that, while this area has recorded below average rainfall over the last 12 years, the ongoing deficiency is not record breaking in historic terms. This contrasts with record breaking droughts across large parts of inland, eastern Australia. The lesser severity of ongoing rainfall deficiencies in the ESB, compared to the rest of the region, is linked to the different impact of observed changes in regional surface pressure and, in particular, changes in the position of the sub-tropical ridge. It is also observed that while tropical modes of variability in the Pacific and Indian oceans are known to influence the climate of eastern Australia, that influence appears very weak and not statistically significant across the ESB. Finally, some issues relevant to future rainfall projections for the ESB are discussed. It is argued that providing reliable climate projections across this climatic region is a difficult challenge.

1. Introduction

The Eastern SeaBoard (ESB) of Australia can be defined as the coastal strip on the eastern side of the Great Dividing Range (GDR) starting in the south in the Victorian Gippsland, and extending north across New South Wales (NSW) and into south-east Queensland (see figure 1). The coastal strip across eastern Australia has long been recognised as a separate climate entity, in the sense that it has a distinct, separate existence: e.g. using rotated Empirical Orthogonal Functions (EOFs) to describe Australian rainfall, the ESB is represented by a separate EOF when as few as 8 EOFs are used [1].

Recently, as part of the South-Eastern Australian Climate Initiative (SEACI), the record breaking, ongoing, protracted drought across South-Eastern Australia (SEA) has been linked [2] to an observed increase in the Sub-Tropical Ridge Intensity (STR-I). However, the spatial extent of the ongoing rainfall deficiency has a marked maximum across the South-West of Eastern Australia (SWEA) decreasing further north in the Northern part of the Murray-Darling Basin (NMDB). It is also evident that the influence of the STR-I, as well as the rainfall deficiency does not appear as strong on the other side of the GDR into the ESB [1]. Using the rainfall-STR-I relationship, three climate entities are defined across SEA (figure 1). The SWEA is defined as a continuous area where in May-June-July (MJJ) the inter-annual variations of the STR-I explain more than 20% of the inter-annual variability of rainfall (correlations 0.45 and above). The NMDB encompasses the rest of the Murray-Darling Basin (MDB), where the influence of the STR-I is less pronounced, and the ESB was defined as the rest of SEA along the east coast.

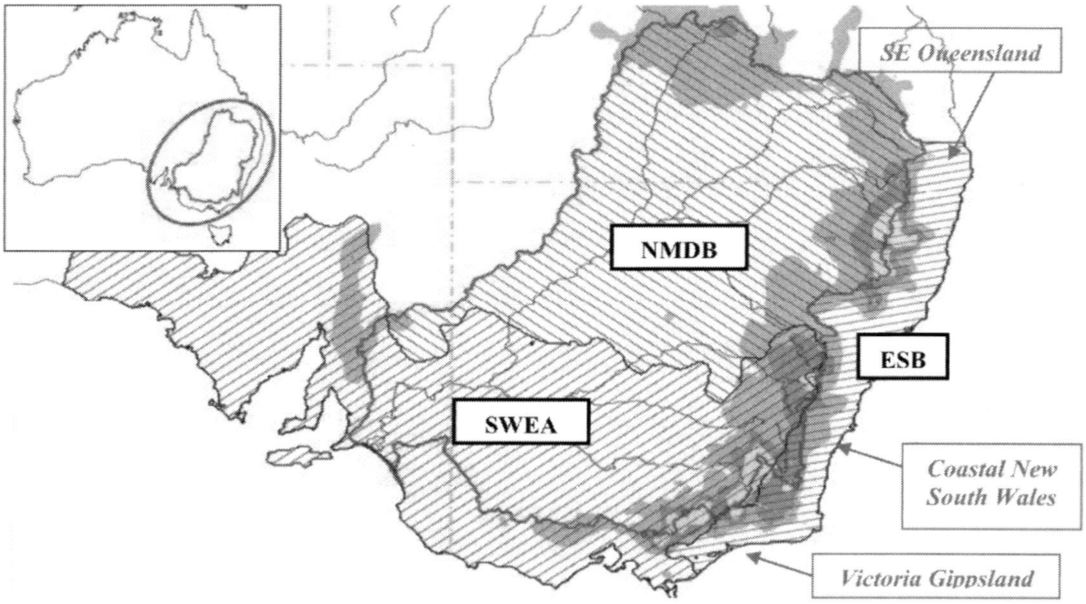

Figure 1. Three climatic regions identified across South-Eastern Australia (SEA): SWEA, NMDB and ESB. Elevations (e.g. the GDR) are shown by colour shading (every 500m). The red contour delimits the MDB; state boundaries are dashed orange lines and other relevant geographical features are indicated in red (see text for details). The position of the area of interest relative to the Australian continent is shown in the top left insert.

The STR-rainfall correlations are calculated in MJJ, since it is the time of the year where the STR-I rainfall relationship is the strongest. In order to obtain a precise boundary, correlations are based on the latest 5km gridded rainfall analysis, generated by the BoM as part of the Australian Water Availability Project (AWAP) [3]. The dataset includes precipitation (amongst other variables) obtained by interpolating surface station measurements onto a 5 km by 5 km grid. The climatological averages are gridded using a three dimensional spline smoothing technique. Analyses of the anomalies are performed using the Barnes successive correction method [3]. The number of stations reporting data varies with time and by variable, with precipitation interpolated from between 5,000 and 7,000 stations across Australia.

The advantage of using gridded rainfall with a fine resolution is the possibility of obtaining small-scale structures, in particular where topography is an important contributor to the rainfall. While the boundary between NMDB and ESB is defined by the highest point of the GDR, the boundary between ESB and SWEA is defined by the STR-rainfall relationship. It is worth noting that this STR-rainfall boundary still corresponds to the topography of the GDR. The definition of the ESB as a distinct climate entity, limited in the west by the GDR, is better captured with finely gridded observations. In the rest of this communication, these three climatic regions (SWEA, NMDB and ESB) are used to (1) analyse the characteristics of the recent dry period compared with the previous historically low rainfall period of 1935-1945 and, (2) analyse the influence of known modes of large-scale variability on the ESB.

2. Comparison of recent drought across the SEA with an historical benchmark

Since 1996, the SEA region, defined as continental Australia, south of 33.5°S and east of 135.5°E, has experienced below average rainfall. Rainfall deficiencies from 1996-2006 [4] were not unprecedented historically, being comparable to the previous lowest 11-year period on record from 1935 to 1945 (widely known as the World War II drought, or WWII drought hereafter). Since 2006 however, the drought has continued, such that it is now the largest on record across SEA [2]. Having now separated the SEA into three different regional entities, the ongoing drought (1997-2008) is compared to the historical benchmark WWII drought (table 1).

Table 1. Rainfall statistics for the 3 climate entities in SEA: long term (1900-2008) mean (annual and seasonal) in mm and anomalies during the WWII dry period and the ongoing protracted drought (as a percent of the long-term climatology). Positive anomalies are shown in blue italics; largest anomalies on record are shown as bold red figures.

Climate entities		Annual	Autumn	Winter	Spring	Summer
	1900-2008	456 mm	103 mm	145 mm	120 mm	87 mm
SWEA	1935-1945	-10.3%	-13.1%	-7.2%	-18.2%	-1.2%
	1997-2008	-10.7%	-26.5%	-12.1%	-4.2%	*+1.2%*
	1900-2008	954 mm	253 mm	191 mm	210 mm	299 mm
ESB	1935-1945	-12.4%	-11.3%	-15.5%	-11.4%	-12.0%
	1997-2008	-6.3%	-16.7%	-7.0%	*+4.1%*	-1.4%
	1900-2008	457 mm	105 mm	89 mm	106 mm	155 mm
NMDB	1935-1945	-15.4%	-17.2%	-6.3%	-29.1%	-10.1%
	1997-2008	-0.5%	-17.9%	-1.7%	*+15.0 %*	*+1.6%*

The climatology of the ESB, with annual rainfall at 954 mm/year, is in clear contrast to both the SWEA and NMDB, which received less than half of this amount. It is also worth noting that while the ESB is dominated by summer rainfall (31% of the annual total) and autumn rainfall, the SWEA region has its largest rainfall in winter (32%) and spring. The annual cycle of rainfall in the NMDB is similar to the ESB. One would expect that, since the ESB has such different background climatology, temporal anomalies are also likely to exhibit significantly different characteristics.

The magnitude of the ongoing decline (1997-2008) in annual rainfall is only highest on record (bold red figures in table 1) in SWEA, about half the magnitude of the WWII drought in the ESB, and insignificant across the NMDB. The dominant feature of the ongoing drought is the decline in autumn rainfall, accounting for about 55% of the total annual decline across SEA [5]. Interestingly, the autumn decline is found to be highest on record across all three regions. However while the autumn rainfall decline is complemented by declines in winter (large) and spring (small) across SWEA, the winter decline is small across the ESB, spring rainfall being above average. This increase in spring rainfall is even more apparent in the NMDB. It is the differences in winter and spring across the three regions that are reflected in the annual total. In general, summer rainfall anomalies are small in the current period. In comparison, the WWII drought, the largest on record in both the NMDB and the ESB, was described by year-round rainfall deficiencies (i.e. no positive anomaly across the four seasons and the three regions). The largest signal is in spring west of the GDR (SWEA and NMDB), and in winter east of the GDR.

3. The role of the sub-tropical ridge

Previous studies have shown that potential changes to the Sub-Tropical Ridge (STR) [6] are a key mechanism in explaining the rainfall deficiency across SEA [5, 6 and 2]. It has been noted [6] that the rainfall decline is primarily related to an intensification of the STR (STR-I). Evidence of a southward shift of the position of the STR (STR-P) appears to play only a secondary role in the rainfall decline.

When linearly combined with the STR-I, changes in the position of the STR do not appear to have any sizeable additional effect on rainfall. Based on the definition of the SWEA and the ESB, it is anticipated that the STR-rainfall relationship between the two regions would be different. Indeed, monthly correlations using regional rainfall averages based on the AWAP data (figure 2) reveal a very contrasting picture. While the relationship with both the STR intensity and position is negative, strong and statistically significant across most months, with a significant rainfall decline in SWEA (April to October), it is not so in the ESB. The relationship between ESB rainfall and the STR-I is mostly insignificant, while the relationship with the STR-P is weak, but marginally significant and of the opposite sign than in SWEA. A shift south of the STR, as observed recently [7], is likely to increase the rainfall in the ESB while reducing rainfall on the other side of the GDR. The observed shift of the STR is consistent with numerous observations of the expansion of the Hadley Circulation [8].

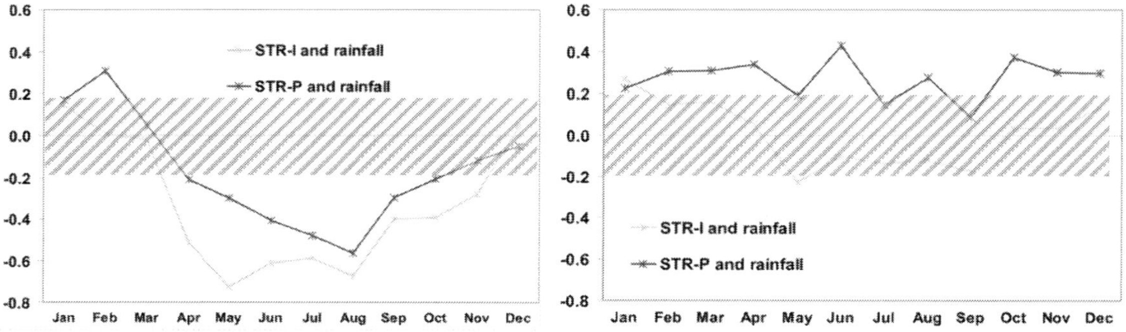

Figure 2. Monthly correlations in SWEA (left) and ESB (right) between regional rainfall and the position and intensity of the Sub-Tropical-Ridge (STR). On both graphs, correlations below the 95% statistical significance level are masked by grey stipples.

4. The role of tropical modes of variability

The contributions of the El Niño-Southern Oscillation (ENSO) and the Indian Ocean Dipole (IOD) [9] in driving inter-annual rainfall and temperature variability across the Australian continent have long been recognised. While both influences are plausible mechanisms to explain the ongoing rainfall deficiency across SEA, their roles are likely to be secondary; since most of the rainfall decline is happening in autumn, when ENSO and IOD influences on rainfall are negligible [10].

The role of the tropical modes of variability on SWEA and ESB rainfall are compared (figure 3) using a tropical tri-pole [11], which attempts to encompass the influence of both ENSO and the IOD. This index was based on the area of tropical SSTs most significant for SEA rainfall variability. The tri-pole is defined as a difference in SSTs. The positive pole is located across the area north of Australia, the Eastern Indian Ocean and the Indonesian maritime continent (0°S -20°S and 90°E to 140°E on the Equator, shifted to 110°E to 160°E at 20°S). Two negative poles are averaged: a region further west in the Indian Ocean (10°N to 20°S and 55°E to 90°E) and a box in the Pacific Ocean (15°N to 15°S and 150°E to 140°W at the northern boundary and 180°E to 140°W at the southern boundary). As expected, the relationships are strongest in winter and spring (autumn and summer not shown). The feature of interest here is the "*white strip*" of weak and statistically insignificant (at the 95% level) correlation between ESB rainfall and the tropical modes of variability. This is not an artefact of using the tri-pole, as white strips are also visible on similar maps calculated using canonical ENSO and IOD indices (not shown).

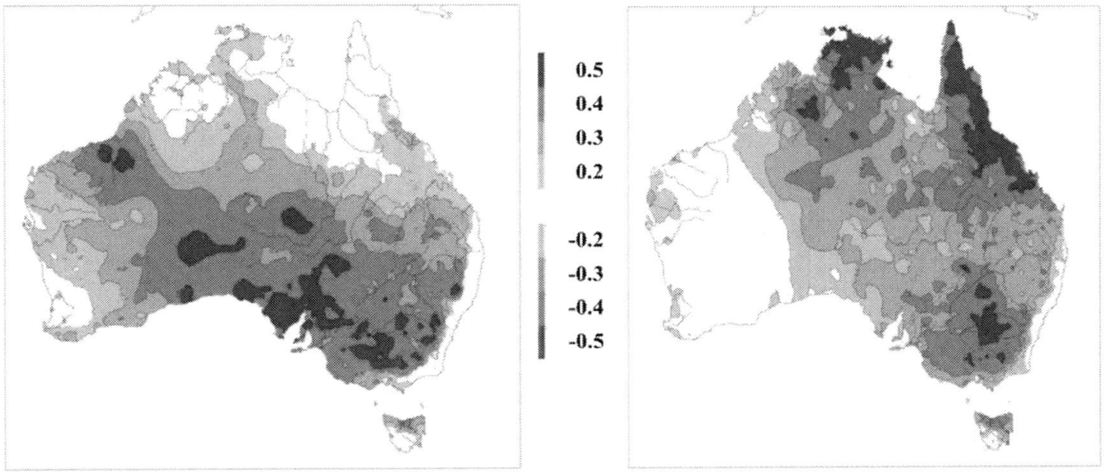

Figure 3. Correlations between detrended values of the annual mean of the Tropical Tri-pole and rainfall across Australia in winter (JJA, left) and spring (SON). Correlations significant at the 95% level and above are colours shaded.

It has been shown previously that tropical modes of variability played an important role during the WWII rainfall deficiency across the Australian continent and in most of SEA [11]. However the absence of a significant relationship between these modes and rainfall in the ESB in the recent record suggests that tropical modes are less likely to be the cause of the WWII rainfall deficiencies for the coastal regions, which were also noted as the largest on record. This is unless a specific mechanism, which does not operate on inter-annual variability, can be established. The reasons for the WWII rainfall deficiencies, and the reasons for the ESB having negligible correlation with tropical modes of variability, require further investigation. The weakening of the ENSO-IOD relationship on the other side of the GDR suggests a mechanism specific to the ESB rainfall (such as moisture transport driven by easterlies) in response to large-scale displacements related to ENSO-IOD variability which across the continent has a negative effect on rainfall.

5. Climate model future projections for ESB

ESB is likely to be a difficult region in which to produce reliable future projections. Current climate projections exhibit very uncertain projections across most of Eastern Australia. In addition, the rainfall projections (available interactively: e.g. http://www.climatechangeinaustralia.gov.au/nswactrain5.php) are very uniform across eastern Australia despite very contrasting background climatology between the coast and in-land regions (http://www.bom.gov.au/jsp/ncc/climate_averages/rainfall/index.jsp).

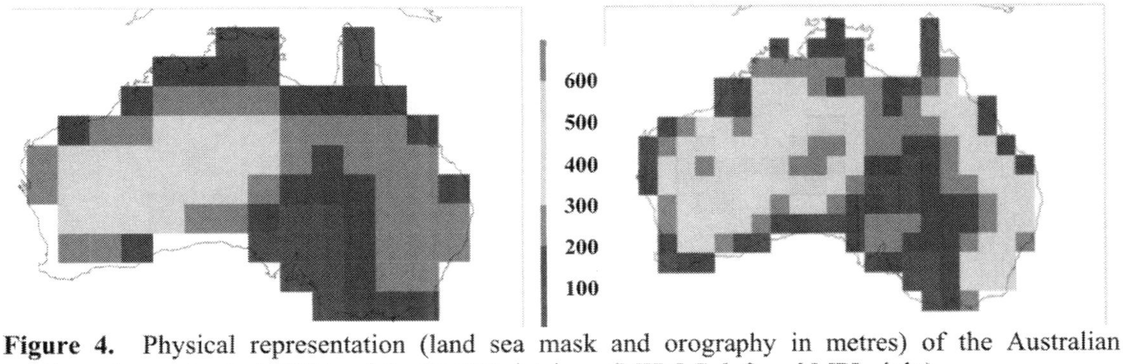

Figure 4. Physical representation (land sea mask and orography in metres) of the Australian continent in two typical GCMs of the CMIP3 database (MIROC, left and MPI, right).

This unlikely behaviour is best explained by plotting the land-sea mask and topography (figure 4) of current General Circulation Models (GCMs). In general, GCMs have insufficient resolution to provide a realistic physical representation of the ESB. The highest elevation points in the GCMs (representative of the GDR, although the altitude does not exceed 500m) are located next to the coast with no (or very few, in the case of the highest resolution models) low elevation grid boxes east of it.

6. Conclusions

In this note, the specificities of the climate of the Eastern SeaBoard (ESB) have been noted, mostly in comparison with other sub-regions in South-East Australia (SEA). The ESB is a small climatic entity, which benefits from twice the rainfall of the rest of SEA. It is dominated by Summer/Autumn rainfall in contrast to the South-West of Eastern Australia (SWEA). Rainfall in the ESB does not suffer significantly from a strengthening of the STR, which is the key driver of the rainfall deficiency in SWEA, while a shift south of the STR is likely to be associated with a rainfall increase in the region. Finally, it does not appear to have the same strong relationship with ENSO and the IOD that is evident across most of Australia. This evidence makes it difficult to understand and attribute the current rainfall deficiencies in the ESB as well as the WWII drought. Besides tropical modes of variability, the largest mode of variability in the Southern Hemisphere is the Southern Annular Mode (SAM). However, it is also unlikely to be the major cause of the rainfall decline, since it exhibits a positive relationship with ESB rainfall [12], and the SAM index has been trending upward during recent decades; however based on [12], it is likely to be a key contribution to the summer anomalies, in addition [12] suggest an increase in cut-off lows related to SAM trend.

The specificities of the ESB climate, and the difficulty of relating its variability and recent trends to important large-scale factors, make it a worthwhile area to explore further [13]. In addition, the inability of GCMs to have a realistic physical representation of the ESB, coupled with their inability to generate realistic looking East Coast Lows (the most significant synoptic event affecting the ESB [14]) means that current GCM projections for future rainfall are likely to be unreliable.

The ESB is perhaps one of the most important socio-economic regions in Australia, containing nearly half of Australia's total population. Understanding the current climate variability and creating reliable future climate scenarios for the region make for a very interesting and challenging scientific endeavour, worthy of a dedicated program of research

Acknowledgements

This communication arises from the involvement of BT in the South-Eastern Australian Climate Initiative (SEACI). The support of the SEACI funding agencies is acknowledged.

References

[1] Drosdowsky W 1993 *Int. J. Climatology*,**13**,1-30
[2] Timbal B 2009 *CAWCR Res. Lett.*, **2**, 8pp
[3] Jones D A, Wang W and Fawcett R 2009, *Aust. Met. Mag.*, submitted
[4] Murphy B and Timbal B 2008 *Int. J. Climatology*, **28**, 859-79
[5] Timbal B *et al* 2007 *SEACI final report for project 1.1.2*, 19 pp
[6] Drosdowsky W 2005 *Int. J. Climatology* **25** 1291-99
[7] Timbal B and Smith I 2010 *SEACI final report for project 1.1.1P* 24 pp
[8] Seidel D J, Fu G, Randel R J and Reichler T J 2008 *Nature Geoscience* **1** 21-24
[9] Nicholls N 1989 *J. Climate* **2** 965-73
[10] Timbal B and Murphy B 2007 *BMRC Res. Lett.,* **6** 6-11
[11] Timbal B and Fernandez E *2010 SEACI final report for project* 1.2.1P 25 pp
[12] Hendon H H, Thompson D W J and Wheeler M C 2007 *J. Climate* **20** 2452-67
[13] Speer M S, Leslie L M and Fierro A O *Clim. Dyn.* DOI10.1007/s00382-009-0726-1
[14] Speer M S, Wiles P and Pepler A 2009 *Aust. Met. Oceanographic Journal* **58** 29-39

IOP Conf. Series: Earth and Environmental Science **11** (2010) 012014 doi:10.1088/1755-1315/11/1/012014

Influence of climate modes on tropical cyclone formation in the Australian region

A Werner
Macquarie University, Department of Environment and Geography, Faculty of Science, NSW 2109 Sydney, Australia.

E-mail: awerner@science.mq.edu.au

Abstract. Relationships between tropical cyclogenesis (TCG) in the Australian region and modes of various atmospheric and oceanic parameters are investigated for the period 1968-2006. Seven thermal and seven dynamic variables describing the state of moisture transport and wind flow patterns in the atmosphere were analysed. In order to identify the dominant modes of each variable, an empirical orthogonal function (EOF) is applied. The three leading modes of each variable were then considered. This study corroborates previous work showing that El Niño-Southern Oscillation (ENSO) is the leading contributor to TCG in the all-Australian, Western and Eastern Australian regions. The ENSO-linked first EOF of the streamfunction at sigma level 0.2101 shows the strongest relationship with TCG variations in the all-Australian and Eastern regions correlating at 0.73 and 0.62, respectively. In the western Australian region, the first EOF of environmental vertical wind shear shows the strongest skill. TCG frequency in the Northern region is mainly affected by the large-scale wind patterns of the subtropics and mid-latitudes such as the subtropical jet and the Rossby wave train.

1. Introduction

Tropical cyclone (TC) formation only occurs when certain atmospheric and oceanic conditions are present. [5] defined basic favourable environmental conditions for tropical cyclone formation to include sea surface temperature (SST), conditional instability in the lower to mid-troposphere, absolute vorticity in the lower troposphere, mid-troposphere relative humidity, relative vorticity in the upper troposphere, and vertical shear of the horizontal winds between the lower and upper troposphere. These parameters were further enhanced in several studies (e.g. [6], [13]), but the general dynamic and thermal requirements on the state of the atmosphere and ocean remains the same. However, understanding the physical processes of TC formation during these favourable conditions is still a subject of ongoing research, particularly with respect to the various climate signals, which also influence the conditions.

In the Australian region, the El Niño Southern-Oscillation (ENSO) signal is known to have a strong impact on TC frequency (e.g. [16], [1], [11]). [17] recently found that next to SST, vertical wind shear of zonal winds between 850 and 200hPa and low-level relative vorticity are the main contributors of ENSO-related variables on TC formation. Other studies have identified atmospheric signals, which influence Australian TCs. The Australian monsoon trough was recognized to have a strong effect on TCG in the Australian region ([14]). Later, [4] suggested that the effect of ENSO-linked dynamics on tropical cyclogenesis (TCG) occurs through a strong relationship between the monsoon trough and ENSO via atmospheric bridge processes. Another study found the equivalent potential temperature gradient between the surface and 500hPa as the most promising variable for prediction [15]. Additionally, the Madden-Julian-Oscillation is known to highly affect TC formation and occurrence on intra-seasonal time-scales (e.g. [7], [8], [12]).

The aim of this study is to identify the most important thermal and dynamic modes for TC development in the Australian region. This will improve our understanding of which parameters are the main contributors to variations in the number of TC formations and which climate signals describe these parameter variations best. Additionally, the lead-time in the correlation is investigated to explore use of the modes as possible predictors of TC frequency.

2. Data and Methods

2.1. Variables

For this study 14 thermal and dynamic parameters are examined (table 1). The ocean temperature and therefore the potentially available moist convection due to evaporation is described by the SST. Relatively moist layers in the lower- and mid-troposphere (relative humidity at 500hPa) are essential, as dry mid levels suppress the continuing development of widespread deep convection. The general distribution of near surface temperature and pressure patterns, are represented by the temperature at 925hPa and sea level pressure, while the potential energy in the troposphere is identified by the geopotential height at 200hPa. The stability of the troposphere, which indicates the support of deep convection, is measured by the convective available potential energy. The gradient of equivalent potential temperature between 100hPa and 500hPa describes the enthalpy between these two layers and hence the likelihood to form cloud clusters.

The general flow pattern is described by zonal and meridional winds at 850 and 200hPa (u 850, v 850, u 200, v 200), as well as by the streamfunction at sigma level 0.2101 (Psi 200). It is imperative, that the environmental vertical wind shear (EVWS) between the lower (850hPa) and upper atmosphere (200hPa) is weak, otherwise the convective system of the eyewall of a TC cannot develop or persist. Also, an existing negative (Southern Hemisphere) low-level relative vorticity anomaly at 850hPa, such as a small cyclonic atmospheric disturbance, a tropical wave or a monsoonal trough with convergence is a necessary initial factor to develop a TC. The attendant strong divergence supports the development of deep convection, which in turn intensifies the disturbance into a low-pressure system.

Table 1. Variables and its abbreviations

	Variables	Abbreviation
Thermal Parameters	sea surface temperateure	SST
	relative humidty at 500hPa	RH 500
	temperature at 925hPa	T 925
	sea level pressure	SLP
	geopotential height at 200hPa	GPH 200
	convective available potential energy	CAPE
	equivalent potential temperature	EPT
Dynamic Parameters	zonal wind at 850hPa	u 850
	zonal wind at 200hPa	u 200
	meridional winds at 850hPa	v 850
	meridional winds at 200hPa	v 200
	streamfunction at sigma level 0.2101	Psi 200
	environmental vertical wind shear	EVWS
	relative vorticity at 850hPa	RV 850

2.2. Data

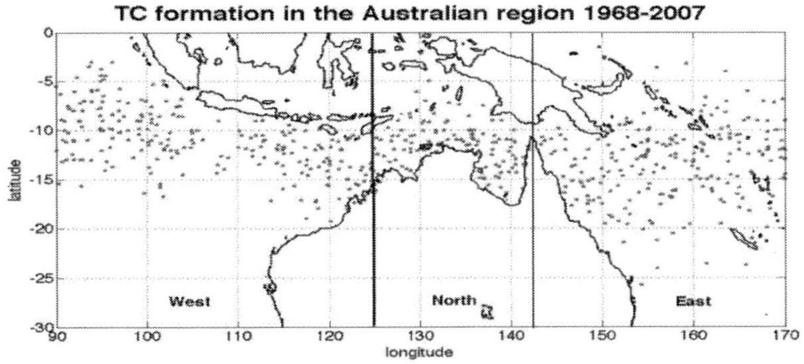

Figure 1 Spatial distribution of TCG occurrence in the Australian region between 1968-2007. The subregions used in this study are also indicated.

This study takes advantage of the global TC best track data set IBTrACS [10] provided by the National Oceanic and Atmospheric Administration (NOAA). TCG is defined as the spatial location where a tropical system exceeding 34 knots is first noted. The Australian region is defined from 0-30°S and 90-170°E. Following [2], this domain is divided into three subregions: a Western, a Northern and an Eastern region with its borders at 125°E and 142.5°E. TCG occurrences identified poleward of 30°S or over land have been removed while TCs which developed exactly on one of the subregional borders at 125° and 142.5°E are assigned the region to which they tracked. Only storms during the Australian TC season from November to April are taken into account. Overall, a total of 564 TCs during the 40-year period from 1968 to 2006 are analysed (figure 1).

The data used to derive the SST indices is the Hadley Centre Global Sea Ice and Sea Surface Temperature (HadISST1) dataset [18], compiled by the UK Met Office Hadley Centre. HadISST1 is a combination of global monthly SST fields and sea ice concentration on a 1° x 1° grid from 1870 to the present. The ENSO signal is described by the NINO4 index and derived from SST anomalies (SSTAs) averaged between 5°S-5°N and 160°E-150°W.

The atmospheric data is the NCEP/NCAR Reanalysis Monthly Means upper-air reanalysis with a 2.5° horizontal resolution on various pressure levels and the spectral reanalysis on a T62 Gaussian grid on various vertical sigma levels [9]. In total, 14 variables, which describe the state of moisture transport and wind flow patterns in the atmosphere are tested. After computing the variables, monthly anomalies of all variables are derived with base period set to 1961-1990.

2.3. Methods

For understanding dominant modes of variability, all potential predictors are decomposed via the empirical orthogonal function (EOF) technique for the tropical Indo-Pacific region. The analysed study domain is 30°N-50°S, 30°E-70°W. For the leading three modes of each predictor, correlations are computed between three-monthly means starting at January-February-March (JFM) prior to the upcoming TC season and ending with March- April-May (MAM) at the end of that TC season, of the principle components (PC) against the regional TCG occurrence in the following season. All results take into account serial correlations in the time series according to [3] and significance is tested against the 95% confidence level. The data is detrended prior to the correlation analysis to focus on interannual variations.

3. Results

The EOF analysis helps to identify the leading modes of each of the investigated variables. Using a serial correlation analysis these modes are linked to ENSO, as well as to estimate the importance of each variable for TC formation in each subregion. In the all-Australian region, the first EOF of the Psi

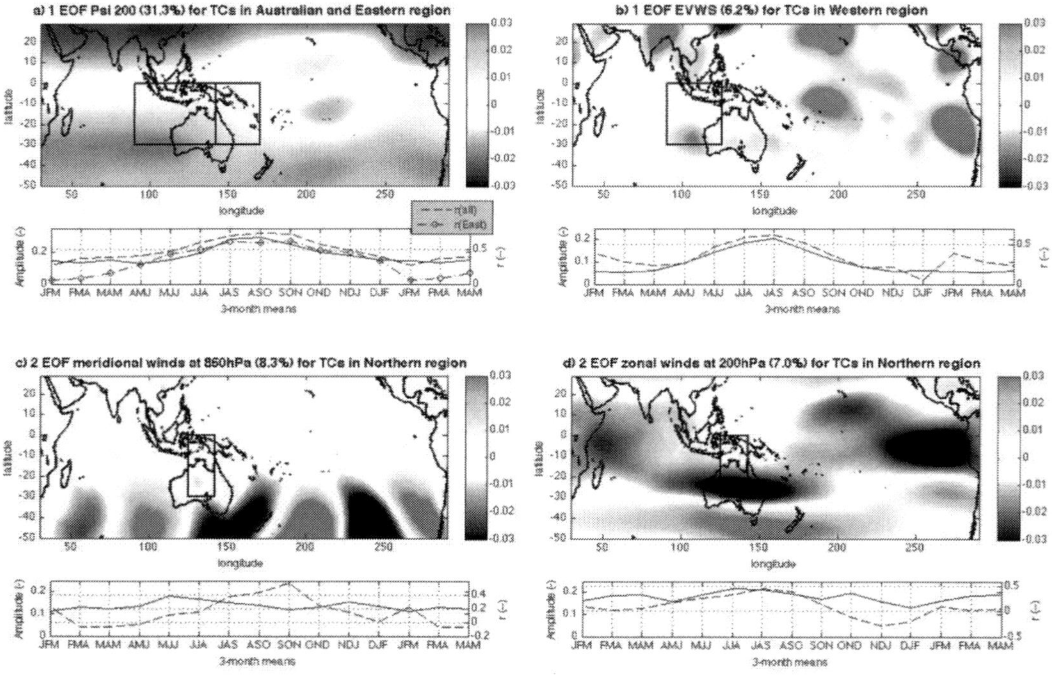

Figure 2: Top: EOF mode with strongest correlation to regional TCG occurrence with the observed region indicated by the box. Bottom: Solid line in plot is the standard deviation of PC during each 3-month period, dashed lines represent the correlation coefficient with regional TC frequency from JFM prior to season till post-seasonal MAM. a) 1st EOF of Psi 200 and TCG in all-Australian and Eastern region, b) 1st EOF of EVWS and Western region, c) 2nd EOF of 850hPa meridional winds and Northern region and d) 2nd EOF of 200hPa zonal winds and Northern region.

200 correlates best with TC frequency. Figure 2a shows the EOF, as well as its amplitude for each 3-month mean and its correlation with seasonal TC variability in the Australian region from pre-seasonal JFM till post-seasonal MAM. The correlation peaks during ASO with r = 0.73 but is already very strong during JAS (r = 0.70). This EOF mode, which explains 31.3% of the variability of Psi 200, shows the ENSO signal and its peak during ASO correlates well with NINO 4 at r = -0.75 during this time. For the subregions also, Psi 200 is most strongly correlated with TC variations in the Eastern region, peaking during SON with r = 0.62 (figure 2a). In the Western region, this correlation reaches r = 0.57 during ASO, while TC formation in the Northern region only correlates at r = 0.26 with Psi 200, peaking during SON (not shown).

The strongest correlation for TC formation in the Western region is with the first EOF of EVWS (r = 0.62; figure 2b). Even though the mode only explains 6.2% of the variance, the pattern is quite distinctive and peaks during JAS. At this time of the year the PC correlates strongly with NINO 4 (r = 0.60). On the other hand, TCG occurrence in the Northern region is dominated by quite different patterns. No link to ENSO was found. The highest correlation coefficient is found between TC frequency and v 850 with 0.57 during SON (figure 2c). This mode is most active during the austral autumn. u 200 also correlates with Northern TC variations at its active phase during JAS at 0.46 (figure 2d).

4. Summary and Discussion

Previous studies have shown high correlations with TCs in the Australian region and NINO4 (e.g. [16], [17]). In this study, we also found ENSO as the most dominant signal affecting the number of seasonal TC formations in the Australian region. However, a different way of explaining variations in TC numbers is provided here, which improves correlations in comparison to the previously chosen ENSO-related SST indices. Additionally, the subdivision into three TC development regions better reflects the regional variability identified in the driving mechanisms prior to the TC season. It was shown, that the first EOF of the streamfunction at sigma level 0.2101 best captures the variations important to TC frequency (figure 2a). The EOF reveals a highly correlated with ENSO dipole in the tropical central Pacific surrounded by the general mean flow pattern peaking during summer. It provides the strongest correlations with annual TC formation in the all-Australian and Eastern subregion and also performs fairly well for TCG occurrence in the Western subregion. For the latter, the first EOF of environmental vertical wind shear between 850 and 200hPa is most strongly correlated with TC variability (figure 2b), showing the greatest variance in the tropical and subtropical in- and out-flow regions of the general circulation. TC formation in the Northern region appears to be driven by different mechanisms, with flow patterns in mid-latitudes playing a more important role. In particular, the variations in the strength and location of the subtropical jet and the mid-latitude Rossby wave train seem to be important. These patterns are seen in the strongest correlated modes with annual TCG occurrence in the Northern region, i.e., in the second EOF modes of the meridional winds at 850hPa and zonal winds at 200hPa. In all regions, the best related predictor for TCG frequency shows consistently high correlations during the JAS 3-month period. This allows for a more reliable forecast of the expected TC variability with a lead time of almost 2 months prior to the start of the Australian TC season.

This study provides a summary of the leading climatic modes in the Indo-Pacific area and how they affect regional TC variability in the Australian region. Strong and robust correlations were found for TC variations in the Australian region, as well as in its subregions. We showed in particular that:

- ENSO is the leading contributor to TC variations in the all-Australian, Western and Eastern region;
- The first EOF of the streamfunction at sigma level 0.2101 is a strong predictor, especially for TC occurrence in the all-Australian and Eastern regions;
- The first EOF of environmental vertical wind shear shows strongest skill in the Western subregion;
- TC variability in the Northern subregion is dominated by different mechanisms. The frequency is mainly affected by the large-scale wind patterns of the subtropics and mid-latitudes such as the subtropical jet and the Rossby wave train;
- The lead time for the highest correlations with TC frequency varies depending on the predictor being examined though a three month lead is usually detected in the modes of interest.

These results advance our knowledge of the interactions of large scale (climate) patterns and how they affect TC variability in the Australian region. It also provides an indication of the possible lead-time for predictors of the TC frequency for the following season, as well as identifying which variables dominate TC formation. This information can be usefully translated to seasonal TC forecast models for the Australian region.

References

[1] Basher R E and Zheng X 1995 Tropical cyclones in the southwest Pacific: Spatial patterns and relationships to Southern Oscillation and sea surface temperature *J. Climate* **8** 1249-60

[2] Dare R A and Davidson N E 2004 Characteristics of tropical cyclones in the Australian region *Mon. Weather Rev.* **132** 3049-65

[3] Davis R E 1976 Predictability of sea surface temperatures and sea level pressure anomalies over the North Pacific Ocean *J. Phys. Oceanogr.* **6** 249-266

[4] Evans J L and Allan R L 1992 El Nino/Southern Oscillation modification to the structure of the monsoon and tropical cyclone activity in the Australasian region *Int. J. Climate* **12** 611-623

[5] Gray W M 1968 Global view of the origin of tropical disturbances and storms *Mon. Wea. Rev.* **96** 669-700

[6] Gray W M 1975 Tropical cyclone genesis *Dept. of Atm. Sci. Paper No. 234, Colorado State University, Fort Collins, CO,* 121 pp [Available from Department of Atmospheric Sciences, Colorado State University, Ft. Collins, CO 80523]

[7] Hall J D, Matthews A J and Karoly D J 2001 The modulation of tropical cyclone activity in the Australian region by the Madden-Jullian Oscillation *Mon. Wea. Rev.* **129** 2970-82

[8] Ho C-H, Kim J-H, Jeong J-H, Kim H-S and Chen D 2006 Variation of tropical cyclone activity in the South Indian Ocean: El Niño –Southern Oscillation and Madden-Julian Oscillation effects *J. Geophys. Res.* **111** D22101, 8PP., doi:10.1029/2006JD007289

[9] Kalnay E and Coauthors 1996 The NCEP/NCAR 40-Year Reanalysis Project *Bull. Amer. Meteor. Soc.* **77** 437–471

[10] Knapp K R, Kruk M C, Levinson D H and Gibney E J 2009 Archive compiles new resource for global tropical cyclone research *Eos Trans. AGU* **90(46)** doi:10.1029/2009EO060002

[11] Kuleshov Y and de Hoedt G 2003 Tropical cyclone activity in the Southern Hemisphere *Bull. Aust. Meteor. Oceanogr. Soc.* **16** 135-137

[12] Leroy A and Wheeler M C 2008 Statistical prediction of weekly tropical cyclone activity in the Southern Hemisphere *Mon. Wea. Rev.* **136** 3637-54

[13] McBride J L and Zehr R 1981 Observational analysis of tropical cyclone formation. Part II: Comparison of non-developing versus developing systems *J. Atmos. Sci.* **38** 1132-51

[14] McBride J L and Keenan T D 1982 Climatology of tropical cyclone genesis in the Australian region *J. Climatology* **1** 13-33

[15] McDonnell K A and Holbrook N J 2004 A Poisson regression model of tropical cyclogenesis for the Australian-southwest Pacific Ocean region *Weather and Forecasting* **19** 440-455

[16] Nicholls N 1984 The southern oscillation, sea-surface temperature, and interannual fluctuations in Australian tropical cyclone activity *J. Climatol.* **4** 661-670

[17] Ramsay H A, Leslie L M, Lamb P J, Richman M B, and Leplastrier M 2008 Interannual variability of tropical cyclones in the Australian region: Role of large-scale environment *J. Climate* **21** 1083-1103

[18] Rayner N A, Parker D E, Horton E B, Folland C K, Alexander L V, Rowell D P, Kent E C and Kaplan A 2003 Global analyses of sea surface temperature, sea ice, and night marine air temperature since the late nineteenth century *J. Geophys. Res.* **108** 4407, doi:0.1029/2002JD002670

IOP Conf. Series: Earth and Environmental Science 11 (2010) 012015 doi:10.1088/1755-1315/11/1/012015

The great Sydney dust event: size-resolved chemical composition and comparison

Michael A Box, Majed Radhi and Gail P Box

School of Physics, University of New South Wales, Sydney, NSW 2052, Australia.

E-mail: m.box@unsw.edu.au

Abstract. In September 2009, a major dust storm crossed eastern Australia, blanketing Sydney on two occasions. We collected size-resolved aerosol samples on both days, and have subjected them to Ion Beam Analysis at ANSTO. In this paper we present these results, and compare them with some of the analysis of similar samples obtained in field trips to the Lake Eyre Basin of central Australia, the primary source region of the dust. In particular, the Fe/Al ratios (~0.9) are similar to LEB values, and higher than northern hemisphere values (~0.55). Salt entrainment indicates a source near dry lakes.

1. Introduction

On 23 September 2009, a major dust storm passed across most of eastern Australia, reaching Sydney in the early hours of the morning. A second, smaller storm occurred on 26 September. We collected aerosol samples during both of these events, as well as the day following, and have analysed them in an attempt to understand their chemical composition, and hence origin. The collection and analysis protocols are the same as we employed in three field trips to sites in the Lake Eyre Basin (LEB) (Radhi et al., 2009; 2010), and it is our intention when all analyses are complete to make a thorough comparison of these data sets.

We used a 12 stage Micro Orifice Uniform Deposition Impactor (MOUDI) (Marple et al., 1991) to collect size-segregated aerosol samples during the two dust storm events, to allow identification of any compositional differences between particles of different size. The sum of the masses on all MOUDI substrates can be considered a good measurement of Total Suspended Particulates in the atmosphere (TSP) for one sampling period, as all particles for different sizes are collected at the one time. (Note that this is a time-averaged mass concentration.)

During the first dust storm the MOUDI was deployed on the roof of the Physics building at the University of the New South Wales (close to the coast), while during the second dust storm the MOUDI was deployed in the backyard of a house in the west of Sydney. Sampling times varied according to ambient dust levels (Table 1).

Table 1. Timings of sample collection and TSP mass concentrations (μg/m^{-3}).

Collection	Start		End		Total time	TSP μg/m^3
	Date	Time	Date	Time		
1st Dust Storm	23/09/2009	9:06:00	23/09/2009	14:32:00	5:26:00	578
Day After	23/09/2009	15:00:00	24/09/2009	13:36:00	22:36:00	33
2nd Dust Storm	26/09/2009	5:50:00	27/09/2009	5:00:00	23:10:00	58
Day After	27/09/2009	8:12:00	28/09/2009	7:25:00	23:13:00	25

2. Gravimetric Mass Distributions

We weighed the substrates before and after sampling at the Australian Nuclear Science and Technology Organization (ANSTO) using a Mettler Toledo MX5 balance with repeatability 0.0008 mg at gross load. This provided the loading on each filter.

During the first dust storm to hit Sydney the TSP was 578 μg/m^3, reducing to 33 μg/m^3 during the day after the dust storm: see Table 1. The second dust storm hit Sydney three days later: the TSP during this period was 58 μg/m^3 and 25 μg/m^3 for dust storm day, and the day after the dust storm, respectively: Table 1. The size-resolved mass concentrations during each dust storm day and the day after are presented in Figure 1. It is clear from this graph that the mass size distribution during dust storm periods has a broad peak in the coarse fraction, reducing somewhat the day after. Nevertheless, the significant coarse fraction indicates that some dust was still present in the Sydney atmosphere the day after each dust storm.

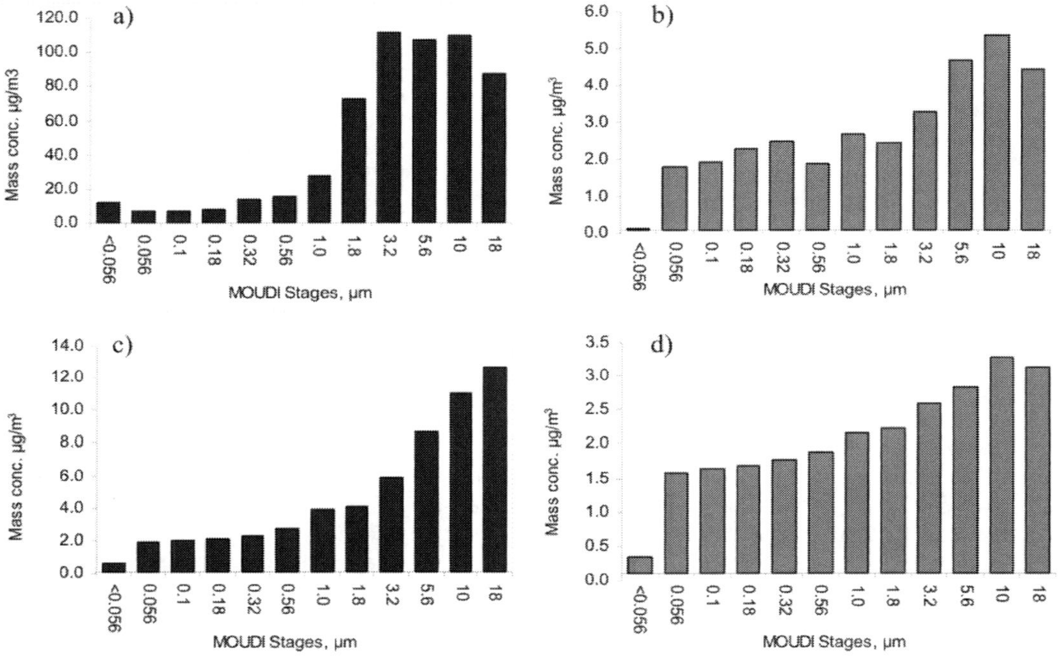

Figure 1. Size-resolved mass distributions of (a) 1st dust storm; (b) Day after 1st dust storm; (c) 2nd dust storm; (d) Day after 2nd dust storm.

The National Environmental Protection Measure (NEPM) for particulate matter is 50 μg/m^3 (24 hour average) for PM_{10} (particulate matter with diameters less than 10μm), although Sydney values are typically 15-25 μg/m^3. We have extracted the PM_{10} values from these 4 data sets, and they are 353, 23, 34 and 19 μg/m^3. Thus we see that only the first storm exceeded the NEPM, although the second storm day was dusty by Sydney norms. (Note that some of the larger particles which remain in the atmosphere the following days will not be included in the PM_{10} standard.)

We have collected similar samples during both dust storm and ambient conditions, at three sites in the Lake Eyre Basin of central Australia: Birdsville Qld. (Radhi et al., 2009); Muloorina SA (near Marree) (Radhi et al., 2010); and Fowlers Gap Station (FGS) NSW : see Figure 2. The LEB is the major dust source region in Australia, and the source of much of this dust storm. To compare the behaviour of the mass size spectrum for these dust storms we applied a smoothing/inversion procedure (Keywood et al., 1999), to the Sydney, Birdsville (a weak storm), Muloorina and FGS data (for both first dust storm day, and the day after), and the results are presented in Figure 3.

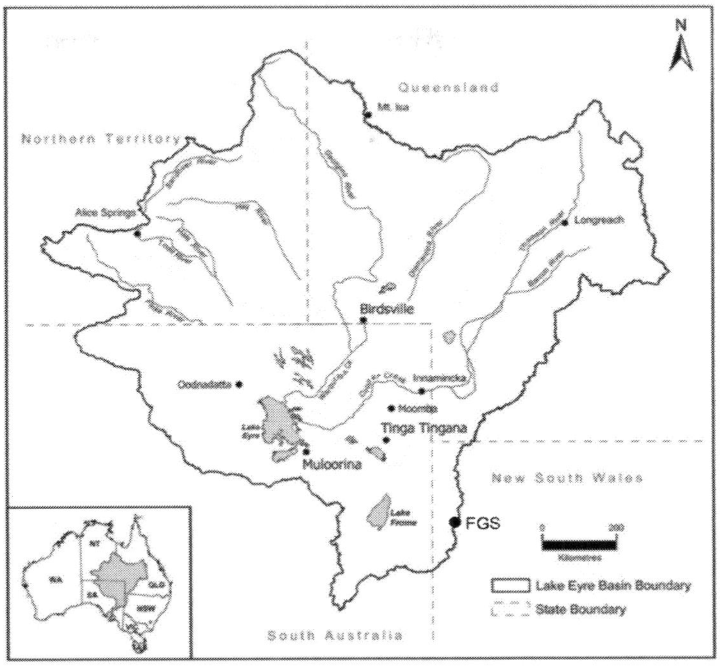

Figure 2. The Lake Eyre Basin, showing sampling sites.

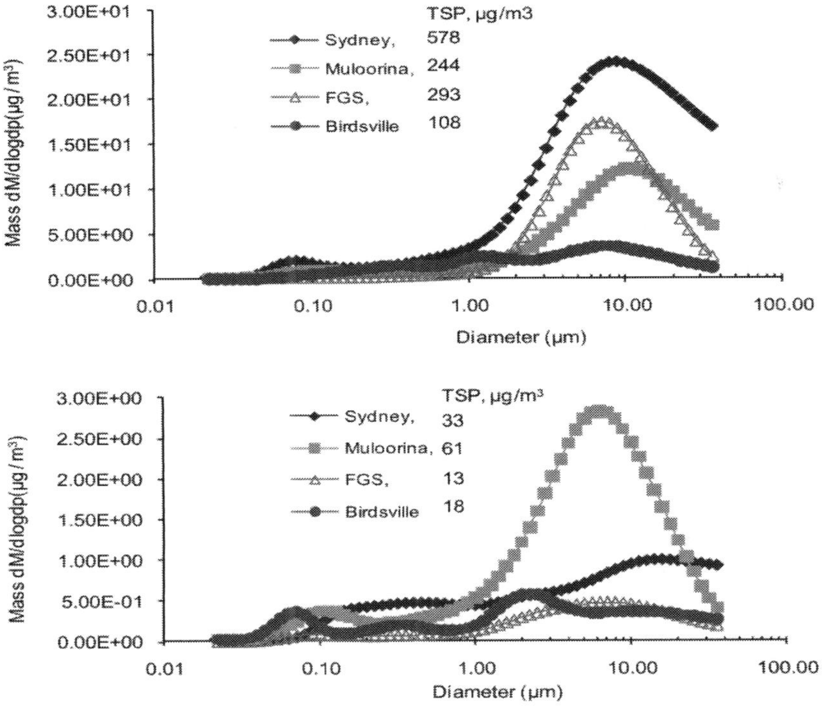

Figure 3. Size-resolved mass concentrations sampled during (top) the 1st dust storms at Sydney, Muloorina, Fowlers Gap Station and Birdsville; (bottom) the day after the dust storms.

For the dust storm days, these results show a very similar spectrum, with a clear peak in the coarse mode between 5 and 10 μm. On the day after the dust storm, the coarse mode peak is significantly reduced for both the Sydney and Fowlers Gap data, but less so for the Muloorina data. This is a clear reflection of the fact that there was still considerable dust in the Muloorina air the day after the dust storm, as the TSP was 61 $\mu g/m^3$ (down from 244 $\mu g/m^3$ during the storm). This, presumably, is a reflection of the proximity of the sampler to some key sources.

3. Elemental Composition

The substrates were then analysed using accelerator-based Ion Beam Analysis (IBA) techniques (PIXE and PIGE) at ANSTO (Cohen, 1993; Cohen et al., 1996). The IBA technique uses an 8 mm diameter beam of 2.6 MeV protons with 10-15 nA target current, and is capable of detecting most elements from Na to Pb. As the spatial distribution of sample on each substrate varied, and the ion beam only sampled the central 10% of the filter, we have used the IBA results primarily to obtain the ratio of elements to the most abundant element in our samples.

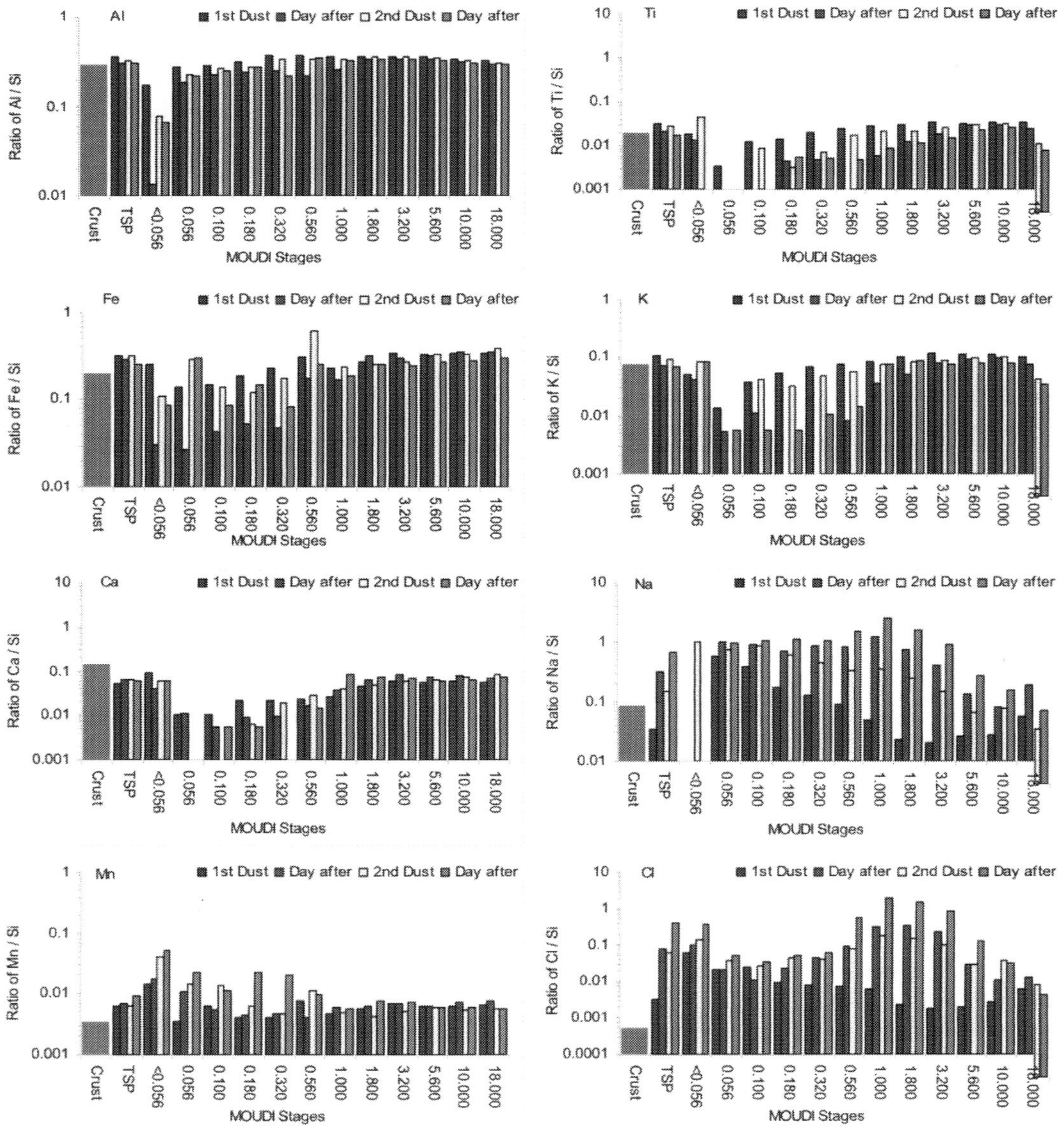

Figure 4. Ratios of some elements to Si over all stages collected by the MOUDI.

Ion beam analysis showed that, as expected, Si is an abundant element in all size fractions, and is used in all our studies as a dust 'indicator' for Australian desert aerosol. The elemental concentrations determined through the IBA are used to calculate the mass ratio of elements to Si. Size-resolved mass ratios for selected elements are presented in Figure 4, along with the mass ratios of these elements in the Earth's crust from CRC Handbook (Lide, 1997).

3.1. Crustal elements

The size-resolved mass ratios in Figure 4 show that Al occurred in most size fractions. By contrast with the crustal value (0.292) the Al/Si mass ratios were moderately higher during dust storm days (0.353 and 0.328) but close to the crustal value during the days after dust storms (0.299 and 0.300).

The Ti/Si mass ratio plot shows that Ti is predominant in the upper fine and coarse fractions in all events. The TSP Ti/Si mass ratios were high during dust storm days, by contrast with the crustal value (0.020), and are consistent with the values found in LEB during dust activities. Nevertheless this ratio is close to the crustal value during the days after the dust storms.

It is clear from the size-resolved mass ratios in Figure 4 that the Fe/Si mass ratio is greatest in the coarse fraction for all events. During the 1st dust storm the Fe/Si TSP mass ratios were 0.312 and 0.287, and during the 2nd dust storm they were 0.320 and 0.247, for the dust storm day and day after, respectively. These ratios are consistent with the values found at various locations within the LEB (Radhi et al. 2009, 2010). When combined with the Al/Si ratios, this gave Al/Fe ratios of 0.884, 0.960, 0.976 and 0.823, respectively. This is consistent with the values found within the LEB – 0.77-0.85 at Birdsville; 0.95-1.02 at Muloorina; and 0.95-1.1 at Fowlers Gap – and are higher than values found during Northern Hemisphere dust storms – 0.55-0.61 (Lafon et al., 2006).

Qin and Mitchell (2009) have performed a cluster analysis of ground-based remote sensing data from a range of sites around Australia, including Birdsville. They have derived 4 classes, which they identify as aged smoke, fresh smoke, coarse dust, and a 'super-absorptive' class of unknown origin. Their coarse dust class has significant absorption in the blue spectral region, which points to a high hematite content. This, again, is consistent with our analysis.

Calcium is mainly in the upper fine and coarse fractions in all events, with the TSP Ca/Si mass ratios lower than the crustal value, which is consistent with what has been found at LEB sites. The Mn/Si mass ratio was dominated by the fine fraction as shown in the size-resolved plot. This ratio was high in all events by contrast with the crustal value but is consistent with the values found at the LEB sites.

The size-resolved K/Si mass ratios were close to crustal in (almost) all size fractions during the dust storm days, but predominated in the coarse fraction during the day after the dust storms. The TSP ratio was higher than the crustal value during the dust storm days but close to this value during the day after. It has been found at FGS during a dust storm that this ratio was 0.12, which is close to the value of 0.11 found during the first dust storm.

3.2. Na and Cl

The size-resolved Na/Si mass ratios show a different structure for each event. During the first dust storm day this pattern was bimodal with a strong peak at 0.056 μm, a minimum in the 1.8-3.2 μm range, and then another peak at 18.0 μm. During the second dust storm this distribution was unimodal with peak in the fine fraction, and reducing steadily but remaining higher than the first storm. The day after the dust storm the Na/Si distribution shows a comparable pattern for both samples, with high values in the fine fraction and a weak peak in the 1.0-3.2 μm range, with lower values in the coarse fraction. The TSP mass ratios of Na/Si are higher than 0.083, the crustal value, during all samples, apart from the first dust storm (0.033) (presumably a reflection of the higher dust concentration on that day): this feature has been reported by Radhi et al. (2010) during dust storm events in Muloorina, and also FGS.

Chlorine occurred in all size fractions for all events. It is clear that the Cl/Si pattern follows the Na/Si pattern during the first dust storm day, while during the rest of the events this pattern has a broad distribution with peak in the 1.0-3.2 μm range. The TSP Cl/Si mass ratios were higher than the crustal value in all events which indicates that non dust sources have contributed to the total Cl measured in all samples. It is worth noting that the Cl/Si mass ratio during the first dust storm is close to the value found at Muloorina and FGS during dust storm activities (Radhi et al., 2010).

The TSP Cl/Na mass ratios were higher than the crustal ratio in all samples; though still lower than 1.8, the value in seawater. During the first dust storm this ratio was very close to the value found at Muloorina site during non-dust days (Radhi et al. 2010), which reflected the influence of dry salt from Lake Eyre. For the other samples the high ratio indicated a clear contribution from another sources (such as sea spray) to total salt found on these samples.

4. Discussion

The Lake Eyre Basin is a region of dry salt lakes, and a major source of aeolian dust (Greene at al., 2009; McTainsh, 1989). Recent drought conditions have increased this dust activity (Mitchell et al., 2010). The dust storms which swept across much of eastern Australia in September 2009 blanketing most of the major population centres in New South Wales and Queensland, including Australia's largest city, Sydney, is a dramatic manifestation of these processes.

We have collected size-resolved aerosol samples in order to investigate their chemical composition. The first key finding is that the Fe/Al ratio of ~0.9 is very similar to this ratio at 3 sites within the Lake Eyre Basin.(~0.8-1.0), and around 50% higher than equivalent northern hemisphere values. This is a direct reflection of the high iron (hematite) content of Australian desert soils, which are far more reddish than northern hemisphere deserts.

The Cl/Si mass ratio during the first dust storm was similar to values measured at Muloorina, close to Lake Eyre, a dry salt lake (although at the time of writing filling with water). This is a clear indication of the entrainment of dry salt along with the dust, giving further support to the suggestion that this region was the source of at least a significant fraction of the dust.

Acknowledgements

This work was supported in part by ARC grant DP0451400. Ion Beam Analysis was performed under AINSE grant AINGRA09057. We thank David Cohen and Ed Stelcer for their assistance.

References

Cohen D D 1993. Applications of simultaneous IBA techniques to aerosol analysis. *Nucl. Instr. and Meth.* **B79** 385-388.

Cohen D D, Bailey G M, and Kondepudi R 1996. Elemental analysis by PIXE and other IBA techniques and their application to source fingerprinting of atmospheric fine particle pollution. *Nucl. Instr. and Meth.* **B109-110** 218-226.

Greene R S B, Cattle S R, and McPherson A A 2009. Role of eolian deposits in landscape development and soil degradation in southeastern Australia, *Aust J Earth Sciences* **56** 55-65.

Keywood M D, Ayers G P, Gras J L, Gillett R W, and Cohen D D 1999. Relationships between size segregated mass concentration data and ultrafine particle number concentrations in urban areas. *Atmos. Environ.* **33** 2907-2913.

Lafon S, Sokolik I N, Rajot J L, Caquineau S, and Gaudichet A 2006. Characterization of iron oxides in mineral dust aerosols: Implications for light absorption. *J. Geophys. Res.* **111** D21207, doi:10.1029/2005JD007016.

Lide D R E 1997. *CRC Handbook of Chemistry and Physics*. (Boca Raton, Fl: CRC Press).

McTainsh G H 1989. Quarternary aeolian dust processes and sediments in the Australian region, *Quarternary Sci. Rev.* **8** 235-253.

Marple V A, Rubow K L, and Behm S M 1991. A microorifice uniform deposit impactor (MOUDI): description, calibration, and use, *Aerosol Sci. Tech.* **14** 434-446.

Mitchell R M, Campbell S K, and Qin Y 2010. Recent increase in aerosol loading over the Australian arid zone, *Atmos. Chem. Phys.* **10** 1689-1699.

Qin Y, and Mitchell R M 2009. Characterisation of episodic aerosol types over the Australian content. *Atmos. Chem. Phys.* **9** 1943-1956.

Radhi M, Box M A, Box G P, Mitchell R M, Cohen D D, Stelcer E, and Keywood M D 2009. Optical, Physical and Chemical Characteristics of Australian Desert Dust Aerosols: Result from a Field Experiment. *Atmos. Chem. Phys. Disc.* **9** 25085-125.

Radhi M, Box M A, Box G P, Mitchell R M, Cohen D D, Stelcer E, and Keywood M D 2010. Size-Resolved Mass and Chemical Properties of Dust Aerosols from Australia's Lake Eyre Basin. Submitted to *Atmos. Environ.*

IOP Conf. Series: Earth and Environmental Science 11 (2010) 012016 doi:10.1088/1755-1315/11/1/012016

Extreme-value time-series analysis of Australian Region A gust wind speeds to examine instrument bias

R P Cechet and L A Sanabria

Risk and Impact Analysis Group, Geoscience Australia
G.P.O. Box 378, Canberra, Australian Capital Territory

E-mail: bob.cechet@ga.gov.au

Abstract. Australian building codes through the Australia/New Zealand Wind Actions Standard as well as the wind engineering community in general rely to a significant extent on the peak gust wind speed observations collected over more than 70 years by the Australian Bureau of Meteorology (BoM). In the mid-1980's BoM commenced a program to replace the aging pressure tube Dines anemometers with cup anemometers. During the replacement procedure, many localities had more than one type of anemometer operating, recording extreme events. Systematic differences between instrument measurements during this overlap period raised serious concerns about the utility of the peak gust wind speed database. This paper presents the results of a reanalysis of the current BoM peak wind gust database for the non-cyclonic region (Region A) of the Australia/New Zealand Wind Actions Standard. The study utilises extreme value distribution analysis and compares estimates of the 500-year return-period (RP) peak gust wind exceedance level derived from segments of the record measured with the Dines and replacement anemometers. Results indicate that the later period appears to have a significant reduction in extreme events; 17 of 31 sites have a mean 500 year RP exceedance level for the replacement anemometer section of the record below the lower 95% confidence limit for the Dines anemometer part of the record. The 3PM mean wind speed time-series observations have also been examined, and they exhibit a similar trend.

1. Introduction

Australian building codes through the Australia/New Zealand Wind Actions Standard [1] as well as the wind engineering community in general rely to a significant extent on the peak gust wind speed observations collected over more than 70 years by the Australian Bureau of Meteorology (BoM). The current wind loading code and the performance of our infrastructure (residential, commercial, industrial and critical infrastructure) is based primarily on the Dines anemometer interpretation of the peak gust wind speed. In the mid-1980's BoM commenced a program to replace the aging pressure tube Dines anemometer with the Synchrotac and Almos cup anemometers. As of 2009 only six Dines anemometers remain in operation, mainly as backup or for high-speed measurement. During the anemometer replacement procedure, many localities had more than one type of anemometer operating, recording extreme events. The passage of Cyclone Vance through Exmouth in 1999 saw Dines and Almos anemometers, separated by 25 metres, recording peak gusts of 144 and 122 knots respectively [2]. A weak cyclone that passed through Townsville in April 2000 recorded a peak gust of 70 knots on the Dines and 59 knots on the Almos anemometer [3]. These systematic differences raise concerns about the consistency and utility of the peak gust wind speed database.

The pressure-tube (Dines) anemometer is described in [4] and [5]. Installation at Australian sites commenced in the 1930's. The Dines anemometers were factory calibrated under steady state conditions (mean wind speeds); their response to transient wind conditions (gusts) was not determined. In practice, a calibration-extrapolation technique was utilised for wind speeds measured outside the calibration wind speed range (i.e. extension of calibration curve). Once installed, the Dines anemometer could not be calibrated due to its large size which incorporated either a 6 or 10 metre long tube, depending on the type of installation. Station intercomparisons were conducted on an ad-hoc basis (mean wind conditions), however it was difficult to undertake this comparison at exactly the same location and height as the head of the Dines anemometer. The BoM anemometer replacement program, which replaced the pressure-tube Dines instruments (paper chart recording) with cup anemometers (digital recording), had the potential to drastically change the characteristics of observed wind speed. It was well known that the Dines had a high minimum start-up speed (i.e. minimum speed before the instrument registers a reading) whilst the cup anemometers suffered from overspeeding (i.e. cups keep rotating even though wind has dropped resulting in higher readings than the actual wind speed) [6]. There was initially a concern that the cup anemometers may read higher than the Dines anemometer during extreme wind conditions due to the overspeeding issue. At the time, the anemometer calibration equipment (wind tunnels) was not able to accurately characterise the transient (gust) response of the cup anemometers. For high (extreme) wind speeds measured by cup anemometers, extrapolation of the wind tunnel calibration continued to be employed.

This paper presents the results of a reanalysis of the current BoM peak gust wind speed database for the non-cyclonic region (Region A) of the Australia/New Zealand Wind Actions Standard AS/NZS 1170.2 [1]. Region A was considered for this initial study as the observed record contains a significant number of extreme events (*synoptic* and *thunderstorm*) over decadal time scales (i.e. extreme events not dominated by one or two tropical cyclone events). A total of 31 BoM recording stations distributed throughout Region A were selected for the analysis. To isolate the issue of anemometer replacement, only observing stations located at airports (consistent exposure) and with more than 30 years of record were considered. Figure 1 shows time-series plots of the daily maximum gust wind speed (above a threshold of 25 m/s) for the Sydney, Melbourne and Adelaide meteorological observing stations. Visual inspection indicates that the early part of the record (Dines anemometer) contains a greater number and also higher amplitude extreme events compared to the later part of the record (replacement anemometer). Other stations considered within the Australian Region A (dominated by synoptic and thunderstorm events) show similar time-series characteristics. The vertical broken line in Figure 1 indicates when the Dines anemometer was replaced.

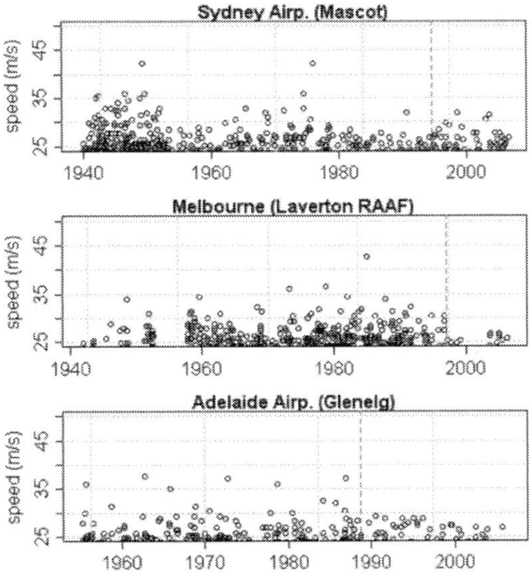

Figure 1.
Time-series plots of the daily maximum gust wind speed (above a threshold of 25 m/s) for the Sydney, Melbourne and Adelaide (airport) meteorological observing stations

2. Dines - Cup Anemometer Intercomparisons

A number of intercomparisons involving Dines and cup anemometers have been undertaken over the last 50 years or so, most primarily concerned with the differences in the mean wind speed; see for example [7]. Logue [8] compared both mean and gust wind speeds measured using a Dines co-located with a standard cup anemometer at the Irish Meteorological Service's Galway observing site (during the year of 1984). Overall the mean wind speeds from the two instruments compared well. However, the cup anemometer significantly underestimated the gust wind speeds when compared to those obtained using the Dines. The cup anemometer used by [8] is similar in design to the Bureau of Meteorology (i.e. heavy construction). These types of anemometers have a large distance constant (length of fluid flow past the sensor required to cause it to respond) compared to lightweight cup anemometers found on automatic weather stations [9]. Since the gust wind speeds recorded by the Dines are higher than those from a cup anemometer, it is clear that the effective gust duration of the Dines gust wind speed measurements must be less than the equivalent cup anemometer values [10].

3. Methods and Data

Three datasets were acquired from BoM for these studies; maximum daily gust speeds, 3PM mean wind speed and weather description. The later contains a three-hour classification of the weather conditions at the recording station. The present weather describes weather conditions during the past hour, whilst the past weather represents the most significant weather conditions within the past three hours of observation but not during the most recent hour. Using this information it is possible to identify whether gusts are due to *synoptic* or *thunderstorm* weather conditions, as explained in [11]. Both maximum daily gust wind speeds and 3PM mean winds are considered to examine the range of wind conditions both within and outside the instrument calibration range. 3PM mean wind speed data was chosen as the mean wind speed reaches a peak at this time in the diurnal cycle. Meta data concerning the replacement of the Dines anemometers was also obtained from BoM.

Wind hazard was assessment by calculating the return period (RP) exceedance levels of maximum gust wind speed for each dataset by using Extreme Value Distributions (EVD). EVD's can determine RP exceedance levels of gust wind speeds well beyond the range of available data [12]. Typically an EVD uses yearly maximum; all other data is discarded. In practice better results are obtained by using the Generalised Pareto Distribution (GPD) which allows the analyst to use all data exceeding a given threshold [11]. For this preliminary analysis we only consider the *combined* (complete) dataset and not the component synoptic and thunderstorm extreme winds. We utilise as a measure of the hazard the 500-year return period (500RP) exceedance level gust wind speed. This hazard level has been chosen as it represents the wind loading design level used for the construction of most of the residential, commercial and industrial buildings in Australia.

One of the problems found in fitting a GPD to given wind speed datasets is the selection of the appropriate threshold value 'u'. High threshold values result in the selection of only a few data points, most likely not enough for a good fitting of the distribution. Low values result in too many samples which are most likely not independent from each other. On the other hand calculation of exceedance levels using GPD distributions is very sensitive to the threshold selection. Although there are methods to help modellers select the appropriate threshold for a given dataset they are mostly visual, subjective techniques, prone to producing inaccurate results and inappropriate for large scale applications. The computer-based, automatic algorithm discussed in [11] was utilised here.

Calculation of exceedance levels of gust wind speeds should be considered incomplete if a confidence interval for the results is not presented. A confidence interval shows the range of values in which the true value of the exceedance levels lies for a given probability. For this study, we have determined confidence intervals with 95% probability; in other words the exceedance levels for 500RP of wind gust speeds with the interval in which the true value can be found in 95% of cases. There are two basic algorithms for calculation of the confidence intervals (CI) of results produced by EVD's: the 'Delta' method and the 'Profile Likelihood' method. Both methods have been implemented in the R environment by Gilleland and Katz based on [12]. Applying the methods to temperature data, it was found that the Profile-likelihood method gave better results because it considers the asymmetry of the data [12]. Since wind speed data is highly asymmetric the Profile-likelihood method as implemented in the R 'extRemes' package [13] was used for this study.

Figure 2 presents exceedance level plots for gust wind speeds for Sydney showing the corresponding curves for the complete dataset *(combined)* and the component *thunderstorm* and *synoptic* winds. Notice that each type of wind poses a different hazard to the environment (dependent on the return period) and hence separation of the types of winds present in the dataset should be considered for realistic modelling over the complete range of return period hazard.

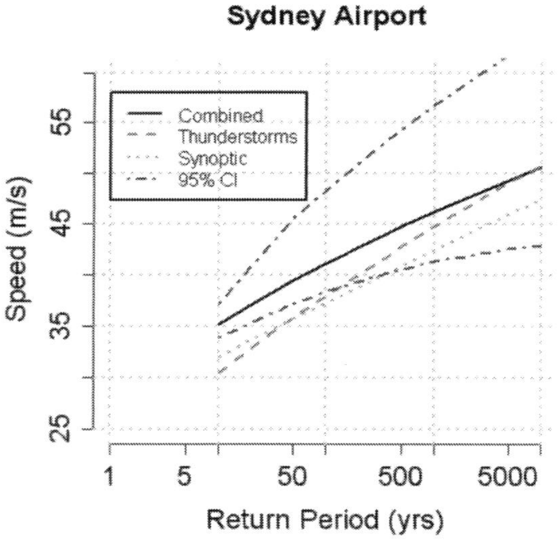

Figure 2.
Return-Period (RP) gust wind speed and 95% confidence interval (CI) for Sydney airport for the "*combined*", "synoptic" and "thunderstorm" gust wind types.

4. Results

The analysis statistically examines the issue of inconsistency in the gust wind speed record by investigating the impact of instrument upgrade within the dataset. The dataset (time-series of measurements) is segmented into the Dines and replacement anemometer observing periods, with the Dines observing period the longest in length in all cases. The latter coincides with the installation of the new generation of cup-anemometers at most observing stations (major Australian airports). All stations considered are currently operational.

We have evaluated both the daily maximum gust and 3PM mean wind speed 500-year return period (500RP) exceedance levels for the total (*combined*) wind speeds for the wind stations selected, where the observations have been segmented into the Dines and replacement anemometer observing periods. The 95% confidence interval (95CI) has been determined for both gust and 3PM mean wind speeds for the Dines anemometer segment of the record. Note that the width of the confidence interval is dependent on the frequency and consistency of the extreme events considered.

For the replacement anemometer 500RP exceedance level gust wind speed estimates, there are 17 observing stations where the replacement anemometer estimates fall below the lower 95% confidence interval (95CI) for the Dines segment of the observing record. For 24 of the 31 observing stations the replacement anemometer estimate fall below the corresponding Dines segment estimate. Considering the 3PM mean wind speeds, there are 18 observing stations where the replacement anemometer estimate falls below the lower 95CI for the Dines segment of the observing record.

Figure 3 displays this information pictorially. The distribution of the Dines 500-year RP exceedance level gust wind speed estimates (data scaled by the mean and standard deviation for each station respectively) is shown. Plotted on the standard normal distribution are blue vertical bars pictorially indicating the fit for each observing station for the section of the record acquired using the upgraded/replacement (cup) anemometer. Estimates for the later part of the record appear biased to the lower tail of the distribution for the early part of the record (Dines anemometer).

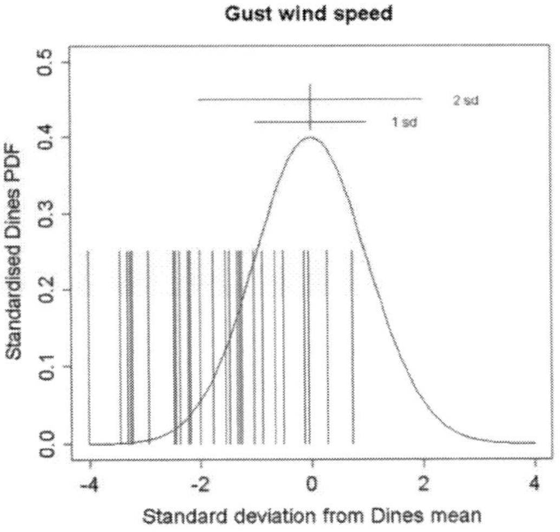

Figure 3.
Distribution of the Dines 500 year RP exceedance level gust wind speed scaled by the mean and standard deviation with the blue vertical bars indicating the fit for the section of the record acquired using the replacement anemometer (also shown is the standard normal distribution curve - provided for guidance).

5. Discussion

One of the main drivers for undertaking this study was to obtain an improved understanding of the historical peak gust wind speed measurements that have been made in the Australian region. Ultimately we would like to have a peak gust wind speed dataset available that could be used to study trends in storminess in the Australian region (climate change detection and attribution studies).

The visual inspection of the peak gust wind speed time-series for most sites considered in this study indicates in general that the early part of the record contains:

(1) a higher frequency of extreme events (2) the largest amplitude extreme events.

A statistical analysis utilising extreme value distributions (EVD's) resulted in more than half of the observing stations considered (later part of the record) having both 500-year RP gust wind speed (17 of 31) and 3PM mean wind speed (18 of 31) exceedance level estimates being in the lower tail of the distribution for the early part of the observing record (period prior to equipment upgrade from Dines to cup anemometer). We should be cautious in interpreting the differences in the return period hazard for the two segments of the record. It is possible that the frequency of extreme wind speeds has actually declined [14]. The most recent part of the record has been dominated by a number of large El Nino events and only one La Nina event, where thunderstorms are known to be more prevalent. The importance of the non-stationary climate needs to be considered when conducting a time series analysis. Circulation changes as discussed in [15] [16], possibly driven by anthropogenic climate change, could also affect the frequency and amplitude of extreme winds. The replacement anemometer part of the record is generally much shorter (20 years) than the Dines segment and it is possible that the very high maximum wind speeds have just not yet been sampled, although the reduced frequency of extreme events in the later part of the record is of concern.

Even considering these issues, this preliminary analysis suggests that the consistency of the peak gust wind speed dataset is suspect, and that the problem requires further examination. This study indicates that it appears unlikely that the later part of the record can be considered consistent with the initial (Dines) segment of the record. Coincident data (where available) needs to be examined.

6. Conclusions and Future Directions

This study compared estimates of the 500-year RP exceedance level peak gust wind speed for 31 BoM airport observing stations, where the observations were segmented into two periods representing the observations from the Dines anemometer and replacement (cup) anemometer. The methodology, which employed the fitting of Extreme Value Distributions to each segment of the dataset, was formulated to explore the consistency of peak gust wind speed measurements due to issues surrounding equipment upgrading. Comparison of results indicated that the later period (cup anemometers) appears to have a reduction in significant events (17 of 31 sites have a mean 500 year RP exceedance level gust wind speed below the 95% confidence limit for the estimated using the Dines anemometer segment of the record. A similar trend was observed for the 3PM mean wind speeds. This analysis indicates that it appears unlikely that the later part of the record can be considered consistent with the initial (Dines) segment of the record. It is possible that circulation changes forced by either climate variability and/or climate change has influenced the data considered.
Further analysis is planned to consider coincident measurements where available. Coincident data from observing stations where the Dines and replacement cup-anemometer instruments were run in parallel for long periods needs to be analysed to determine whether a systematic bias is evident for extreme events. At the Townsville RAAF Base, Dines and cup-anemometer instruments have been operating side-by-side for 18 years (Dines observations are only on paper record).

The Cyclone Testing Station (CTS) at James Cook University has recently acquired a prototype Pressure Load Actuator (PLA) from the University of Western Ontario, Canada. A detailed field investigation of the characteristics of the Dines is now possible. The PLA is designed to apply fluctuating positive and negative pressure loads to the structure of a house at frequencies of up to 10 Hz, and can easily be adapted to apply a fluctuating positive pressure to the open tube of the head of the Dines (10 metres above the ground using a lifting device). Since the PLA is specifically designed to take some form of input pressure time history, this feature can also be implemented as part of the calibration. A study following the above procedure has been funded by the Federal Government (Department of Climate Change). This study should provide sufficient information to fully specify the dynamic response of the Dines over the range of operating conditions and hence determine the resulting effective gust duration and its variation with wind speed.

References

[1] AS/NZS 1170.2 2002 *Structural design actions, Part 2: Wind actions* Australian/New Zealand Standard
[2] Reardon G F Henderson D and Ginger J D 1999 *A structural assessment of the effects of Cyclone Vance on houses in Exmouth WA*, (Technical report, James Cook University of North Queensland. Cyclone Testing Station) No. 48
[3] Reardon G F 2000 *Anemometers: Dines vs. AWS* (Australasian Wind Engineering Society Newsletter, May 2000)
[4] Jacobson M Z *2005 Fundamentals of Atmospheric Modeling* (New York: Cambridge University Press, 2nd ed.) 828pp
[5] Knowles W E K and Spilhaus A F 1953 *Meteorological Instruments* (University of Toronto Press – Third Edition)
[6] Gorman J 2004 *The conversion equation of the Synchrotac 706 anemometer* (Bureau of Meteorology Instrument Test Report) No. 677
[7] Smith S G 1981 Comparison of wind speeds recorded by pressure-tube and Meteorological Office electrical cup generator anemographs *Meteorol. Mag., 110* 288-300

[8] Logue J J 1986 Comparison of wind speeds recorded simultaneously by a pressure tube anemograph and a cup-generator anemograph *Meteorol. Mag., 115* 178-185

[9] Sparks W 1997 *Equations of Motion for Munro Anemometers* (Meteorological Office, Observations - Logistics and Automation) Technical Report No.11

[10] Miller C 2007 Defining the effective duration of a gust (*Proceedings of the International Conf. on Wind Engineering, ICWE12, Cairns, Australia*) 759-766

[11] Sanabria L A and Cechet R P 2007 *A Statistical Model of Severe Winds* (Geoscience Australia record 2007/12) http://www.ga.gov.au/hazards/publications.jsp#severewinds

[12] Coles S 2001 *An Introduction to Statistical Modeling of Extreme Values* (Springer series in statistics. London)

[13] Gillelland E and Katz R W 2005. Extremes Toolkit: Weather and Climate Applications of Extreme Value Statistics. *National Center for Atmospheric Research (NCAR). Boulder CO, USA*

[14] Smits A Klein Tank A M G and Konnen G P 2005 Trends in storminess over the Netherlands, 1962-2002 *Int J Climatology, 25* 1331-1344

[15] Plummer N Salinger M J Nicholls N Suppiah R Hennessy K J Leighton R M Trewin B Page C M and Lough J M 1999 Changes in Climate Extremes Over the Australian Region and New Zealand During the Twentieth Century *Climate Change, 42* 183-202

[16] Alexander L V Uotila P Nicholls N and Lynch A 2010 A New Daily Pressure Dataset for Australia and Its Application to the Assessment of Changes in Synoptic Patterns during the Last Century *Journal of Climate 23:5* 1111-1126

IOP Conf. Series: Earth and Environmental Science **11** (2010) 012017 doi:10.1088/1755-1315/11/1/012017

Development of methods for assessing the vulnerability of Australian residential building stock to severe wind

Martin Wehner[1], John Ginger[2], John Holmes[3], Carl Sandland[1] and Mark Edwards[1]

[1]Geoscience Australia, GPO Box 378 Canberra, ACT 2601 Australia.
[2]James Cook University, Townsville, QLD 4811 Australia
[3]JDH Consulting, PO Box 269 Mentone, VIC 3194 Australia

Email: Martin.Wehner@ga.gov.au, John.Ginger@jcu.edu.au, jdholmes@bigpond.net.au, Carl.Sandland@ga.gov.au, Mark.Edwards@ga.gov.au

Abstract. Knowledge of the degree of damage to residential structures expected from severe wind is used to study the benefits from adaptation strategies developed in response to expected changes in wind severity due to climate change. This study will inform government, the insurance industry and provide emergency services with estimates of expected damage. A series of heuristic wind vulnerability curves for Australian residential structures has been developed. In order to provide rigor to the heuristic curves and to enable quantitative assessment to be made of adaptation strategies, work has commenced to produce a simulation tool to quantitatively assess damage to buildings from severe wind. The simulation tool accounts for variability in wind profile, shielding, structural strength, pressure coefficients, building orientation, component self weights, debris damage and water ingress via a Monte Carlo approach. The software takes a component-based approach to modelling building vulnerability. It is based on the premise that overall building damage is strongly related to the failure of key components (i.e. connections). If these failures can be ascertained, and associated damage from debris and water penetration reliably estimated, scenarios of complete building damage can be assessed. This approach has been developed with varying degrees of rigor by researchers around the world and is best practice for the insurance industry.

1. Introduction

The vulnerability of buildings to damage caused by severe wind can be described by an S shaped curve (for example [1,2]) which relates damage index (defined as repair cost / building replacement cost) to the 3 second gust wind speed at 10m height ($V_{10m,3sec}$) at the building's location. The curve is termed a vulnerability curve and is described in [3] by an equation of the following form:

$$DI = 1 - \exp\left[-\left(\frac{V_{10\,m,3\,sec}}{e^{\beta}}\right)^{\frac{1}{\alpha}}\right] \qquad (1)$$

where DI is the damage index, α and β are parameters describing the curve's shape and position.

This paper describes the development of a set of heuristic vulnerability curves for the Australian residential building stock and the development of a simulation tool used to estimate the vulnerability curve for generic housing types based on calculations of component level damage. As the simulation tool provides information on the range of damage indices across a population of houses, it can generate fragility curves that show the proportion of houses in different damage states with increasing wind speeds.

2. Development of heuristic vulnerability curves

The Australian residential building stock comprises a wide variety of housing of different forms, construction and age. Each of these varieties will respond differently in a severe wind event. In order to be able to estimate damage from severe wind nationally, the range of vulnerabilities exhibited in the building stock needs to be estimated.

In order to develop a suite of heuristic vulnerability curves that adequately covered the variety of housing, a study was undertaken whereby members of the Australian wind engineering community assigned vulnerability curves from a suite of standard curves to a detailed categorization of Australian housing types.

The Australian residential building stock was categorized on the basis of region, age, roof material and wall material. For the modern housing types, a design wind speed was assigned from AS 4055 [4]. This produced a possible 306 categories of which 203 were considered valid types. For example, no houses would be expected for the category "*Queensland Region C / 1891 - 1913 / sheet metal roof / reinforced masonry walls*".

A series of six vulnerability curves representative of typical Australian construction types were developed drawing on the outcomes of a workshop attended by members of the Australian wind engineering community (Figure 1) [3].

Responses were subsequently sought from eight recognized vulnerability experts in the Australian wind engineering community who could draw upon their experience in post-event survey activity. Each of the respondents were given a blank copy of the categorization and the suite of six vulnerability curves and asked to independently assign a curve (or an intermediate curve) to each of the available fields in the categorization.

For each available field in the schema, the responses were analyzed to produce a heuristic vulnerability curve for each categorization. The process took account of the average curve (i.e. position left to right) and also the variability in responses that is reflected in the steepness of the curve (Figure 2). The results are expressed as an alpha and beta value for each field that describe the vulnerability curve by Equation (1).

Although responses were sought from the insurance industry, none were received. An anecdotal response to Geoscience Australia indicated that the lower sections of the suite of six curves overestimated the damage at low gust wind speeds.

The vulnerability curves generated require calibration against observed damage from severe wind events. Such data is scarce due to the rare occurrence of high wind speed events (especially at the higher end) and as post-storm surveys of wind damage are expensive and time consuming. Furthermore, for surveyed damage to be useful, an estimate of the spatial nature of gust wind speed is required which can be difficult to ascertain due to limited available measurements.

3. Development of a Wind Vulnerability Simulation Tool

Work has commenced on developing a software tool to simulate damage to buildings from damaging wind (expressed as the peak 3s gust wind speed in the event observed at a 10 metre height) to assess the heuristic curves and to enable a quantitative assessment of the extent of damage and repair or the beneficial effect of structural upgrading of residential buildings.

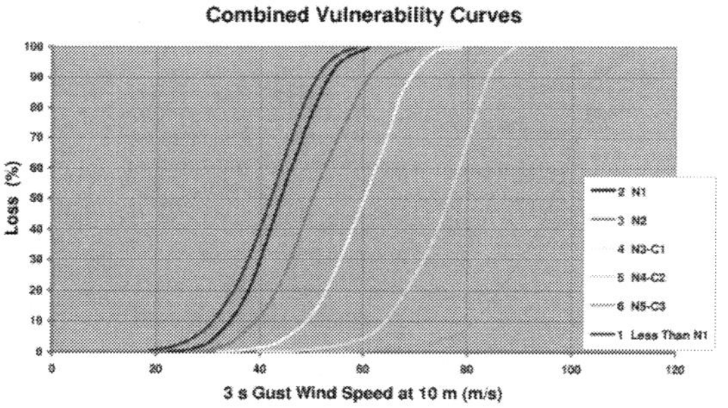

Figure 1. Suite of vulnerability curves for selection. The N and C numbers refer to design wind speeds as defined in [4].

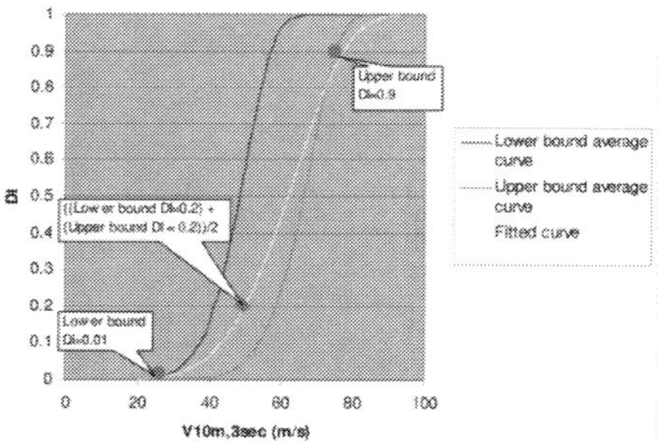

Figure 2. Method of calculating vulnerability curve from lower and upper bound curves. Note that the fitted curve will not pass precisely through all three points.

3.1. Simulation tool logic

The tool takes a component-based approach to modelling building vulnerability. It is based on the premise that overall building damage is strongly related to the failure of key components (i.e. connections).

The tool generates a house with values for component/connection strengths, external and internal pressure coefficients, shielding coefficients, wind speed profile, building orientation, debris damage parameters, and component weights randomly generated from predetermined probability distributions. Then, for successive wind speed increments, it calculates the forces in all critical connections using structural analysis methods, assesses which connections have failed and translates these into a damage scenario which it costs and thus calculates a damage index for that wind speed. Loads on connections are determined using influence coefficients. The building envelope is divided into zones to which wind pressures are applied. Each connection – zone pair has an influence coefficient that specifies the force generated in that connection from a unit pressure on that zone.

Failures of roof sheeting or roof batten connections trigger a redistribution of load to adjacent intact connections thus making allowance for member continuity in these elements. Failure of roof structure connections revises the influence coefficients for the roof structure load effects depending on the specific connection that failed thus making allowance for redistribution of forces within the roof structure. Connections that have failed and the effects of redistribution are preserved for successive wind speed increments thus ensuring that increasing wind loads act on the damaged structure rather than beginning anew with an intact structure.

The effect of high level structural (i.e. cladding) failures removing load from lower level structure (i.e. batten) is modelled by a hierarchical approach to assessing component failures at each wind speed. Roof sheeting connections are assessed first and any necessary distribution from failed roof sheeting connections is undertaken before loads are calculated on batten connections which are then assessed and redistributed before loads on roof structure connections are calculated and the connections assessed. Hence, if an area of roof sheeting were to fail (lift off), the underlying battens and rafters would not receive any load from that area.

Once the tool has determined which connections have failed, it relates the extent of damage in a group of connections to a percentage damage for a damage scenario such as loss of roof sheeting, damage to wall cladding, loss of roof structure, etc. A costing module then gives a total repair cost for the accumulation of damage scenarios at a particular wind speed level. After the set maximum wind speed is reached, a new, intact, house is generated with new randomly selected parameters.

At the completion of the simulation of a user specified number of houses of the same basic type, each at a number of wind speeds, a vulnerability curve can be plotted and fragility curves calculated.

3.2. Probability distributions for an example residential building

Various types of probability distributions are used in the tool to generate wind loading and house component strength parameters.

3.2.1. *Wind speed profiles.* Variation in the profile of wind speed with height is captured by the random selection of a profile from a choice of 10 profiles for a given terrain category [5]. Figure 3 shows an example of a set of profiles from reference [5]. The wind speed used for vulnerability curves is the 3 second gust at 10m height however calculation of wind loads on a house requires the wind speed at mid-roof height. The appropriate factor to account for the change in wind speed from 10m to the house's mid-roof height is selected from the chosen gust profile.

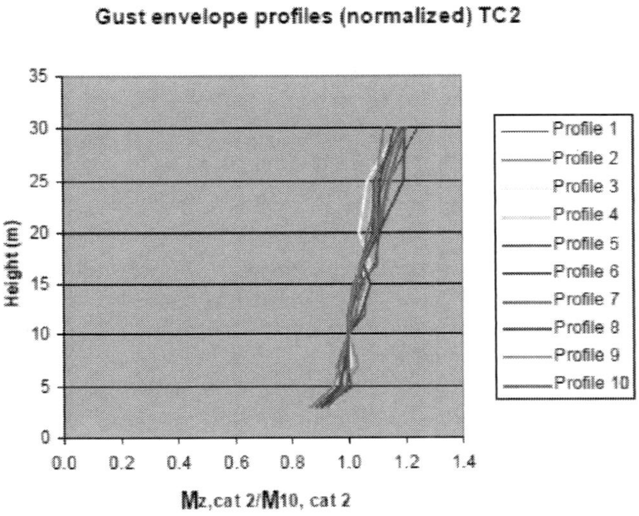

Figure 3. Example of a set of wind speed profiles for Terrain Category 2, reproduced from [5]. The ratio Mz,cat 2 / M10,cat 2 is the ratio of the terrain, height multiplier at height z to the terrain, height multiplier at 10m height. TC 2 denotes Terrain Category 2 as defined in [7].

3.2.2. *Shielding coefficients.* Variability in shielding to a house within its surroundings is accounted for by modifying the incident wind speed. The modifying factor is randomly chosen from a population of shielding coefficients for suburban houses developed from suburban wind speed field observations [6]. Variability in the degree of shielding applied to different house surfaces is accounted for by modifying the external pressures using factors obtained from wind tunnel experiments [6].

3.2.3. External and internal pressure coefficients. Pressure coefficients for different zones of the house surface are randomly chosen from a Type III extreme value distribution with specified statistical means and specified coefficients of variation for different zones of the house surface for different load effects: cladding, structural member, internal pressure (no openings), and internal pressure (with openings) [8]. The tool samples pressure coefficients for every zone of each house surface for each wind direction and each load effect; approximately 6,500 coefficients.

3.2.4. Building orientation. For each simulated house, its orientation with respect to the wind is chosen from the eight cardinal directions.

3.2.5. *Connection strengths.* For each connection strengths are sampled for each house generated from log-normal probability distributions for each type of connection. The mean and coefficient of variation (CoV) values [9] have been developed from testing, for example [10], and damage survey, for example [11].

3.3. Results

Examples of the simulation tool output are given in Figures 4 and 5 for a high-set, fibro clad Queenslander type house with corrugated steel roof cladding, timber battens at 900 centres and timber framed roof structure consisting of timber rafter pairs at 900 centres with collar ties to every second pair. The roof structure connections modelled are rafter to top plate, rafter to ridge board and collar tie to rafter. The outputs are given in two parts:

- Graphical view of results of Damage Index versus wind speed with a fitted vulnerability curve Results from simulations on 20 houses (simulations at each wind speed increment approaching from a single direction) are shown in Figure 4. The small black dots are individual simulation - wind speed combination results. The large red dots are mean DI at each wind speed increment. The curve conforms with that described by Equation (1) fitted to the mean DI values.
- Graphical view of calculated Fragility Curves for slight, medium, severe and total damage shown in Figure 5. The curves capture the spread of results about the mean vulnerability curve and show the proportion of houses expected to suffer no damage, slight damage, medium damage and total damage at each wind speed. The stepped nature of the curves results from the low number of simulations (20 in this case). The DI at each level of damage is arbitrarily set.

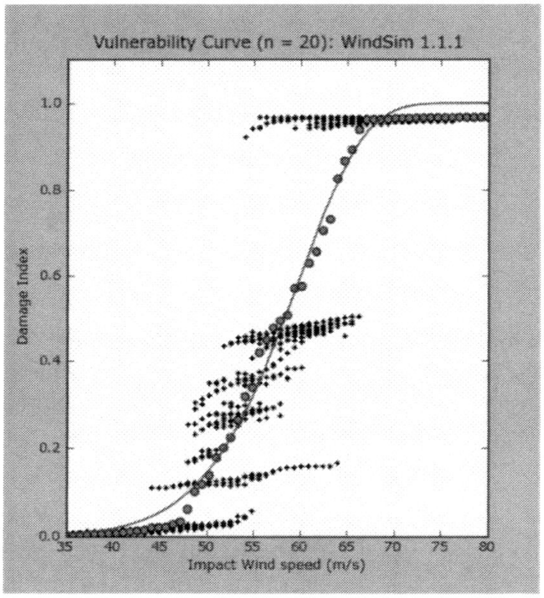

Figure 4. Output of simulation tool showing the DI versus $V_{10m,3sec}$ (m/s) vulnerability curve. The jump in damage index from about 0.5 to about 0.95 is caused by the failure of the lower storey structure.

3.4. Current and future work

To date the tool has been developed to model the damage to roof sheeting, roof battens, roof structure, lower storey structure, damage from wind-borne debris and damage from water ingress. Current and future work involves extending the scope of the tool to include: damage to wall cladding and damage to wall structure, as well as calibrating results against observed damage.

Data has been developed for a single house type: a high-set, fibro clad Queenslander type house dominant in residential building structures in the 1960's and early 1970's from south-east Queensland to Darwin. Current and future work involves assembling data sets for two further building types: a portal frame industrial shed typical of north Queensland and a two storey tiled roof house with brick veneer walls typical of western Sydney.

4. Conclusions

An extensive set of heuristic vulnerability curves has been developed for the Australian residential building stock. The set of curves is currently being used by Geoscience Australia for a National Wind Risk Assessment of Australia's residential building stock. In order to quantitatively verify these curves and to enable quantitative assessment of building adaptation a wind vulnerability simulation tool is under development.

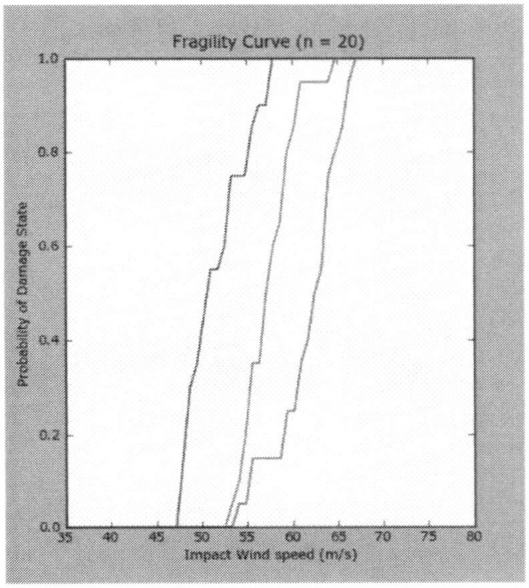

Figure 5. Output of simulation tool showing fragility curves calculated from the results shown in Figure 4.

Acknowledgements

The authors wish to acknowledge that the work reported in this paper was undertaken as part of a program of work co-funded by the Australian Federal Department of Climate Change and Energy Efficiency. Geoscience Australia gratefully acknowledges the work of their collaborators James Cook University and JDH Consulting.

References

[1] Ellingwood B R, Rosowsky D V, Li Y, Kim J H 2004 *J. Struct. Eng.,ASCE.* **130** 1921
[2] Spence R, Brown A, Cooper P 2004 Whether to strengthen? Risk Analysis for Strengthening Decision Making *Henderson Colloquium, Magdalene College, Cambridge*
[3] Timber Ed Services 2006 *Report on Wind Vulnerability Research Workshop 14,15 March, 2006*
[4] AS4055 Wind loads for housing, 2006 *(Sydney, NSW: Standards Australia)*
[5] JDH Consulting 2009 *Interim report to Geoscience Australia on Randomized wind gust profiles for vulnerability modelling*
[6] JDH Consulting 2009 *Interim report to Geoscience Australia on Variability of shielding effects for vulnerability modelling*
[7] AS1170.2 Structural design actions Part 2: Wind actions, 2002 *(Sydney, NSW: Standards Australia*
[8] JDH Consulting 2007 *Effective wind pressure coefficients for two house types – report prepared for Geoscience Australia*
[9] Cyclone Testing Station, James Cook University, 2009 *Wind Vulnerability Tool Simulation – JCU Group 4 House Update Report Version 2, December 2009*
[10] Cyclone Testing Station, James Cook University, 1983 *Report TR19, Testing a High Set House Designed for 42m/s Winds, Part 1- Preliminary Results*
[11] Cyclone Testing Station, James Cook University, 2009 *Report TR55, Investigation of Peformance of Housing in Brisbane Following Storms on 16 and 19 November, 2008*

IOP Conf. Series: Earth and Environmental Science **11** (2010) 012018 doi:10.1088/1755-1315/11/1/012018

Rapid Inventory Collection System (RICS) and the 2009 Victorian Bushfires impact assessment

N Habili, N Corby and R P Cechet

Risk and Impact Analysis Group, Geoscience Australia
G.P.O. Box 378, Canberra, Australian Capital Territory

E-mail: nariman.habili@ga.gov.au

Abstract. The Rapid Inventory Collection System (RICS) is a vehicular data collection system (image and GPS) used for building/infrastructure damage and inventory assessment. The system consists of Ethernet cameras attached to a tripod mounted on a motor vehicle, a GPS receiver and software written in C++. RICS was deployed following the 2009 Victorian Bushfires and the collected data was used by the Victorian Bushfires Royal Commission for the impact assessment (field survey) which quantified the extent and severity of the damage caused to residential buildings by the fire-storm. The regions of Kilmore East, Murrindindi, Churchill, Maiden Gully – Bendigo and Bunyip had nearly 5400 residential structures within the fire perimeter. Analysis utilising both aerial and vehicular (RICS) imagery indicates that just over 2100 homes were destroyed and an additional 800 received minor damage.

1. Introduction

The Rapid Inventory Collection System (RICS) is a roving vehicular data collection system (image and GPS) that offers time and cost saving benefits to Geoscience Australia (GA) in the efficient and effective assessment of building and infrastructure damage following natural disasters. Its chief purpose is for building inventory assessment (residential, commercial and industrial) in support of the National EXposure Information System (NEXIS), which is a buildings information database being developed by GA. In particular, RICS:

- complements and augments the detailed field damage assessments (house-to-house, structure-to-structure) currently undertaken during post-event impact assessments using hand-held personal digital assistants (PDAs);
- provides a quick "first-look" of the damage impact area to prioritise the worst-hit regions to be surveyed first;
- allows for 100% coverage of building damage in a disaster-affected area;
- collects data focusing on "population coverage" and in disaster situations quickly collects information on undamaged structures allowing key engineering and GIS staff to focus on damaged structures; and
- allows fieldwork to be undertaken more efficiently.

The software was written in C++ using an iterative development process. The software has been extensively tested in building inventory applications where the purpose is to characterise structural types and building details. RICS was deployed following the 2009 Victorian Bushfires to collect post-disaster imagery in order to quantify the extent and severity of the damage caused by the fire-storm.

2. RICS Equipment

First generation RICS utilised a desktop PC and three Firewire Pixelink cameras. An external petrol generator was used to power the PC and monitor located inside the vehicle. During field trials we found the use of the PC and monitor to be cumbersome requiring significant time for set up with limited portability. The Pixelink cameras captured images containing distortion (skew) due to their use of *rolling shutter* technology. Firewire connectivity between cameras and the PC also proved troublesome due to software and hardware incompatibilities. There were occupational health and safety issues with the petrol generator such as excessive noise and fumes. The transportation of the generator on aircraft (rapid deployment) was also a challenge as it is heavy and contains flammable liquids.

In order to overcome the problems of the first generation RICS, the design of the second generation system revolved around the use of a high-end laptop, Prosilica cameras [1] with global shutters and Gigabit Ethernet connectivity, powered by an inverter connected to a vehicle's battery. The laptop allows rapid setup while also adding to the portability of the system and provides greater comfort for the passenger. RICS requires a high end laptop because it compresses (in JPEG format) and saves images from up to four cameras at four frames per second which is computationally intensive. RICS captures GPS data and adds the data to an SQLite database [2]. Moreover, ESRI's ArcPad, used to generate a tracklog, may also run alongside RICS. As RICS is multithreaded, a multi-core system is necessary for optimal performance. We currently use a quad-core laptop which has proven satisfactory when utilising four cameras.

The Prosilica cameras employ global shutters, not rolling shutters, therefore image distortion is minimised. Gigabit Ethernet connectivity, common on all modern laptops, allows a more reliable data transfer through the use of multiple cameras through an Ethernet switch. The Ethernet switch uses jumbo-packet capable hardware to maximize network throughput.

3. RICS Software

3.1 Components

RICS was implemented in C++ using Microsoft Visual C++ Express 2008. After reviewing a number of Graphical User Interface (GUI) libraries, wxWidgets [3] was chosen because of its use of native controls, well written documentation, large user-base, licensing scheme and its cross-platform nature (i.e., can be easily ported to Linux). wxThread, included in wxWidgets, was used to implement software threading. SQLite is used to store image file names and their corresponding GPS data, enabling images to be geo-referenced. Besides the Prosilica SDK, the MarshallSoft GPS library is the only other commercial component used in the software. The MarshallSoft GPS library is used to receive and decode standard NMEA 183 sentences from a GPS receiver. It is likely that in the future either a library will be written or a third-party open-source library will be used to serve the same purpose. It is envisioned that RICS will be released as open-source software at some stage in the future.

3.2 Graphical User Interface

The GUI was designed to be user-friendly, consistent and intuitive. The main window consists of three sections (see Figure 1). Streamed images are displayed in the top section. The middle section displays the collected GPS data, specifically time (UTC), latitude, longitude, bearing, number of satellites, fix quality and speed. Based on user feedback the speed, in kilometres per hour, is displayed prominently in a larger font. The bottom section is a notepad, where users can type notes during a field session. If a user wishes, the current time, latitude and longitude can be added to the end of each line, therefore the notes are time and geo-referenced. The notepad text is saved as a "comma separated value" file, and can therefore be opened in a database or spreadsheet program.

Figure 1:RICS Graphical User Interface (GUI)

4. 2009 Victorian Bushfires impact assessment

The scale of the Victorian bushfires of 7 February 2009 in terms of the loss of life and assets is unprecedented in Australia. This presented significant challenges to agencies tasked with providing recommendations for changes to building design, planning and management regulations. In response to the 2009 Victorian fires, a group of researchers from various states, fire agencies and research organisations were assembled to provide fire and land management agencies and the 2009 Victorian Bushfires Royal Commission (VBRC) with an objective scientific analysis and dataset of the factors surrounding these series of fires. The VBRC recommendations (due July 2010) will in time take into consideration the impact of fire at a range of spatial scales, from broad regional assessments of fire extent and severity, to detailed assessments of the incidents related to single properties. A physical survey was performed on five regions affected by fire to understand the magnitude and nature of the impact of the February 7th bushfires. The survey had two major objectives:

- To provide a dataset that facilitates ongoing reform of policies, regulation and education initiatives.
- More specifically, to provide a dataset and data assessment methodology to address a range of issues such as house vulnerability (in relation to observed intensity, fire attack mechanisms and related winds), effectiveness of various prevention measures (such as defendable space, sprinkler systems), and impact of land-use planning and building controls.

Damage assessment (fire impact) for residential structures was undertaken using a combination of Google Street View (limited coverage in region), aerial photography (before and after imagery available) and RICS (total coverage of residential structures within 200 metres of roads and tracks).

Figures 2 & 3 illustrate the use of RICS for the Victorian Bushfires field deployment, where only three cameras were used. RICS imagery collection was undertaken over a period of three weeks and more than 500,000 images were obtained involving about 8000 structures. Figure 4 provides examples of the imagery obtained using RICS in the Kinglake region.

Figure 2. RICS (3 cameras) deployed on the tray of the mapping vehicle during the 2009 Victorian Bushfires impact assessment.

Figure 3. Laptop set-up inside the mapping vehicle.

A large number of fires were reported on 7 February 2009. The regions of Kilmore East, Murrindindi, Churchill, Maiden Gully – Bendigo and Bunyip had nearly 5400 residential structures within the fire perimeter. Due to the extreme nature of the fires, houses that were affected were generally severely damaged. The impact assessment of the Victorian fires therefore employed just three categories:

- DESTROYED - defined as a structure that was not habitable without major repairs.
- MINOR DAMAGE - defined as a structure that was habitable but which still required repair.
- NO DAMAGE - no visible damage (hi-res images) although smoke damage possible.

Figure 4. Images showing residential building damage captured by RICS.

The tabulated results for the five regions surveyed are shown in Table 1. They indicate that just over 2100 homes were destroyed and an additional 800 received minor damage.

Table 1. Summary of house damage for February 7th *Black Saturday* bushfires

Fires	Number of houses within fire perimeter	Houses Damage **DESTROYED**	House Damage. **MINOR**	Houses Damage. **NONE**	**Fatalities** [1]	**Forest Area burnt** [2]
Bunyip	240	35	21	184	0	180,000+ ha
Churchill	359	133	86	140	11	150,000+ ha
Kilmore East	3540	1244	530	1766	121	32,800+ ha
Maiden Gully	172	48	21	103	1	500+ ha
Murrindindi	1064	590	74	400	38	24,500+ha
Total	5375	2118	832	2593	171	390,000+ ha

1. Source: Victoria Police
2. Source: Victorian State Government, Department of Sustainability and Environment (DSE)

The assessment utilised locations of existing structures based on the GNAF (Geocoded National Address File; geocoded address index for the whole country) and Geoscience Australia's NatMap (National Mapping) structure database in conjunction with structures identified in either before and after aerial photography. The RICS and Google Street View images were essential for delineating differences between houses and other buildings such as sheds, commercial premises and animal shelters. Structures that were ruins prior to the fire were removed from the database. It is possible that some houses have been classified as sheds (in general corrugated shell construction) and some sheds as houses (old houses being used as storage facilities).

5. Discussion and Future Developments

The accuracy of the latitude and longitude ascribed to each image depends on a number of factors, such as the speed of the vehicle, frame rate, GPS capture rate, network throughput and the way threading is implemented in software. Field trials suggested that four consecutive frames are necessary to capture an entire house with an average frontage of 20m. With four cameras attached to the laptop, images were captured, compressed and saved at approximately two frames per second. This suggests a maximum vehicle speed of 36 km/h. With threading, the challenge was to determine the sequence of operations (i.e., GPS and camera data fetch, display and compression) to allow an accurate match between the position of the vehicle (i.e., latitude and longitude) and the captured image. The camera thread captures and compresses an image from a camera, whereas the GPS thread captures the GPS data. The main GUI thread displays the captured images and GPS data. There are relatively large lags involved in both image capture (because of network throughput) and compression. It was determined that the following sequence of operations (in software) would provide the most accurate geo-located imagery: GPS data capture; image capture; image and GPS data display and image compression. As GPS data capture has a relatively small lag, the GPS data is ascribed quickly to the captured image. Using Google Street View as ground-truth, we determined that RICS captures geo-located imagery with a high degree of accuracy (less than a 5m misalignment from a buildings central frontage).

Following the Victorian Bushfires, a number of changes were made to the RICS software. Firstly, vibrations inside a vehicle, pronounced on rough roads, made mouse clicks for camera "play" and "stop" challenging. Therefore, keyboard shortcuts were implemented for a number of critical operations. In addition, if a camera became detached from the Ethernet switch, the software would "crash" and would sometimes require the laptop to be restarted. This was annoying and time consuming. A call-back function was implemented so that the software would exit gracefully instead of crashing.

6. Conclusions

A vehicular data (image and GPS) collection system called RICS has been developed at Geoscience Australia, using software written in C++ and an iterative development process. With the aim of releasing the software as open-source, most of its third party dependencies are open-source. Deployment of RICS following the 2009 Victorian Bushfires to collect geo-located imagery on bushfire affected areas and structures (impact assessment) allowed the quantification of the extent and severity of the damage caused to residential buildings by the fire-storm [6]. The regions of Kilmore East, Murrindindi, Churchill, Maiden Gully – Bendigo and Bunyip had nearly 5400 residential structures within the fire perimeter. Analysis utilising both aerial and vehicular (RICS) imagery indicates that just over 2100 homes were destroyed and an additional 800 received minor damage. A number of improvements to RICS were made to the software following the bushfires field survey, namely key-board shortcuts for critical operations and better handling of hardware failures. Using Google Street View as ground-truth (where available), it was determined that RICS captures geo-located imagery with a high degree of accuracy. A number of avenues for the further development of the software, such as automatic testing and JPEG compression speedup, are being investigated.

References

[1] Prosilica Inc 2009 www.prosilica.com
[2] SQLite 2009 www.sqlite.org
[3] wxWidgets 2009 www.wxwidgets.org
[4] MarshallSoft GPS Component for C/C++ 2009 www.marshallsoft.com/mgc4c.htm
[5] Subversion 2009 subversion.tigris.org
[6] Bushfire Cooperative Research Centre 2009 *Victorian 2009 Bushfire Research Response; Chapter 3, Buildings and Land Use* (available at http://www.bushfirecrc.com/research/taskforce2009.htm)

IOP Conf. Series: Earth and Environmental Science **11** (2010) 012019 doi:10.1088/1755-1315/11/1/012019

An initial assessment of the impact of Australian aerosols on surface ultraviolet radiation and implications for human health

C Y Chee[1,2] and F P Mills[3]

[1] Medical School, Australian National University, Canberra, ACT, 0200, Australia
[3] Research School of Physics and Engineering and The Fenner School of Environment and Society, Australian National University, Canberra, ACT, 0200, Australia

E-mail: Frank.Mills@anu.edu.au

Abstract. Aerosols can have significant influence on surface radiation, and the intense surface ultraviolet radiation Australia experiences contributes to Australia's high incidence rates for related human diseases. Aerosol properties, such as total column aerosol optical depth, have been measured over several years for varying lengths of time at sites across Australia using sunphotometers. Statistical analysis of the average daily aerosol optical depth over sites near Alice Springs, Canberra, Darwin, and Perth provides one measure of the annual atmospheric loading of aerosols over these sites. The sunphotometers used at these sites do not make measurements in the UV-B spectral region and have only one channel in the UV-A spectral region, the regions of most interest for assessing human health impact. Consequently, model calculations using standard aerosol types have been used to make an initial estimate of the impact of the aerosols found over these four sites on surface ultraviolet radiation. The aerosol loading is at times sufficient to significantly reduce the surface ultraviolet radiation, but few such days occur each year. The annual average effect of aerosols on surface ultraviolet radiation, thus, appears to be small compared to lifestyle factors, such as clothing and use of sunscreen.

1. Introduction

Human exposure to surface ultraviolet (UV) radiation is an important public health concern and significant changes in surface UV intensities, which are likely due to global environmental changes, have been reported, such as an 11.3% per decade increase in clear sky solar irradiance at 324 nm for 1991–2006 at Thessaloniki, Greece [1]. This issue is particularly important in Australia, where UV intensities are significantly greater than in European countries [2; 3].

Surface ultraviolet radiation is commonly classified by wavelength as UV-A (315–400 nm), UV-B (280–315 nm), and UV-C (100–280 nm). Oxygen and ozone in the atmosphere completely absorb UV-C and the highest energy UV-B, allowing only UV-A and a small fraction of lower energy UV-B to reach the surface. The amount and energy of UV-B photons reaching the surface is a strong function of overhead ozone column. Public health officials raise concerns over exposure to both UV-A and UV-B. With a predominantly fair-skin population and an outdoor lifestyle, skin-cancer rates in Australia are among the highest in the world [4]. Excess UV exposure may also contribute to posterior subcapsular cataracts, currently the leading cause of blindness

[2] Present address: Canberra Hospital, Canberra, ACT, Australia

worldwide [5]. High UV can also harm terrestrial and oceanic vegetation, degrade man-made objects, and change the chemistry of the lower atmosphere. Conversely, insufficient exposure to UV has been postulated as a risk factor for development of autoimmune diseases (such as Type 1 diabetes, rheumatoid arthritis, and eczema) via suppression of T-helper 1 lymphocytes [6; 7]. Consequently, understanding the factors that affect Australian surface UV radiation levels has significant scientific and social interest.

Surface UV is determined primarily by the exoatmospheric solar irradiance, surface elevation, solar zenith angle, surface cover, ozone, aerosols, and clouds. The impact of many of these factors is reasonably well known. The effects of aerosols and clouds on surface UV, however, can be significant but highly variable and poorly constrained. Usually particles reduce surface UV, but scattering by non-UV-absorbing particles can increase the UV intensity on non-horizontal surfaces [8]. A previous study on the effects of aerosols on surface UV radiation found tropospheric aerosols can reduce surface UV by $\sim 50\%$ [9], but the amount of the reduction is dependent on the chemical and physical properties of the aerosols [10]. Thus, the effect of aerosols on surface UV may vary greatly for different regions. Over Australia, two major types of aerosols are smoke from fires and mineral dust. Australia's low level of anthropogenic pollution provides a clean environment for studying their effects on surface UV, and the only previous examination of the impact of Australian aerosols on surface UV found smoke from fires near Darwin reduced surface UV by as much as 40–50% at 290–300 nm and 20–25% at 320–400 nm [11]. Results for mineral dust over Alice Springs were inconclusive because no day during the study period had a high loading of mineral dust aerosols [11]. A low aerosol abundance has been noted for other Australian sites, such as Cape Grim in Tasmania [12]. No study to date, however, has investigated the annual average effect of Australian aerosols on surface UV. This study reports an initial assessment.

2. Observations

Four sites were chosen for this study based on proximity to moderate-size population centres and availability of data: Alice Springs (23.80S, 133.89E, 546 m elevation), Canberra (35.271S, 149.111E, 600 m), Darwin (12.42S, 130.89E, 29 m), and Rottnest Island (32.000S, 115.502E, 70 m), which is near Perth. Aerosol optical depth (AOD) measurements at 500 nm were obtained from two sources: the Bureau of Meteorology (BoM) for Alice Springs and Darwin [13] and AERONET for Canberra, Darwin, and Rottnest Island [14]. The AERONET data were level 2.0 daily averages of data obtained and processed using version 2 of the direct sun algorithm. They had been pre- and post-field calibrated, automatically cloud cleared, and manually inspected. The BoM data were obtained from Dr. Bruce Forgan as half-daily averages of data that had been calibrated, cloud cleared, and inspected. The BoM data were screened for clouds by applying the Baseline Surface Radiation Network (BSRN) algorithm [15] to the standard deviation of the direct solar irradiance over a minute. 500 nm (501 nm at Alice Springs) AODs were used because it was common to all instruments at all sites. An earlier intercomparison of sunphotometers at Alice Springs found negligible bias (95% uncertainty interval < 0.007 at 500 nm) [13]. Probability density functions for all four sites, constructed by dividing histograms of the data by the total count, are shown in figure 1. Data were available for Alice Springs (BoM) for April 2002 to June 2006, Canberra (AERONET) for October 2003 to April 2004 and September 2005 to June 2006, Darwin (BoM) for June to October 2003 and January 2004 to March 2005 and August 2005 to June 2006, Darwin (AERONET) for April to September 2004 and May to November 2005, and Rottnest Island (AERONET) for September 2003 to March 2004.

3. Approach and Methodology

Based on the limited scope and human health focus for this assessment, a simplified approach was adopted. Monthly mean AOD was computed from the daily and half-daily averages. For

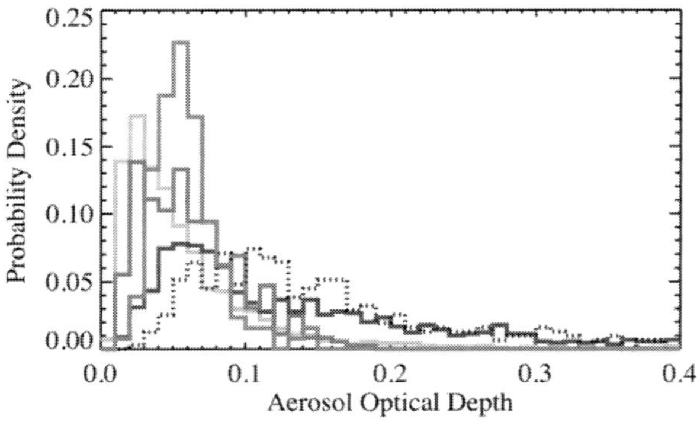

Figure 1. Probability density functions (histogram normalized by total count) for measured 500 nm AOD (501 nm at Alice Springs). Alice Springs (BoM half-daily average) in green, Canberra (AERONET daily average) red, Darwin (BoM half-daily average) blue, Darwin (AERONET daily average) dotted black, and Rottnest Island (AERONET daily average) grey. 13 of 2482 and 27 of 899 half-daily averages for Alice Springs and Darwin, respectively, had AOD ≥ 0.4.

each site, the month with the largest monthly mean AOD was identified. The monthly mean ozone, pressure, etc, for these months were used as input to calculations using libRadtran v. 1.1β [16], which determined the effective sun protection factor (SPF) due to aerosols as a function of AOD. A linear fit to these calculations was used to convert the probability distribution functions in figure 1 into probability distribution functions for the effective SPF due to aerosols over each site. These effective aerosol SPFs could then be compared with other SPFs.

The basic model atmosphere profiles used were tropical (Darwin) and mid-latitude summer (Alice Springs, Canberra, and Rottnest Island). Extraterrestrial solar flux was from ATLAS3 ($\lambda \leq 407.8$ nm), ATLAS2 ($407.8 < \lambda \leq 419.9$ nm), and MODTRAN3.5 ($\lambda > 419.9$ nm) [16]. Surface albedo was that for grass [17]. The DISORT algorithm [18] with delta-M scaling and 32 streams was used. Monthly mean total column ozone was computed by averaging the Earth Probe (EP) Total Ozone Mapping Spectrometer (TOMS) v. 8 daily total column ozone [19] for dates prior to the end of 2005. Starting from 2006, the Ozone Monitoring Instrument (OMI) TOMS v. 8.5 total column ozone data [20] were used. Ozone over Perth was used for Rottnest Island, and the mean of Sydney and Melbourne was used for Canberra. The latter overestimates the Canberra ozone column by $< 4\%$ if the vertical profiles are the same. Monthly mean surface pressures for Alice Springs were derived from half-daily averages included in the BoM AOD data files. 3-hourly pressures for Canberra Airport, Darwin Airport, and Rottnest Island from the BoM climate data service were averaged for the other sites. Surface elevations were included with the AERONET AOD data, and elevations for the co-located BoM airport weather stations were used for the BoM AOD sites. The solar zenith angle at local noon for the middle of each month was used. A previous study found the impact of uncertainties in surface pressure and ozone on AOD retrievals is negligible when surface pressure is measured on-site and monthly mean ozone from stations of commensurate latitude is used [13].

For the aerosol calculations, the AOD at 550 nm, the reference wavelength for aerosol calculations in libRadtran, was set to the monthly mean measured 500 nm AOD. The urban aerosol type in libRadtran [21] gave the greatest effective SPF, so it was used for all sites.

Calculated direct and diffuse downward irradiances at the surface were combined and then summed over 280–400 nm with relative spectral weights from an erythemal action spectrum [22] to determine the effective UV radiation intensity, UVR_{eff}, equation 1 [2],

$$\mathrm{UVR}_{eff} = \sum_{280}^{400} \mathrm{E}_\lambda \mathrm{S}_\lambda \Delta_\lambda \qquad (1)$$

where E_λ is calculated surface irradiance (W m^{-2} nm^{-1}), S_λ is relative erythemal spectral effectiveness [22], and Δ_λ is the width of each interval. The effective SPF due to aerosols is,

Table 1. Site-specific information for calculations.

Site	Peak AOD Month Year	Mean	Ozone (DU)	Pressure (hPa)	Fit Parameters Slope	Intercept
Alice Springs	Nov 2002	0.25	283	951	0.81	0.99
Canberra	Jan 2006	0.09	265	947	0.63	1.0
Darwin	Oct 2004	0.38	261	1007	0.88	0.97
Rottnest Island	Sep 2003	0.07	321	1010	0.68	1.0

thus, the ratio of the calculated UVR_{eff} for a clear sky case divided by the calculated UVR_{eff} with aerosols present, equation 2,

$$SPF_{eff}(\text{aerosol}) = \frac{UVR_{eff}(\text{clear sky})}{UVR_{eff}(\text{aerosol})} \tag{2}$$

4. Results

The monthly mean AODs over each site are shown in figure 2. The months with the highest mean AOD for each site and the site-specific information used for the radiative transfer calculations are shown in table 1. The slope and intercept for the best linear least-squares fits to the effective aerosol sun protection factor (SPF) as a function of AOD for each site, figure 3, are given in table 1. The linear fits in table 1 were used to convert the probability density functions from AOD, figure 1, to effective aerosol SPF, figure 4.

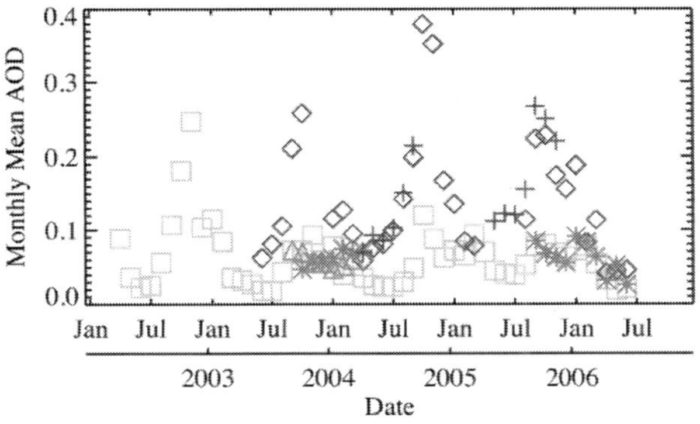

Figure 2. Monthly mean 500 nm AOD (501 nm at Alice Springs) for April 2003 to June 2006. Alice Springs (BoM) in green squares, Canberra (AERONET) red asterisks, Darwin (BoM) blue diamonds, Darwin (AERONET) blue crosses, and Rottnest Island (AERONET) grey triangles. The monthly means were computed as unweighted means over the days (half-days) on which AOD was measured.

5. Discussion

The results shown in figure 4 indicate the effective sun protection factor (SPF) due to aerosols over the four sites was less than 1.1 for half of the days (half-days) on which measurements were made and less than 1.4 on 95% of the days. A previous study on the impact of smoke over Darwin on surface UV found the smoke reduced surface UV by as much as 50% [11], an effective SPF of two. This is a larger maximum effective SPF than we found in the present study, but both studies suggest the longer-term average impact of aerosols on surface UV may be minimal from the viewpoint of human health. This is particularly true when compared to the strong influence that human behaviour has in determining exposure to UV.

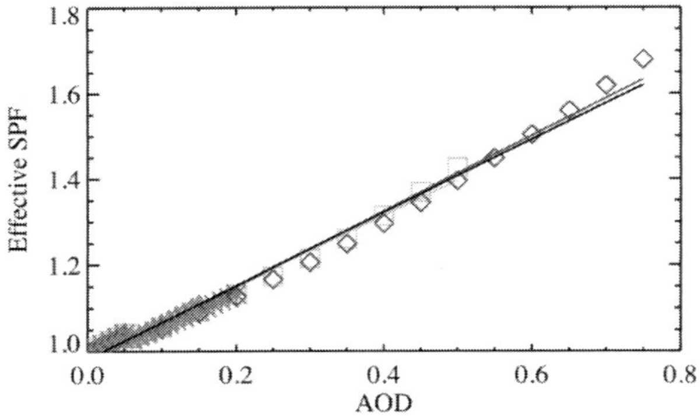

Figure 3. Calculated effective sun protection factor (SPF) due to urban aerosols over Alice Springs (green squares), Canberra (red asterisks), Darwin (blue diamonds), and Rottnest Island (grey triangles) for the conditions in table 1. Linear least-squares fits for each site are the corresponding solid lines and for all sites is the black solid line.

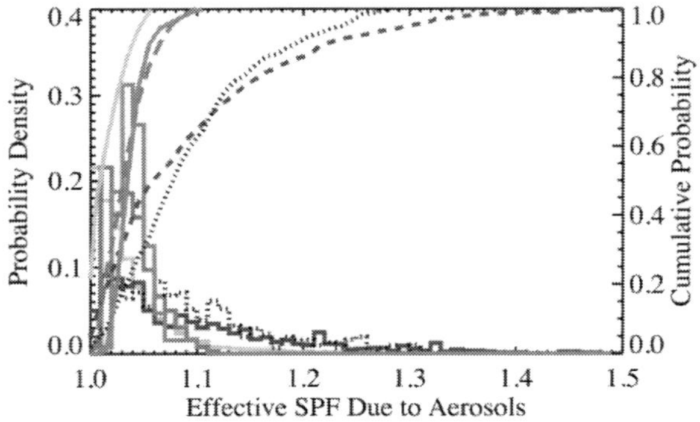

Figure 4. Probability density functions for effective aerosol SPF over Alice Springs (green), Canberra (red), Darwin (BoM blue, AERONET dotted black), and Rottnest Island (grey), assuming "urban" aerosol at all sites. Corresponding cumulative probabilities: Alice Springs (green), Canberra (red), Darwin (short-dashed blue–BoM–and dotted black–AERONET), and Rottnest Island (grey). One of 2482 and three of 899 half-daily averages for Alice Springs and Darwin, respectively, had effective SPF > 1.5.

The present study provides an initial assessment of the potential importance of aerosols for human exposure to UV, but this study has a number of significant limitations. A standard (northern hemisphere) "urban" aerosol type was assumed for calculating the effective SPF. A better procedure would be either to fit the multispectral observations of AOD at each site with a polynomial or to determine the Angstrom exponent and then use either the polynomial fit or the Angstrom exponent to extrapolate into the UV. Ideally, this would be done for each day or half-day so as to identify, and potentially classify, different types of aerosols over each site. Another significant limitation is that the present study computed the effective SPF due to aerosols at local noon for the middle of the month with the highest AOD. The effective SPF for a given AOD will vary through a day and throughout the year and is likely to increase at higher solar zenith angles. Finally, the time periods for which AOD observations were available, particularly for Canberra and Rottnest Island, were limited and provided irregular coverage. A more detailed study is being undertaken that is intended to address these issues.

6. Conclusions

An initial assessment of the impact of Australian aerosols on surface UV over four sites near moderate-size population centres has been undertaken. Australian aerosols can substantially attenuate surface UV but this occurs on only a few days each year. Thus, the time-average effect of aerosols on surface UV appears to be much smaller than the influence of human behaviour.

Acknowledgments

We thank Drs R. Mitchell, B. Forgan, and B. Holben for establishing, operating, and maintaining the sunphotometers that provided the data for this project. We thank Dr. M. Kimlin and two anonymous referees for helpful comments. This was originally a MBBS student project by CYC.

References

[1] Bais A F, Lubin D, Arola A, Bernhard G, Blumthaler M, Chubarova N, Erlick C, Gies H P, Krotkov N, Lantz K, Mayer B, McKenzie R L, Piacentini R D, Seckmeyer G, Slusser J R and Zerefos C S 2007 *Scientific Assessment of Ozone Depletion: 2006* (Geneva: World Meteorological Organization) chap 7, pp 7.1–7.54 Global Ozone Research and Monitoring Project–Report No. 50

[2] Gies P, Roy C, Javorniczky J, Handerson S, Lemus-Deschamps L and Driscoll C 2004 *Photochem. Photobiol.* **79** 32–9

[3] McKenzie R L, Aucamp P J, Bais A F, Björn L O and Ilyas M 2007 *Photochem. Photobiol. Sci.* **6** 218–231

[4] Slevin T, Clarkson J and English D 2000 *Radiat. Prot. Dosim.* **91** 303–6

[5] Andrady A *et al.* 2004 *Photchem. Photobio. Sci.* **3** 1–5

[6] van der Mei I A F, Ponsonby A L, Blizzard L and Dwyer T 2001 *Neuroepidemiology* **20** 168–74

[7] Lucas R M and Ponsonby A L 2002 *Med. J. Australia* **177** 594–8

[8] Dickerson R, Kondragunta S, Stenchikov G, Civerolo K, Doddridge B and Holben B 1997 *Science* **278** 827–30

[9] Krotkov N A, Bhartia P K, Herman J R, Fioletov V and Kerr J 1998 *J. Geophys. Res.* **103** 8779–8793

[10] Balis D, Amiridis V, Zerefos C, Kazantzidis A, Kazadzis S, Bais A, Meleti C, Gerasopoulos E, Papayannis A, Matthias V, Dier H and Andreae M 2004 *Atmos. Chem. Phys.* **4** 302–321

[11] Kalashnikova O V, Mills F P, Eldering A and Anderson D 2007 *Remote Sens. Environ.* **107** 65–80

[12] Wilson S R and Forgan B W 2002 *J. Geophys. Res.* **107** 4068

[13] Mitchell R M and Forgan B W 2003 *J. Atmos. Ocean. Tech.* **20** 54–66

[14] Dubovik O, Smirnov A, Holben B N, King M D, Kaufman Y J, Eck T F and Slutsker I 2000 *J. Geophys. Res.* **105** 9791–806

[15] Alexandrov M D, Marshak A, Cairns B, Lacis A A and Carlson B E 2004 *Geophys. Res. Lett.* **31** L04118

[16] Mayer B and Kylling A 2005 *Atmos. Chem. Phys.* **5** 1855–77

[17] Feister U and Grewe R 1995 *Photochem. Photobiol.* **62** 736–744

[18] Stamnes K, Tsay S C, Wiscombe W and Jayaweera K 1988 *Appl. Opt.* **27** 2502–2509

[19] McPeters R D *et al.* 1998 *Earth Probe Total Ozone Mapping Spectrometer (TOMS) Data Products User's Guide* NASA Technical Publication 1998-206895 NASA Goddard Space Flight Center

[20] Bhartia P K and Wellemeyer C W 2002 *OMI Algorithm Theoretical Basis Document vol. II OMI Ozone Products* ed Bhartia P K (Greenbelt, Maryland: NASA Goddard Space Flight Center) chap 2, pp 15–32 ATBD-OMI-02, version 2.0

[21] Shettle E P 1989 *Models of aerosols, clouds and precipitation for atmospheric propagation studies*, paper presented at Conference on Atmospheric Propagation in the UV, Visible, IR, and MM-Region and Related System Aspects, NATO Adv. Group for Aerosp. Res. and Dev., Copenhagen

[22] McKinlay A F and Diffey B L 1987 *Human exposure to ultraviolet radiation Risks and Regulations* ed Passchier W F and Bosnjakovic B F M (Amsterdam: Elsevier) chap 11, pp 83–88 proceedings of a seminar held in Amsterdam, 23–25 March 1987

Climate change and the effects of dengue upon Australia: An analysis of health impacts and costs

D Newth and D Gunasekera

CSIRO Centre for Complex Systems Science, CSIRO Marine and Atmospheric Research, CSIRO, GPO Box 3023, Canberra ACT 2601, Australia.

E-mail: david.newth@csiro.au

Abstract. Projected regional warming and climate change analysis and health impact studies suggest that Australia is potentially vulnerable to increased occurrence of vector borne diseases such as dengue fever. Expansion of the dengue fever host, *Aedes aegypti* could potentially pose a significant public health risk. To manage such health risks, there is a growing need to focus on adaptive risk management strategies. In this paper, we combine analyses from climate, bio-physical and economic models with a high resolution population model for disease spread, the EpiCast model to analyse the health impacts and costs of spread of dengue fever. We demonstrate the applicability of EpiCast as a decision support tool to evaluate mitigation strategies to manage the public health risks associated with shifts in the distribution of dengue fever in Australia.

1. Introduction

Australia continues to be at risk from tropical infectious mosquito-borne diseases such as dengue fever, yellow fever, Japanese encephalitis and malaria. Regional warming and climate change projections and health impact analysis suggest that this risk will increase considerably over the coming decades. Of these diseases, dengue fever poses a significant threat to Australia due to the potential expansion of the home range of the disease host vector *Aedes aegypti* [1, 2, 3, 4]. At present, locally acquired dengue infections have been reported only from urban areas in the north-east of Queensland, where *Ae. aegypti* is most abundant [5]. The potential for epidemics of dengue fever in north Queensland has increased in recent years [6]. Between 1991 and 2008, there were a total of 3385 clinical cases of dengue fever reported [2]. Human infections caused by dengue fever can result in fatalities, reduction in quality of life, loss of productivity and increased public health costs. It has been estimated that the economic cost of dengue in major developing countries in Asia and Latin America is around $1.8 billion annually [7]. It is important to recognize that dengue fever is a rapidly growing public health problem in tropical and sub-tropical regions of the world. Dengue fever is endemic in many developing countries in our region. Around 2.5 billion people, two-fifths of the global population, live in parts of the world affected by dengue. An additional 120 million people travel to dengue-affected regions annually. Between 50 and 100 million people are infected by dengue each year. The specific economic costs of dengue fever include lost productivity, school and work place absenteeism and unpaid time from caregivers [7]. Assessing the potential risk presented by infectious and vector borne diseases and their associated social, economic and epidemiological impacts, along with formulating and evaluating mitigation strategies is a key issue for policy makers and health care providers.

In this paper, we demonstrate how a high resolution population model for disease spread (EpiCast) can be used in combination with analyses from climate, bio-physical and economic models to quantify the health impacts and costs associated with the potential spread of vector borne diseases such as dengue fever. The high resolution population-based nature of the EpiCast modelling framework enables us to better represent the dynamics of the spread of dengue fever when large numbers of individuals are potentially infected. We regard this as a useful approach which complements other recent studies [1, 2, 3, 4, 8] of potential spread of dengue. In the following section, we outline the potential risk that dengue fever presents to Australia, and the likely shift in *Ae. aegypti* home range due to climate change. A short overview of the EpiCast modelling framework is presented in section 3. We then outline illustrative scenarios analysed in this study. Finally, we provide a discussion of the results of our analysis and make some closing remarks.

2. Climate change and dengue risk factors

Recent studies on the effect of climate change on *Ae. aegypti* suggest that within Australia, the mosquito will spread across many parts of the country, and there will be an increase in the annual number of potential generations [6, 8, 9]. These studies argue that human adaptation to climate variability and change —through the use of inappropriately maintained water storage tanks, swimming pools and other standing bodies of water, which provide breeding sites for *Ae. aegypti*— may heighten the risk posed by dengue fever to the Australian population [8, 9]. The ongoing dry conditions in many parts of Australia, particularly in the southeast of the country, have seen an increase in the number of domestic water storage containers. Government subsidies have encouraged the uptake of rainwater tanks throughout the country. The proportion of households having rainwater tanks has steadily increased over the past two decades from 15% in 1994 to 17% in 2004 and on to 19% in 2007 [10]. This steady increase in domestic collection and storage of water has been flagged as a matter of potential concern, as it provides an ideal breeding habitat for *Ae. aegypti* in urban areas [1].

In the present analysis, we use the key findings from recent climate change projections and bio-physical and ecological studies that generate projections of the potential home range of *Ae. aegypti* to identify regions that are likely to be affected by dengue fever. These studies highlight their challenges such as the limitations of use of point occurrence data and environmental parameters to map the potential distribution of *Ae. aegypti* and the difficulties in understanding the non-linear relationships between the daily survival of *Ae. aegypti*, the extrinsic incubation period of the dengue virus, temperature and humidity [8]. Figure 1 which is extracted from reference [8] shows the potential distribution of *Ae. aegypti* in 2030 under the IPCC SRES A1B (a mid range scenario for economic growth and greenhouse gas emissions) climate change scenario. In our analysis these projections are included as a static scenario which is incorporated with regional household water storage tank data [10] in the EpiCast Model. This provides a snapshot in 2030 of a situation whereby climate change is projected to expand the range of *Ae. aegypti* to include much of northern, eastern and southeast Australia in the presence of an increase in uptake of water storage tanks.

3. The EpiCast modelling framework

EpiCast [11] is a stochastic individual-based discrete-time simulation model. It is a high-resolution model that takes a "whole of population" approach to modeling the spread of infectious diseases and population demographics. EpiCast has been used to develop US policy in response to potential infectious disease spread, such as avian influenza and swine influenza, as well as bioterrorism. We have calibrated EpiCast for the Australian population and expanded the model to include human-vector interactions. EpiCast has four basic components: population demographics, contact groups, disease natural history and long distance travel. Each of these components is described in more detail below (for further information and details about the EpiCast model structure, the reader is directed to reference [11]):

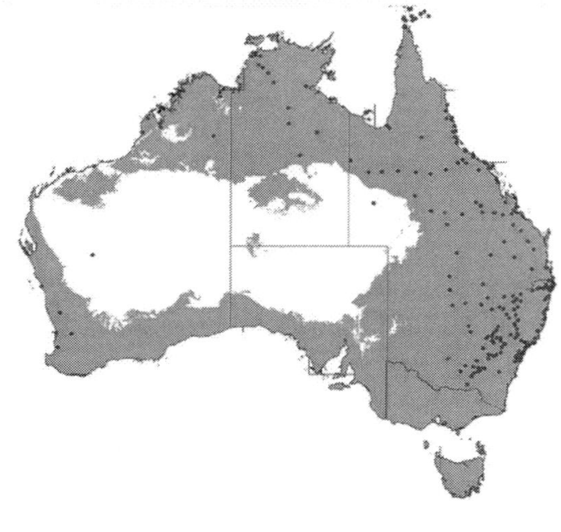

Figure 1. Projected distribution of *Ae. aegypti* in 2030 due to climate change. The dots represent collection sites of *Ae. aegypti* used in the projections *cf.* [8]

1. *Population demographics*. The model accurately depicts a synthetic population of 21 million individuals, derived from the Australian Census [12]. This representation includes age, sex, family structure, household structure, occupation, sector of employment, income, travel to work and place of work;
2. *Contact groups*. The model runs in two cycles (day and night), during which we identify seven contact groups. In five of the contact groups (households, household clusters, preschools and playgroups, schools and work groups) relatively close person-vector association can occur. The remaining two groups (neighborhoods and communities) provide unspecified contact groups (e.g. shopping centers or parks) within which occasional casual person-vector interactions can occur. The household contact group contains information about the water bodies such as water tanks. For the purposes of this study, households are the only breeding sites for *Ae. aegypti*;
3. *Disease natural history*. The dengue fever disease transmission dynamics and natural history used in the current analysis is based on those presented in [13]. Specifically, after being bitten by an infected mosquito, the disease has an incubation period of 5-8 days and a symptomatic period of 4-7 days. Also, 76% of all cases will be subclinical and of the 24% of clinical cases, 0.8% will result in death. Once exposed to an infectious individual, the host vector, *Ae. aegypti* requires 7 days for viral load to become high enough to infect humans [13]. The life span of an *Ae. aegypti* mosquito is 14 to 31 days [14]; and
4. *Long distance travel*. This captures the long distance travel habits of the Australian population. This information is drawn from the time usage survey, travel to work data and long distance travel data [12].

The combination of these components results in a high resolution population model that can be used to simulate vector borne disease spread. The model allows the policy maker to examine the effect of an outbreak at the level of the individual, household, special sub-group (e.g. age group), economic class, sector of employment, local government area, state or nationally. The structure of the model allows policy makers to test various mitigation and intervention strategies.

4. Scenarios and analysis

4.1. Scenarios

We examine three illustrative scenarios. In all scenarios it is assumed that *Ae. aegypti* has colonized all key regions of the country as projected by [8] (see figure 1). Further, it is assumed that an important breeding ground of *Ae. aegypti* is unsecured bodies of water such as uncovered or ineffectively screened rain water tanks. For the purpose of this study we assume that different types of water bodies are uniformly distributed across regions. Analyses of these illustrative scenarios help ascertain the potential social and economic impacts of dengue fever within Australia. No medical or public health interventions are considered here. Also, it is important to recognise that household environment is not the only breeding site for *Ae. aegypti*. The illustrative scenarios analysed here include:

1. *Baseline scenario*. This is the reference case under which there are no mitigation response strategies in place. Dengue can originate from any breeding site. The baseline scenario is based on one of the recently published [8] projections of the potential expansion of the *Ae. aegypti* in Australia due to climate change adaptation and climate change. It corresponds to the IPCC global warming SRES A1B scenario [8]. The baseline scenario represents the potential combined effects of two key factors. First, a situation whereby climate change is projected to expand the spread of *Ae. aegypti* to much of northern, eastern and southeast Australia by 2030 (see figure 1) and second, the uptake of water storage tanks that is expected to increase in response to reduced rainfall over that period;
2. *Scenario A (Low response scenario)*. This scenario examines how the impact of dengue fever changes with a minor mitigation strategy. We assume that 25% of the breeding sites for *Ae. aegypti* are properly secured and maintained, and the vector is eliminated from these sites; and
3. *Scenario B (High response scenario)*. This scenario examines how the impact of dengue fever changes with a major mitigation strategy. We assume that 75% of the breeding sites for *Ae. aegypti* are properly secured and maintained, and the vector is eliminated.

All scenarios are seeded with a small outbreak of dengue-carrying people in Queensland. For each scenario and parameter configuration 100 simulation runs were carried out. Each run simulated 100 days, effectively a summer outbreak period.

4.2. Basic measures

To measure the various epidemiological, social and economic impacts of dengue upon the Australian society, we make use of the following statistics and measures: (1) *Basic Reproductive Number, R_0*, which is defined as the number of secondary infections generated by a primary infectious case via the vectors within an entirely susceptible population. Studies of dengue outbreaks estimate R_0 to be between $0.49 \leq R_0 \leq 6$ [15]. Each scenario examines three R_0 values of 1.5, 2.0 and 2.5; (2) *Disability-Adjusted Life Year (DALY)*. DALY is a measure of overall disease burden upon a population. It combines mortality and morbidity into a single common metric; (3) *Exposed*. This is the number of people currently residing within a community that could be infected by the dengue virus; (4) *Potential Health Cost*. This measure combines direct costs incurred by the health-care system and costs incurred for public health surveillance, prevention and control activities. For dengue fever, the public health cost per year, on average, in Australia is conservatively estimated to be around $2.80 per exposed person [2]; and (5) *Labour Absenteeism*. This measures the number of days of lost productivity for a particular sector. Conservative estimates for Australia are that a subclinical dengue case will be absent from work for two days and a clinical dengue case will be absent for 10 days [2].

4.3. Results

The key results of our analysis are shown in tables 1–3 and they highlight the following. First, the assumed mitigation response measures (Scenarios A and B) help to lower (relative to the baseline scenario) the estimated DALYs, hence reducing the overall burden of dengue fever across the country. For example, the estimated DALYs under the high mitigation response scenario vary, on average, between 20 and 37 depending on the assumed R_0 values. Our results illustrate that even a low mitigation response strategy (Scenario A) has the potential to reduce the overall size of an outbreak of dengue fever and hence lower the associated adverse socio-economic impacts. To put our estimated DALYs in perspective, it is worth noting that the regional estimates of DALYs (in 2000) ranged from 15–40 in Western Pacific and Eastern Mediterranean to 240 in Southeast Asia [16].

Second, the estimated absenteeism in key labor intensive sectors such as construction, manufacturing, health and retail activities indicate that the average number of days of lost productivity per dengue outbreak tends to decline with mitigation response measures (relative to the baseline scenario). This finding highlights the importance of the mitigation response measures in minimising the disease burden on working population and hence the associated adverse economic implications.

Third, relative to the baseline scenario, the estimated public health costs tend to decline with the mitigation response measures. For example, intervention measures help to lower the potential public health costs (in 2000 dollars) – from around $1.9m (baseline scenario) to about $0.9m (Scenario B) (see table 3) i.e. halving the public health costs on average. Again, to put our public health cost estimates in perspective, it is important to note that the comparable dengue related health costs under different climate change scenarios analysed for the Garnaut Review vary from $1.3m (in 2020) to $1.8m (in 2050) (in 2000 dollars) [2].

Table 1. Characteristics of simulated dengue fever in Australia (R_0=1.5)

	Baseline	**Scenario A**	**Scenario B**
Exposed	320,101	298,316	279,985
Public Health Cost ($)	902,686	789,305	789,305
DALYs	29 (242)	26 (133)	20 (72)
Sector Absenteeism (days):			
Construction	7.46 (62)	5.50 (60)	3.76 (26)
Health	9.02 (151)	10.72 (61)	5.18 (22)
Manufacturing	10.73 (116)	10.73 (62)	8.21 (34)
Retail	9.42 (60)	8.16 (48)	5.97 (26)

Numbers in parentheses are the maximum values

Table 2. Characteristics of simulated dengue fever in Australia (R_0=2.0)

	Baseline	**Scenario A**	**Scenario B**
Exposed	427,087	342,720	275,008
Public Health Cost ($)	1,204,305	966,471	775,524
DALYs	67 (1072)	39 (333)	25 (84)
Sector Absenteeism (days)			
Construction	12.22 (272)	9.60 (75)	6.07 (32)
Health	18.35 (354)	10.54 (103)	6.60 (31)
Manufacturing	23.83 (529)	12.43 (122)	8.65 (30)
Retail	20.99 (487)	12.29 (119)	8.81 (31)

Numbers in parentheses are the maximum values

Table 3. Characteristics of simulated dengue fever in Australia (R_0 =2.5)

	Baseline	**Scenario A**	**Scenario B**
Exposed	672,499	484,260	322,947
Public Health Cost ($)	1,896,450	1,365,614	910,711
DALYs	155 (2,251)	88 (1,950)	37 (171)
Sector Absenteeism (days)			
Construction	30.62 (504)	17.68 (385)	8.14 (36)
Health	45.57 (746)	26.25 (563)	9.73 (60)
Manufacturing	52.46 (707)	30.06 (535)	12.69 (77)
Retail	29.64 (685)	17.28 (441)	10.90 (81)

Numbers in parentheses are the maximum values

5. Concluding remarks

The continuing dry conditions and ongoing water restrictions in many parts of urban Australia have resulted in an increasing number of households collecting rainwater. But inappropriately maintained water tanks with poor water storage hygiene and stagnant water carries the risk of mosquito borne diseases such as dengue fever. There is a continuing need to eradicate mosquito breeding sites on a regular basis including proper maintenance of rainwater tanks and other household water containers. Our illustrative analysis indicate that if potential breeding sites such as rainwater tanks around homes are properly maintained, the dengue related public health costs could be reduced by 50% relative to what would otherwise be.

Acknowledgements

The authors would like to thank John Finnigan, Roger Bradbury, Paul De Barro, Clem Davis and the two anonymous reviewers for their valuable comments.

References

[1] Prime Minister's Science, Engineering and Innovation Council 2009 Epidemics in a changing world, Report prepared by the Expert Working Group on Epidemics in a Changing World, Canberra, June.
[2] Bambrick H, Dear K, Woodruff R, Hanigan I and Mc Michael A 2008 The impacts of climate change on three health outcomes. Garnaut Climate Change Review. http://www.garnautreview.org.au/CA25734E0016A131/pages/all-reports--resources-commissioned-reports.html
[3] McBride W J H, Mullner H, Muller R, Labrooy J and Wronski I 1998 *Am J Epidemiol* **148** 1111–16
[4] Murray-Smith S, Weinstein P and Skelly C 1996 *Aust NZ J Public Health* **20** 545–7
[5] Russell R C, Currie B J, Lindsay M D,Mackenzie J S, Ritchie S A and Whelan P I 2009 *Med J Aust* **190** 265–8
[6] Woodruff R, Hales S, Butler C and McMichael A 2005 Climate change health impacts in Australia: Effects of dramatic CO_2 emission reductions, Australian Medical Association and Australian Conservation Foundation
[7] Suaya J A, Shepard D S, Siqueira J B, Martelli C T, Lum L C S, Tan L H, Kongsin S, Jiamton S, Garrido F, Montoya R, Armien B, Huy R, Castiio L, Caram M, Sah B K, Sughayyar R, Tyo K R, and Halstead S B 2009 *Am J Trop. Med. Hyg.* **80** 846–55
[8] Beebe N W, Cooper R B, Mottram P and Sweeney A W 2009 *PLoS Negl Trop Dis* **3** e429 doi:10.1371/journal.pntd.0000429
[9] Kearney M, Porter W P, Williams C, Ritchie S and Hoffman A A 2009 *Funct Ecol* **23** 528–538
[10] Australian Bureau of Statistics 2007 Environmental issues: peoples views and practices, Canberra, Cat No. 4602.0
[11] Germann T C, Kadau K, Longini I M and Macken C A 2006 *Proc Nat. Acad Sci* **103** 5935–40
[12] Australian Bureau of Statistics 2007 Census of population and housing, Cat no. 2914.0.55.002
[13] Shepard D S, Suaya J A, Halstead S B, Nathan M B, Gubler D J, Mahoney R T, Wang D N C and Meltzer, M I 2004 *Vaccine* **22** 1275–80
[14] Tabachnick W J, Munstermann L E and Powell J R 1978 *Evolution* **33** 287–95
[15] Chowell G, Diaz-Duenas P, Miler J C, Alcazar-Velazco A, Hyman J M, Fenimore P W and Castillo-Chavez C 2007 *Mathematical Biosciences* **208** 571–89
[16] Clark D V, Mammen Jr M P, Nisalak A, Puthimethee V and Endy T P 2005 *Am J Trop Med* **72** 786–91

IOP Conf. Series: Earth and Environmental Science **11** (2010) 012021 doi:10.1088/1755-1315/11/1/012021

Severe wind hazard using dynamically downscaled climate simulations

L A Sanabria and R P Cechet

Risk and Impact Analysis Group, Geoscience Australia
G.P.O. Box 378, Canberra, Australian Capital Territory

E-mail: augusto.sanabria@ga.gov.au; bob.cechet@ga.gov.au

Abstract. Tropical cyclones, thunderstorms and sub-tropical storms can generate extreme winds that can cause significant economic loss. Severe wind is one of the major natural hazards in Australia. In this study, regional return period gust wind hazard (10 meter height over open terrain) is determined using a new methodology developed by Geoscience Australia. The methodology involves a combination of three models: A Statistical Model (ie. data-based model) to quantify wind hazard using extreme value distributions. A methodology to extract gridded hourly maximum mean (time-step) wind speed and direction fields from a high-resolution regional climate model (RCM). Area-averaged measurements from the RCM are "corrected" for point measurement exposure by applying a regional factor to the RCM values. And a Monte Carlo method to calculate gust wind from RCM mean wind, the former is the value of interest in wind hazard.

To assess accuracy, model results were compared against three wind recording stations in Tasmania. These sites were selected because their weather stations and anemometer measurements are located at airports, avoiding the problem of houses or trees affecting the instruments, and also due to these sites having wind gust records. The model works only with synoptic gust wind speeds. A methodology to calculate severe wind gusts associated with thunderstorm down-burst winds is under development.

1. Introduction

Regional and local wind gust hazard estimates are utilised in the Australian building codes through the Australia/New Zealand Wind Actions Standard [1]. The wind engineering community relies to a significant extent on the peak gust wind speed estimates derived from observations collected over more than 70 years by the Australian Bureau of Meteorology (BoM). The current wind loading code and the performance of our infrastructure (residential, commercial, industrial and critical infrastructure) is based primarily on hazard estimates from a small dataset, consisting of mainly airport sites. In this paper we present a fast algorithm for the calculation of gust wind hazard from gridded climate-simulated hourly maximum mean wind speeds for a region (island of Tasmania). Currently we deal only with gusts associated with synoptic winds as the climate model only provides mean winds at a resolution of 14 km, which does not resolve thunderstorms.

The algorithm involves two very computationally demanding processes:

- Calculation of return period (RP) for gust wind speed using extreme value distributions; and
- A Monte Carlo method to generate synthetic gust speeds by performing a numerical convolution of mean wind speeds and regional *gust factors*.

A detailed Monte Carlo process to calculate gust hazard at a wind recording station was developed in 2007 [2]. However the process is too computer intensive and cannot be used for large-scale gridded applications, that is, calculation of wind hazard over a region. A region consists of thousands of cells each with many years of simulated data. Clearly a more computationally efficient methodology is necessary for gridded analysis.

2. Extreme Value Distributions

The core of statistical model is the fitting of extreme value (EV) distributions to point-specific anemometer observations. EV distributions allow wind analysts to make inferences about the magnitude and frequency of extreme events beyond the range of years available in the observed record.

There are two basic functions to model extreme values; one is the Generalised Extreme Value distribution (GEV) and the other is the Generalised Pareto distribution (GPD) [3]. The former utilises annual maximum (i.e. using annual observations). This is a limitation of the distribution because it is possible that the highest values below the maximum in a single year exceed the maximum of another year, and yet they are not included in the model. The GPD works with all dataset exceeding a given threshold. This distribution not only uses more data but by selecting the appropriate threshold, samples can be better distributed in time guaranteeing their independence (one of the conditions of extreme value distributions). In recent years most wind researchers have adopted the GPD for extreme value analysis due to its superior performance [4,5]. The statistical model uses the GPD to calculate the return period of wind hazard. Its main feature is an automatic algorithm to calculate the appropriate threshold to fit the GDP to given wind speeds [6].

A second important feature of the statistical model is its capacity to separate the gust wind components from the BoM-provided datasets. In the non-cyclonic regions of Australia the main gust components are from thunderstorm and synoptic winds [6]. In this paper, we only deal with *synoptic* winds. A paper on the calculation of the corresponding spatial gust hazard from thunderstorm winds is in preparation.

3. Monte Carlo (MC) generation of synthetic *gust* speeds

In the MC process the physics of wind gust generation is simulated. Surface wind gusts originate from the deflection of air parcels flowing higher in the boundary layer, which are brought down by turbulent eddies [7]. The method separately takes into account the mean wind and the turbulent structure of the atmosphere. Turbulence is represented by the gust to mean ratio, referred to as the gust factor in wind engineering terms. The process consists of the numerical convolution of daily maximum mean wind speed and the gust factor to produce synoptic gust wind speeds.

As explained in [2] the implementation of the full MC involves three steps:

1. Extraction of wind speeds from climate simulations at the location of BoM recording stations, and the calculation of the corresponding RP "daily maximum mean" wind speed exceedance level (daily maximum assessing each modelled time-step);
2. Comparison of climate model simulated mean wind hazard with the observed hazard. Area-averaged speeds from the climate simulations are "corrected" to observed station (single point) measurements using linear regression; and
3. The mean speeds are used to generate synthetic gust speed RP curves using MC.

3.1 Gust Factor

The gust factor (GF) is defined as the ratio of maximum wind speed (gust) and mean wind speed for the same time period. In order to capture the regional characteristics of Tasmania, BoM-provided half-hourly datasets for the three selected stations (Hobart, Launceston and Wynyard Airports) were used. The datasets were joined to calculate a *regional* GF. This GF was used for the sampling process in the Monte Carlo simulation.

3.2 Stratified sampling

To guarantee that enough samples are generated in the tail of the MC distribution, the most important part of the distribution for wind hazard studies, a simple stratified sampling process was implemented as follows. The numerical convolution of the daily maximum mean wind speed and gust factor was carried out for wind speeds from the distribution's mean (m_1) value onwards to higher values in order to generate the same sample size as in the original vector (mean wind speed). Mean wind speed values below m_1 are of no concern in wind hazard studies. The process is then repeated to generate the 'peak' samples: one fifth of the number of samples of the original vector is drawn from the 98% quantile onwards. The wind gust used in the calculation of the RP is the sum of both vectors. It was observed that the model is very sensitive to the percentage of samples used in each step: too many peak samples result in a high, flat curve. Too few peak samples result in a curve with very low values at high RP. The methodology explained above was developed empirically; it provides simulated RP curves which have a similar shape to those of the three selected stations.

4. High Resolution Climate Model

The climate simulation data used for this project was obtained from CSIRO's Conformal-Cubic Atmospheric Model (CCAM). Simulations focusing on Tasmania, using IPCC scenario A2, for the period 1960 to 2100, were used. Five coupled general circulation models were used to drive CCAM (dynamic downscaling) as explained in [8]: CSIRO mark 3.5; ECHAM 5; GFDL_CM 2.1; MIROC 3.2 and UK Hadley_CM3. We extracted maximum hourly mean wind speeds (10 meter height over open terrain) from each of the five simulations using NCO tools [9]. The mean hourly speeds were then transformed to maximum daily mean speeds using the R package 'zoo' [10]. The final wind speeds utilised are the average of the 5 simulations considered.

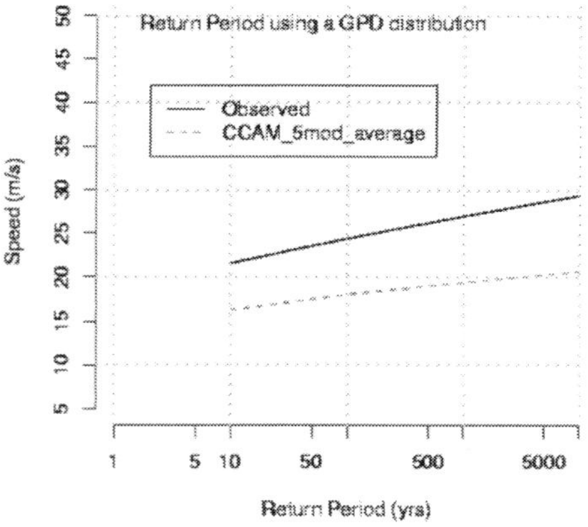

Figure 1. RPs of observed and modelled mean wind speeds (Hobart Airport).

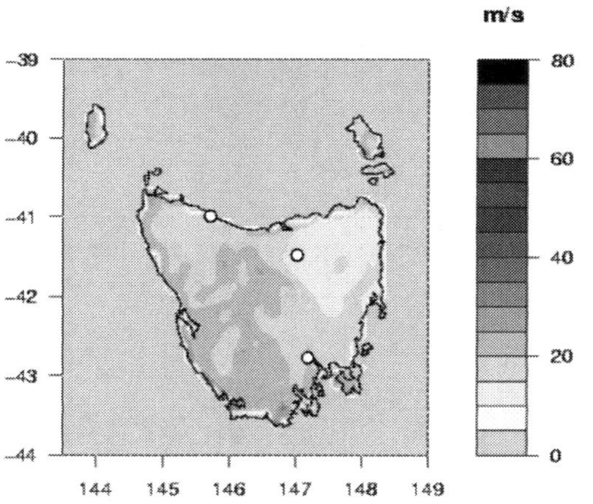

Figure 2. CCAM-modelled exceedance levels of 500-yr average mean wind speeds.

4.1 Bias correction

The bias introduced by the fact that CCAM calculates area-average values whilst the recording stations give location-based winds has to be corrected otherwise meaningless results could be obtained. To correct this bias consider the observed and CCAM-modelled RP of mean wind speeds at Hobart Airport as shown in Figure 1. A linear regression expression (LR) between observed and CCAM-modelled mean wind speeds can be found to correct the CCAM-modelled RP and the same expression can be used to correct the bias in the curves of gust speed RP to make the current climate results to coincide with the observed results.

For gridded results we combined the data from the Launceston and Wynyard sites to undertake the correction because both stations are located in the north of the state. The LR correction at 500-yr RP for these stations is 4.2 m/s. The correction for Hobart calculated from Figure 1 is 7.2 m/s. Since this correction is different from the correction calculated for Wynyard/Launceston, the Hobart correction was applied to all cells south of latitude 42.3S, whilst the correction calculated for Launceston/Wynyard was applied to all cells north of 42.3S.

5. Fast MC simulation for gridded data

In the MC simulation wind gusts are generated by selecting a sample from the CCAM-generated mean wind vector in an orderly fashion and multiplying it with a randomly selected sample from the regional gust factor distribution. A uniform distribution is used for the random selection. As explained in Sect. 3.2 the process is carried out twice to generate enough samples in the distribution's tail.

6. Calculation of gust speed over gridded data

6.1 Current Climate

The methodology was applied to the gridded data from Tasmanian to obtain RP of gust wind speeds. As a reference the 500-year RP exceedance levels for five-model average mean wind speed is shown in Figure 2. The black squares show the location of the three selected meteorological recording stations. Representative gust wind speeds in each cell were then generated from these mean wind speeds using the MC process discussed above in order to calculate the corresponding RP.

Figure 3 shows the map of (synoptic) gust speed exceedance levels for 500-year RP under current climate conditions. As expected the gust wind speeds follow closely the texture of the mean values (Figure 2), in particular note the increase of synoptic wind speeds due to the effect of the mountain slope in both the west and the north-east parts of Tasmania. This is consistent with detailed studies of the effect of mountain slopes on wind speeds [11].

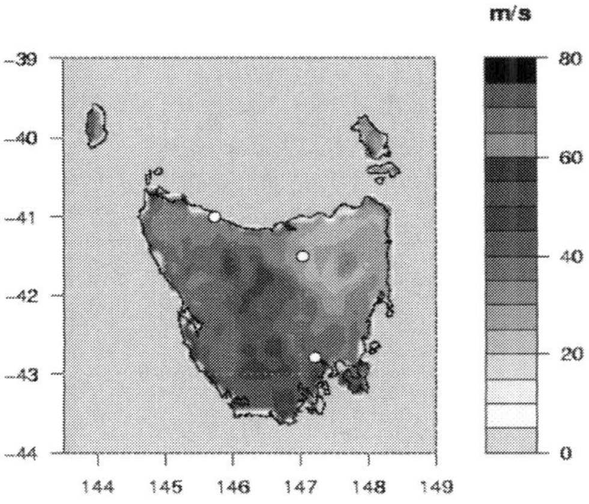

Figure 3. 500-yr RP exceedance levels of gust wind speeds (current climate).

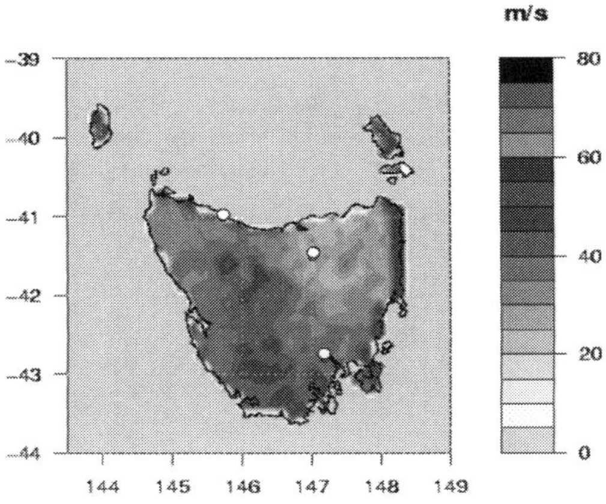

Figure 4. 500-yr RP exceedance levels of gust wind speeds around 2030.

6.2 Future Climate

To gain some insight into possible future extremes with regards to gust wind speeds, simulated daily maximum mean wind speeds from two windows of 20 years around the years 2030 and 2070 (IPCC scenario A2), were extracted from the 5 CCAM climate simulations. Their corresponding gust wind speeds were calculated using the MC procedure and RP curves for these 20-year windows using the statistical model. Figure 4 shows the 500-yr RP exceedance levels of gust wind speed around 2030 and Figure 5 shows the corresponding 500-yr RP exceedance levels around 2070.

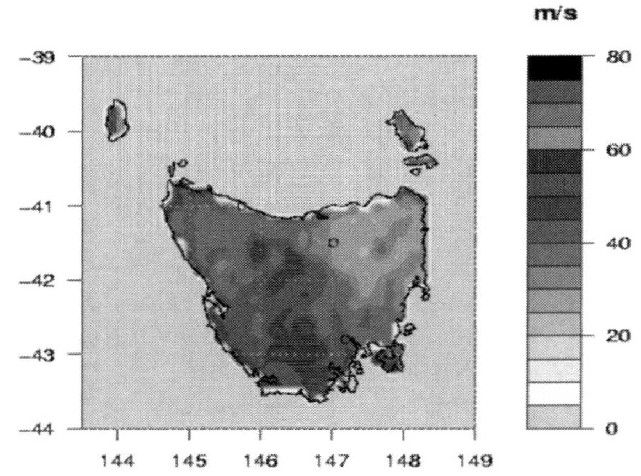

Figure 5. 500-yr RP exceedance level of gust wind speed around 2070.

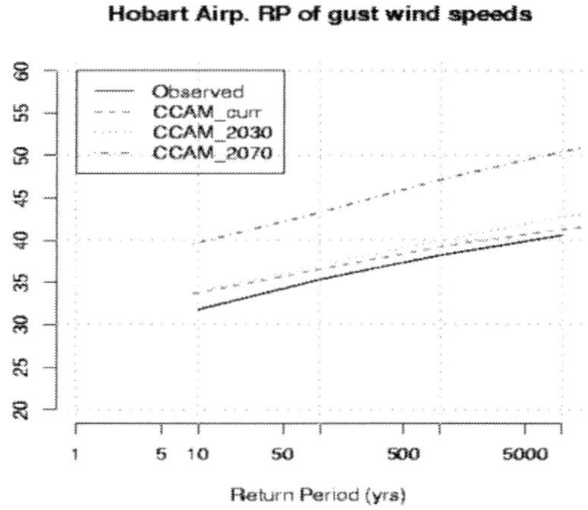

Figure 6. Corrected RP of gust wind speeds for Hobart Airport.

Comparing Figure 3 with Figures 4 and 5, it is possible to visually observe a small increase in gust wind speed hazard by 2030 compared with current climate. The increase is more noticeable in the north-east coastal areas of the state and the mountainous region of the north-west. It is also possible to see a greater increase in wind hazard by 2070 with respect to the current climate, especially in the north-western coastal areas and in the elevated regions of the South. In general, the hazard increases from the current climate through 2030 to 2070. Only the Northeast of the Tasmania experiences a decrease between 2030 and 2070 (ensemble average), and this decrease is evident in three of the five models considered. We believe that this is due to a change in the synoptic circulation possibly connected with intense east-coast low activity, and are currently undertaking a synoptic typing study to investigate this issue.

6.3 Future climate gust wind hazard at the Tasmanian BoM meteorological stations.

Plotting both the future-simulated and observed RP exceedance levels at the meteorological stations allows the determination of the trend in gust wind speed for a changing climate. Figure 6 shows the observed and corrected CCAM-modelled RP of wind gust for the Hobart Airport. Similar results were obtained for the other two stations but are not presented here because of space limitations; however Table 1 presents the corrected exceedance levels of gust wind speeds generated by the model for the selected stations.

Table 1. Corrected 500-yr RP exceedance levels of gust wind speed at stations

	Hobart	Launceston	Wynyard
Observed	37.5	37.1	36.3
CCAM_current	39.1	36.1	32.7
CCAM_2030	39.8	38.7	34.7
CCAM_2070	47.0	38.9	38.4

The station-location results confirm the trend observed in the RP maps: there is a small increase in wind gust hazard by 2030 for high RP; and a substantial increase by 2070 with respect to current climate conditions. These results agree with similar studies carried out for Victoria [12].

For location-based gust speeds, where observational mean wind speeds are available, the full MC algorithm should be used as discussed in [2]. The Fast MC algorithm presented in this paper does not give the same accuracy of the full MC model; however it allows wind analysts to observe variability in severe wind hazard over a given region. The results are highly dependent on the quality of the mean speeds provided by the climate-simulation models, and also the quality and representativeness of the observations (i.e. is the record length representative of a long-term climatology for extreme wind gusts?).

7. Conclusions

A new methodology to assess severe wind hazard for the Australian region has been developed at Geoscience Australia. The model uses simulated daily-maximum mean wind speeds produced by a regional climate model to generate the corresponding gust wind speeds via a Monte Carlo process. Wind hazard is calculated by generating curves of RP exceedance levels over the given region by using extreme value distributions. The methodology has been applied to the Tasmanian region utilising dynamical downscaling from five coupled general circulation model simulations, which provide a 14km grid of simulated data. A major characteristic of the methodology is that it provides some guidance as to the change in synoptic wind hazard under future climate conditions at the regional level.

The methodology currently only considers synoptic winds; however work is underway to also include thunderstorm downburst gust winds in order to produce a *combined* understanding of RP exceedance levels for gust wind speeds.

References

[1] AS/NZS 1170.2 2002 *Structural design actions Part 2: Wind actions* Australian/New Zealand Standard

[2] Sanabria L A and Cechet R P 2010 Severe Wind Hazard Assessment using Monte Carlo Simulation *Environ. Model. and Assessment* Vol. **15** 147-154

[3] Coles S 2001 *An Introduction to Statistical Modeling of Extreme Values* London:Springer

[4] Holmes J D and Moriarty W W 1999 Application of the generalized Pareto distribution to extreme value analysis in wind engineering *J. of Wind Eng. and Industrial Aerodynamics* **83** 1-10

[5] Brabson B B and Palutikof J P 2000 Test of the Generalized Pareto Distribution for Predicting Extreme Wind Speeds *J. of Applied Meteorology* **39** 1627-1640

[6] Sanabria L A and Cechet R P 2007 *A Statistical Model of Severe Winds* Record 2007/12 Canberra:Geoscience Australia

[7] Brasseur O 2001 Development and Application of a Physical Approach to Estimating Wind Gusts *Monthly Weather Review* **129** 5-25

[8] Corney S, J J Katzfey, J L McGregor, M Grose, J Bennett, C J White, G Holz and N L Bindoff, Climate Futures for Tasmania technical report: methods and results on climate modelling, Antarctic Climate and Ecosystems Cooperative Research Centre, Hobart, Tasmania, in press (due 2010)

[9] NetCDF Operator (NCO Tools) http://nco.sourceforge.net

[10] Zeileis A and Grothendieck G 2005 Infrastructure for Regular and Irregular Time Series *J. of Stat. Software* **14** 1-27

[11] Holmes J D 2007 *Wind Loading of Structures* 2nd Ed. UK:Taylor & Francis

[12] McInnes K L, Abbs D J and Bathols J A 2005 *Climate Change in Eastern Victoria.The effect of climate change on coastal wind and weather patterns* Report to Gippsland Coastal Board

IOP Conf. Series: Earth and Environmental Science **11** (2010) 012022 doi:10.1088/1755-1315/11/1/012022

Role of exposure in projected residential building cyclone risk for the Australian region

Denis Waters, Bob Cechet and Craig Arthur
Risk and Impact Analysis Group, Geoscience Australia, Canberra, Australia

E-mail: denis.waters@ga.gov.au

Abstract. The paper presents a methodology to analyse the direct impact of tropical cyclone hazard on communities in northern Australia. The study focuses on the *maximum potential intensity* (MPI) of the cyclonic wind hazard, and location. Storm surge impacts were developed using a simple relationship between intensity and storm surge height and mid-point sea-level rise projections. The impact on residential building stock of severe wind and storm surge hazards associated with IPCC climate change scenarios is considered. Changes in residential building stock, for over 500 coastal statistical local areas (SLAs) from Southeast Queensland anticlockwise to Perth, were estimated using Australian Bureau of Statistics population projections through to 2100. A Probable Maximum Loss (PML) curve was derived, and the average annual cost across a 5000 year period (or "annualised loss") was evaluated for each region. The projected population growth and the drift to coastal locations are significant elements in determining the damage associated with possible future cyclone threat.

1. Introduction

During 2009, the Prime Minister made a number of announcements that the likely population of Australia in 2050 might be about 35 million; an increase of some 6 million on 2005 projections. The drift to the north, especially to coastal Queensland is expected to continue. The House of Representatives Standing Committee on Climate Change, Water, Environment and the Arts inquiry [1], identified "*Sea Change*" as an important feature of the future. The Committee notes the pressure on ecologically sustainable development with social, economic and environmental consequences (p.169), and concludes that greater understanding of the impacts of population is needed (p.249). Further, the Committee recommends better monitoring of coastal demographic and population growth (p.260). This echoes the Queensland Government submission which recognises the risks faced by coastal communities resulting from population growth coupled with climate change (p.264). In this context the Committee notes that the coastal zone is especially exposed to sea-level rise, increased cyclone intensity, ocean acidification and ocean temperature rise. Against this background the purpose is to:

- present details of the *Hazard, Exposure, Vulnerability and Impact* paradigm used by Geoscience Australia for the Garnaut Climate Change Review; and
- provide greater understanding of the relative importance of population growth and climate change in the projected loss of residential buildings across the period to 2100.

2. Contribution to the Garnaut Climate Change Review

The Garnaut Climate Change Review examined the impacts of climate change on the Australian economy and possible policy responses. Technical Paper Number 4: *Methodology for modelling climate change impacts* [2], discusses key areas of measurable impacts that have enough certainty to be modelled within the Monash Multi Regional Forecasting (MMRF) model. Tropical cyclones were an area meeting this certainty criterion. The **Hazard→Exposure→Vulnerability→Impact** methodology applied by Geoscience Australia to this problem is set out below. Each element of the paradigm is required to ensure a comprehensive, consistent and integrated assessment.

3. Methodology for Analysing Tropical Cyclone Impacts

3.1. Hazard

Changes in the activity of tropical cyclones may be among the more potentially damaging results of global climate change. Climate models; CSIRO MK 3.0, GFDL CM 2.1, and NCAR CCSM 3.0 were selected on the basis of their performance for temperature and rainfall patterns between 1961 and 1990 for the Australian region [3]. The lack of horizontal resolution in general circulation models necessary to resolve the inner core of tropical cyclones requires a thermodynamic approach to identify changes in large-scale environmental factors known to affect cyclone intensity (similar to [4]). Scenarios A1FI (the so-called "business as usual") and A1B and B1 which stabilise greenhouse gas emissions at 750ppm and 550ppm CO_2 equivalents are considered. The potential riverine and/or flash flooding, the potential pole-wards movement and changed frequency in cyclone activity and the interaction with La Niña and El Niño [5, 6] are not assessed.

Maximum potential intensity (MPI) obtained from a given vertical temperature and humidity profile, was utilised. This sets an upper limit for the distribution of tropical cyclone intensity at a given point (see [7, 8]). The methodology suggests higher maximum wind speeds and storm surge levels across northern Australia in line with [9, 10].

Tropical cyclone intensification, since satellite measurements, are consistent with the observed increases in sea surface temperature [11, 12]. Frequency changes are less certain [13]. Overall, more intense cyclonic weather implies additional coastal impacts. In addition, rising sea levels will threaten low-lying areas and accelerate coastal retreat [14]. The distribution of normalised intensity of cyclones (following [15]) is based on distributions derived from the Pacific and North Atlantic experiences.

Return-period hazard levels for severe wind and storm surge (covering 50, 100, 200, 500, 1000, 2000 and 5000 years) are used to assess impact and annualised loss. However, it is difficult to separate decadal variability from human induced forcing for the early decades of the 21st century.

A cumulative distribution function of cyclone intensity [15] is combined with the annual frequency to determine the wind speed for a range of return period annual exceedence probabilities (AEP). The AEP value of MPI at the closest coastal point to the centroid of the statistical local area (SLA) is used and then reduced for the distance inland to allow for the decay of the cyclone over land.

3.1.1. Maximum wind gust speed Regional wind speeds from tropical cyclones are modelled over flat terrain at a 10 metre height. Differences in terrain, building height, the surrounding structures and topographic features affect wind speed. Wind multiplier formulae, according to [16, 17], are used to qualify local wind conditions. All multipliers were aggregated to the SLA level for a sample of 85 SLAs, and generalised to the entire coastline considering the slope, direction of slope, building density, vegetation and wind regions [16].

3.1.2. Storm Tide Hazard Due to a paucity of storm-tide studies where quality Digital Elevation Models (DEMs) exist, return-period storm tide hazard estimates were obtained only for Brisbane, Gold Coast, Sunshine Coast, Cairns, Innisfail, Darwin, Perth and Port Hedland.

Climate change influence on storm tide was assessed as proportional to the increase in cyclone intensity. The "bathtub" inundation method was adapted from [21]. By simple addition, estimates of sea-level rise, tidal range and storm surge were combined and applied to a spatial grid giving the location of likely inundation. Sea-level rise projections were from mid-point IPPC AR4 [9]. Riverine (freshwater) flooding and event timing and clustering interactions are not assessed.

3.2. Exposure

Understanding the exposure of the built environment to hazards is fundamental to risk assessment. Building type, roof and wall construction, age, number of storeys, business activity and replacement value are critical parameters. The number and type of businesses, critical infrastructure and people exposed are also of interest though not analysed here.

3.2.1. National EXposure Information System NEXIS, a nationally consistent building level exposure catalogue, provides the location and value of residential buildings "exposed" to the hazards [22]. "Present-day" NEXIS at SLA level was projected into the future using the Australian Bureau of Statistics (ABS) *Series B* population projection.

All post year 2000 residential structures are assumed to be built to the present-day building standard in the same spatial footprint, resulting in increased building density. Further, the 2006 ratio of population per building from the 2006 ABS census has been maintained. No allowance is made for the continued increase in floor area (evident in recent decades). All values are in 2007 Australian dollars.

3.2.2. Population Projections Population projections covering the period 2004 to 2101 were released by the ABS on 25 November 2005 [23]. *Series B*, which largely reflects 2004 trends in fertility, life expectancy at birth, net overseas migration and net interstate migration is the primary population projection used here. The *A* and *C Series* (high and low projection) are incorporated by scaling the *Series B* results. State metropolitan and non-metropolitan projections out to 2054 are used to identify trends in regional population growth. The drift to the north and the coastal regions is driven by this. Beyond 2054, the Australia wide projections are used.

Under these projections Australia's population is projected to rise from current levels of 21 million to around 44 million by 2101 under *Series A* assumptions, or only to 23 million under *Series C* values.

3.3. Vulnerability

Vulnerability, the third leg of the paradigm, assesses through engineering analyses the capacity of buildings, infrastructure and people to withstand hazards.

3.3.1. Wind vulnerability Relationships derived from insurance data by Dr. George Walker of Aon Re for North Queensland structures were used [24]. Walker's pre-1980 curve represents North Queensland vulnerability before the lessons of cyclones Althea (1971) and Tracy (1974) influenced construction practice. His post-1980 curve significantly reduces vulnerability due to changes to the Queensland Building Act. These curves are considered the best vulnerability relationships available and have been used by other researchers [25] and are applied to Perth, Brisbane and the Gold Coast. A correspondence between contemporary brick veneer damage losses and Walker's pre-1980 curve has been shown [26]. For Cairns both curves were used as a building survey is available [27].

3.3.2. Storm Surge vulnerability Storm tide damage was determined at the buildings level by employing a "stage curve" derived from insurance data that relates loss to water depth. Floor heights for buildings were determined from digital elevation models mostly obtained from LiDAR, and assuming a 30 centimetre difference between the ground and floor level for all buildings considered.

3.4. Impact Assessment

The direct damage to residential structures is calculated from the interaction of wind speed and storm surge, building exposure and vulnerability. The value of the loss was calculated as the full cost of repair or replacement ("new for old"). This is significantly greater than the insured value where payout is often on a "like for like" basis. Contents value was assessed using empirical insurance information.

The costs of damage were based on the return period of exceedence loss levels (50 to 5000 years) across 520 coastal SLAs from the Queensland/NSW border to southern Western Australia. In turn the SLA losses were aggregated to Queensland, Western Australia and the Northern Territory. Losses were regressed to obtain a Probable Maximum Loss (PML) curve for each region. The losses range from frequent minor losses through to those associated with catastrophic events. Annualised loss, which is evaluated by integrating the area under the PML curve, represents the average annual cost to the region due to exposure to the hazard viewed through a very wide 5000 year window.

In examining only residential building and contents losses from wind and storm surge, many areas of hazard and impact are not considered. Flood damage and other climate change induced hazards are not addressed. Other direct losses incurred by commercial and industrial businesses, critical infrastructure owners and rural/agricultural are not counted. Agricultural damage can be significant as evidenced by Tropical Cyclones Justin (1997) and Larry (2006) which caused losses of approximately $150 million and $500 million. Damage to environmental resources such as forests, reefs, fisheries, was not calculated and indirect business losses due to life line disruption were not considered. Nor were costs of clean up, death and injury, emergency response and relief efforts considered.

3.5. Findings

The projected loss of residential buildings in Australia's northern regions, from cyclonic winds and storm surge, differ according to the exposure and the hazard. Using the ABS *Series B* population projections, the projected annualised loss of residential building stock and contents across Australia is presented in Figure 1.

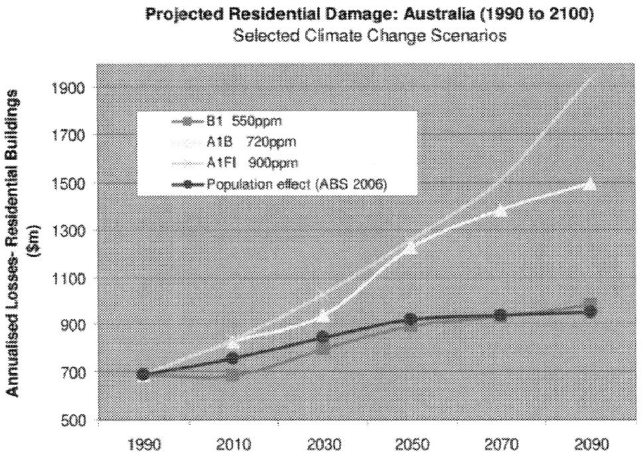

Figure 1. Projected residential damage associated with selected climate change scenarios: Australia: 1990 to 2100 (Annualised loss $m) for climate scenarios: A1FI (Business as Usual), selected CO2 stabilisation scenarios (A1B, 720ppm and B1, 550ppm) and the climate of the 20th Century.

The level of projected damage depends on the quantum of building stock exposed. Wind and storm surge losses projected from unrestrained 'business as usual' emissions (A1FI) are almost three times the level of damage in 1980s and 1990s. A1FI losses more than double the loss projected for the stabilisation scenario A1B where the emissions are stabilised at about 720ppm. Furthermore, vulnerability plays a part in as much as the new building stock meets the higher wind standards established in 1980. For scenario B1, where there is a significant stabilisation of CO_2 equivalent emissions, projected residential property damage is lower still. The 'business as usual' (A1FI) and A1B projections demonstrate considerable and increasing annualised residential and contents losses.

The greatest level of annualised loss, presented in the Table 1, is projected to occur in Queensland, While not shown here, the *Series A* population projection increases the annualised losses. By contrast the *Series C* population projections lead to significantly lessened wind and storm surge losses.

Table 1. Projected residential damage associated with selected climate change scenarios: Queensland: 1990 to 2100 (Annualised loss $m)

Centre of 20 year periods	A1FI 900ppm Business as Usual	A1B 720ppm	B1 550ppm	Climate of 20th Century
1990	482	482	482	482
2010	581	570	535	523
2030	703	632	628	586
2050	841	832	688	632
2070	1029	926	727	646
2090	1333	982	748	655

4. Role of Exposure in Tropical Cyclone Impacts under Climate Change Scenarios

The Garnaut Review results can be disaggregated into exposure and the greater cyclone intensity .

To develop the annualised loss associated only with the change in population, a scenario of the climate of the 20th Century was run. The intensity of the simulated cyclones was that of the period 1981-2000. Table 2 presents only the 2005 *Series B* population projections. Under a continuation of the modestly intensive cyclones of the late 20th century, annualised losses from cyclone hazard increase solely due to the increased location of people in the cyclonic regions of Australia. The net impact of the exposure is determined by removing the pre-2000 exposure impact (the '1990' observation in Table 2) from the future projected loss associated only with exposure.

Table 2. Projected residential damage (Total and Net): Continuation of the climate of the 20th Century and 2005 ABS Series B population projection: 1990 to 2100 (Annualised loss $m)

Centre of 20 year periods	Australia		Queensland		Western Australia		Northern Territory	
	Total	Net	Total	Net	Total	Net	Total	Net
1990	688	0	482	0	17	0	188	0
2010	754	67	523	41	21	3	210	22
2030	845	157	586	104	25	8	233	45
2050	919	231	632	150	28	11	259	70
2070	939	252	646	164	29	12	264	76
2090	952	265	655	173	30	12	268	79

The exposure effect dominates the climate change effect where there has been significant mitigation. Under conditions where mitigation is less effective, the climate change effects dominate the exposure effects by a relatively wide margin. These shaded cells in Table 3 show where the climate change effect exceeds the exposure effect. Table 3 also shows the percentage loss caused by climate change.

Under scenario A1FI net annualised loss from climate change induced cyclones is greater than the exposure component; similarly for the A1B scenario. Only where there is significant success in stabilising emissions to less than 550ppm, does the exposure effect exceed the hazard effect. This has important implications for policy makers in that even if the greenhouse gas concentration can be stabilised to produce low increases in hazard levels, the projected population increase will continue to deliver increased annualised loss due to the continuing growth in exposure through population growth in the northern coastal communities.

Table 3. Net projected residential damage: selected climate change scenarios (2005 ABS Series B population projection): Queensland: 1990 to 2100 (Annualised loss $m)

Centre of 20 year periods	A1FI 900ppm Business as Usual		A1B 720ppm		B1 550ppm		Climate of the 20th Century
1990	0		0		0		0
2010	58	(59%)	47	(53%)	13	(24%)	41
2030	116	(53%)	46	(31%)	42	(29%)	104
2050	209	(58%)	200	(57%)	56	(27%)	150
2070	383	(70%)	280	(63%)	81	(33%)	164
2090	678	(80%)	327	(65%)	93	(35%)	173

Shaded cells indicate where climate change effects dominate the population effects

4.1. 2008 ABS Population Projections

Since the Garnaut Review, the ABS released further population projections on 4 September 2008. *Series A* is projected to grow to 62 million in 2101, *Series B* to 45 million and *Series C* to 34 million. The mid century projection, from *Series B* under these more rapid growth conditions, suggests an extra 6 million persons (similar to *Series A* from 2006 projections). The new *Series A* projection suggests a further 7 million by around 2050 or a total population of some 40 million.

As demonstrated, population growth can be translated into an increase in exposure to tropical cyclones. Should these new population projections be realised, the losses associated with the 2008 *Series B* projection would exceed the earlier 2005 *Series A* projection. The implication of this is that there is a growing exposure to cyclone hazard irrespective of the climate change policy settings and there will thus need to be serious consideration in policy circles about the likely future population of Australia and its regional distribution.

5. Conclusions and policy implications

The disaggregation of the Garnaut Review results into population increases and climate change (wind and storm surge effects) provides insights into the relative contribution of exposure and hazard to the extent of tropical cyclone damage through to 2100. It can be concluded that a more rapid rate of population increase around the coast in northern Australia produces a much greater exposure to the potential cyclone hazard and hence introduces a much greater risk to the community. Exposure must not be ignored in considering policy settings, including climate change mitigation policies. Practical steps to reduce both exposure (planning) and vulnerability (building standards) of coastal communities to today's extreme weather would alleviate the impacts under any future climate threat [28].

This research also identifies many areas where further research is warranted to ensure a complete picture of the risk of tropical cyclone under future climate conditions is drawn. The economic consequences extend well beyond the residential damage considered here.

Acknowledgements

The authors wish to acknowledge the valuable input from Risk and Impact Analysis Group colleagues: Krishna Nadimpali, Nariman Habili, Ken Dale and Tina Yang, and the two anonymous reviewers.

References

[1] House of Representatives Standing Committee on Climate Change, Water, Environment and the Arts 2009 *Managing our coastal zone in a changing climate: The time to act is now* Available at http://www.aph.gov.au/house/committee/ccwea/coastalzone/report.htm

[2] Garnaut Review 2008 *Methodology for modelling climate change impacts* Technical Paper Number 4 Available at http://www.garnautreport.org.au/

[3] Suppiah, R, Hennessy K J, Whetton P H, McInnes K L, Macadam I, Bathols J, Ricketts J and Page C M, 2007 Australian climate change projections derived from simulations performed for the IPCC 4th Assessment Report *Aust. Met. Mag.* **56** 131-52

[4] Vecchi G A and Soden B J 2007 Increased tropical Atlantic wind shear in model projections of global warming *Geophysical Research Letters* **34** *(L08702)* doi:10.1029/2006GL028905

[5] Plummer N, Salinger M J, Nicholls N, Suppiah R, Hennessy K J, Leighton R M, Trewin B, Page C M and Lough J M 1999 Changes in climate extremes over the Australian region and New Zealand during the Twentieth Century *Climate Change* **42** 183–202

[6] Kuleshov Y and de Hoedt G 2003 Tropical cyclone activity in the Southern Hemisphere *Bull. Aust. Met. Oceanogr. Soc.* **16** 135–7

[7] Emanuel K A 1999 Thermodynamic control of hurricane intensity *Nature* **401** 665-9

[8] Holland G J 1997 The Maximum Potential Intensity of Tropical Cyclones *J. Atmos. Sci.* **54** 2519–41

[9] IPCC 2007 *Climate Change 2007: The Physical Science Basis. Summary for Policy Makers,* Contribution of Working Group I to the Fourth Assessment Report of the Intergovernmental Panel on Climate Change, Cambridge Uni. Press

[10] Backlund P, Janetos A and Schimel D 2008 *The Effects of Climate Change on Agriculture, Land Resources, Water Resources, and Biodiversity in the United States. Synthesis and Assessment Product 4.3* (U.S. Climate Change Science Program and the Subcommittee on Global Change Research) 193pp

[11] Emanuel K 2005 Increasing destructiveness of tropical cyclones over the past 30 years *Nature* **436** 686-8

[12] Webster P J, Holland G J, Curry J A and Chang H R 2005 Changes in tropical cyclone number, duration, and intensity in a warming environment *Science* **309** 1844-46

[13] Meehl G A *et al.* 2007 Global Climate Projections in *Climate Change 2007: The Physical Science Basis. Contribution of Working Group I to the Fourth Assessment Report of the Intergovernmental Panel on Climate Change* [Solomon S, Qin D, Manning M, Chen Z, Marquis M, Averyt K B, Tignor M and Miller H L (eds)] Cambridge University Press, Cambridge, United Kingdom and New York, NY, USA

[14] Nicholls R J and Tol R S J 2005 Impacts and responses to sea-level rise: A global analysis of the SRES scenarios over the 21st Century *Philosophical Transaction of the Royal Society A–Mathematical, Physical and Engineering Sciences*, **361** 1073-95

[15] Emanuel K A 2000 A Statistical Analysis of Tropical Cyclone Intensity *Monthly Weather Review* **128** 1139-52

[16] AS/NZS 1170.2 2002 *Structural design actions. Part 2: Wind actions* Australian/New Zealand Standard

[17] AS/NZS 1170.2 Supp 1, 2002: *Structural design actions - Wind actions- Commentary*, AUS/NZ Standard

[18] Leslie L M , Karoly D, Leplastrier M and Buckley B W 2007 Variability of tropical cyclones over the southwest Pacific Ocean using a high-resolution climate model *Meterol. Atm. Physics* **97** 171-80

[19] Walsh K J E and Ryan B F 2000 Tropical Cyclone Intensity Increase near Australia as a Result of Climate Change *J. Climate* **13** 3029-36

[20] Kaplan J and DeMaria M 1995 A simple empirical model for predicting the decay of tropical cyclone winds after landfall *J. Appl. Met.* **34** 2499-512

[21] Eastman J R, Kyam P A K, Toledano J and Jin W 1993 *GIS and Decision Making* (Exploration in Geographical Information Systems Technology, **4**, United Nations Institute for Training and Research, UNITAR), 111pp

[22] Nadimpalli K, Edwards M and Mullaly D 2007 National Exposure Information System (NEXIS) for Australia: Risk Assessment Opportunities (MODSIM 2007; International Congress on Modelling and Simulation, Oxley L and Kulasiri D (Eds)) 1674-80

[23] Australian Bureau of Statistics 2006 *Population projections for the period 2004 to 2101 for Australia* (released on 14 June 2006) http://www.abs.gov.au/Ausstats/abs@.nsf/mf/3222.0

[24] Walker G R 1995 *Wind Vulnerability Curves for Queensland Houses* Alexander Howden Reinsurance Brokers

[25] Harper B A 1999 Numerical modelling of extreme tropical cyclone winds *J. Wind Eng. Ind. Aerodyn.* **83** 35-47

[26] Edwards M R, Lin X G and Corby N 2004 Observations on tornado damage to residential structures in Bendigo *Eleventh Australasian Wind Engineering Society Workshop* Darwin

[27] Granger K, Jones T, Leiba M and Scott G 1999 Community risk in Cairns – A multi-hazard risk assessment (*AGSO publication Geo Cat No. 33548*) Available at Geoscience Australia http:/www.ga.gov.au/

[28] Arthur W C, Schofield A and Cechet R P 2008 Assessing the impacts of tropical cyclones *Aust. J. E. Mgt.* **23** 14-20

Formaldehyde and nitrogen dioxide in smoke plumes from Australia's Black Saturday fires

E L Young and C Paton-Walsh

School of Chemistry, University of Wollongong, Wollongong, NSW, Australia

E-mail: ely677@uow.edu.au, clarem@uow.edu.au

Abstract. The 'Black Saturday' fires were a series of devastating bushfires that burned across Victoria, Australia, during February 2009. The smoke plume from Saturday the 7th February, the worst day of the fires, separated from subsequent emissions and persisted for several weeks, providing the opportunity to track the changing composition of the smoke plume as it aged. In this study we have used satellite data from the Ozone Monitoring Instrument (OMI) and the Atmospheric InfraRed Sounder (AIRS) to characterise the emissions of formaldehyde and nitrogen dioxide from the fires. Emission ratios with respect to carbon monoxide are determined for formaldehyde (0.017 ± 0.004 $mol.mol^{-1}$) and nitrogen dioxide (0.004 ± 0.001 $mol.mol^{-1}$). Additionally OMI UV Aerosol Index is used to track the smoke plume and infer how the concentration of these gases changes as the smoke ages. Our study suggests that formaldehyde concentrations within the smoke plume increase during the first day before declining to background levels within 2 days after emission. Nitrogen dioxide concentrations show a monotonic decrease reaching background levels about 1 day after emission.

1. Introduction

Biomass burning is a major contributor to the release of trace gases and aerosols into the atmosphere. Some of the trace gases emitted by biomass burning contribute directly to global warming as "greenhouse" gases [1], others indirectly by influencing the abundance of hydroxyl radicals [2] and thus the atmospheric lifetimes of other "greenhouse" gases that react with them [3]. Several of the trace gases emitted by biomass burning can influence the acidity of precipitation [4] and both trace gases and aerosols can degrade air quality and pose a threat to human health [5]. Aerosols can also influence radiative transfer processes in the atmosphere through direct absorption and scattering, or indirectly as cloud condensation nuclei [4, 5].

Models of atmospheric composition require some measurement or estimate of the emissions from biomass burning [6]. Total emissions of a particular gas from fires are normally calculated as the product of the area burned, the average fuel load, the efficiency of combustion and the emission factor for the gas of interest, where the emission factor is the mass of a compound released per unit of dry matter consumed [6]. To determine emission factors directly, detailed field studies that include determination of the carbon content of the fuel are required. Alternatively an emission factor may be determined by measuring the emission ratio of the gas to a reference gas such as carbon monoxide (CO) or carbon dioxide (CO_2) and then multiplying by a previously measured emission factor for the reference gas [6]. Emission ratios are obtained by dividing the enhanced concentration of a trace gas by the enhanced concentration of a reference gas. "Enhanced" concentrations are obtained by subtracting ambient concentrations of the gases from those concentrations measured in the smoke plume. The emission ratio can also be considered as the slope of the concentration of the species plotted against the concentration of the reference species [6].

The 'Black Saturday' bushfires were a series of devastating fires burning across Victoria on Saturday the 7th February 2009. At least 400 individual fires were ignited on the day and these continued to burn throughout the state until mid-March, when favourable weather conditions helped to extinguish them. The fires claimed 173 lives [7], which is Australia's largest ever death toll from a bushfire. The smoke plumes from the Black Saturday bushfires eluded the ground-based remote sensing spectrometers at the University of Wollongong, but were observed by numerous satellite-based sensors. In this study satellite data was used to determine emission ratios for formaldehyde (H_2CO) and nitrogen dioxide (NO_2) with respect to CO from the Black Saturday fires and examine how the ratio of these gases changed as the smoke plume aged. Descriptions of the satellite platforms and sensors included in the survey are given below.

The Ozone Monitoring Instrument (OMI) is one of four instruments aboard the NASA satellite Aura and records daily global measurements of backscattered solar radiation in the visible and ultra-violet. OMI can distinguish between aerosol types such as smoke, dust and sulfates based upon differences in spectral absorbance and geographical location [8]. The UV Aerosol Index is derived from OMI measurements and is an indicator of near-UV absorbing aerosols such as smoke [9]. The index measures the deviation of an observed spectrum from the ideal molecular atmosphere as influenced by the presence of UV absorbing aerosols [10]. The mapped index over the period from the 7th February 2009 to the 3rd March 2009 shows the smoke plume from the original Victorian Black Saturday bushfires form and then separate from subsequent emissions. The plume then appears to be transported to the north of New Zealand where it lingered for over a week before it travelled west over northern Australia, and eventually dissipated completely over the Indian Ocean nearly a month later. The UV Aerosol Index can therefore be useful as a tracer of regional aerosol transport.

OMI also gathers spectral information that can be used to determine column amounts of trace gases such as H_2CO (measured at spectral resolution of ~0.45 nm in the UV-2: 310-365nm) [11, 12] and NO_2 (measured at spectral resolution of ~0.63 nm in the visible: 350-500nm) [11, 12] that are used in this study [13]. OMI differentiates trace gases by measuring backscattered solar radiation at known optimal wavelengths for individual gas species [11]. A combination of processing algorithms are used to extract OMI data products including TOMS Version 8, DOAS (Differential Optical Absorption Spectroscopy), Hyperspectral BUV retrievals and forward modelling [14].

The Atmospheric InfraRed Sounder (AIRS) is carried aboard NASA's Aqua satellite. AIRS measures the infrared energy from the Earth's surface, separating wavelengths which are each sensitive to variants such as temperature, water vapour and some trace gases [15, 16]. Of particular interest to this study, is the instrument's ability to measure the abundance of CO in the atmosphere.

2. Deriving Emission Ratios from OMI and AIRS data

Maps visualizing daily OMI UV Aerosol Index from 7th February 2009- 3rd March 2009 were obtained from the Giovanni online data system [17]. H_2CO and NO_2 total column amounts were obtained from the OMI sensor onboard the satellite Aura (Level 2 data products were used). For CO, the total column amounts from the AIRS instrument onboard the satellite Aqua were used (ascending swaths – Level 3 data products).

The daily mapped column amounts for the reference gas, CO, were used to define the region of the active fires by limiting analyses to areas where CO concentration was enhanced on each individual day. The column amount data for CO from the 7th February showed insufficient enhancements for an emission ratio to be determined. The lack of enhancement in CO concentration could be attributed to the possibility that the smoke plume had yet to develop strongly on the 7th February, as most of the major fires had only just begun on this day. Total column amount averages of H_2CO and NO_2 from the 8th, 9th and 10th of February 2009 for each grid space (1° latitude by 1° longitude) in the defined areas were plotted against the co-located total column amounts of CO. The gradient of the best straight line fit (using a linear regression) to these points gave the emission ratio of each gas relative to CO. The standard deviation of the gradient was taken as an estimate of the uncertainty in the calculated emission ratio. It should be noted that the quoted uncertainties are from statistical analysis that assumes that the uncertainties in each point are uncorrelated. Since this is not the case the true uncertainty is likely to be larger. In addition, firestorms can cause cloud to develop and this smoke and cloud could degrade the instruments' abilities to measure below the height of the top of the plume adding to the uncertainty of the measurements.

There were significantly enhanced concentrations of CO, H_2CO and NO_2 over the active fire region on the 8th February 2009. Data from this day yielded emission ratios with respect to CO of 0.004 ± 0.001 mol.mol^{-1} for NO_2 with a correlation coefficient of 0.81; and 0.017 ± 0.004 mol.mol^{-1} for H_2CO with a correlation coefficient of 0.81 (see Figure 1 for the plots from which these emission ratios were derived). OMI H_2CO and NO_2 total column amounts were poorly correlated with AIRS CO total columns on the 9th and 10th February 2009. There appeared to be a significant fraction of OMI data missing over the relevant areas, especially on the 9th February, leading to increased uncertainties when averaging over the 1° latitude by 1° longitude grid spaces.

There are no previous measurements of emission ratios or emission factors for NO_2 from Australian temperate forest fires in the literature and results for different types of vegetation and in different climatic regions of the world vary by several orders of magnitude [6, 18].

The emission ratio for H_2CO with respect to CO from the 8th February 2009 is in broad agreement with emission ratios from *Paton-Walsh et al* (0.019 ± 0.008 mol.mol^{-1} [19] and 0.023 ± 0.007 mol.mol^{-1} [20]) and with *Andreae and Merlot's* emission factor (2.2 ± 0.5 g/kg) assuming their mean emission factor for CO for extratropical forests of 107 ± 37 g/kg [6].

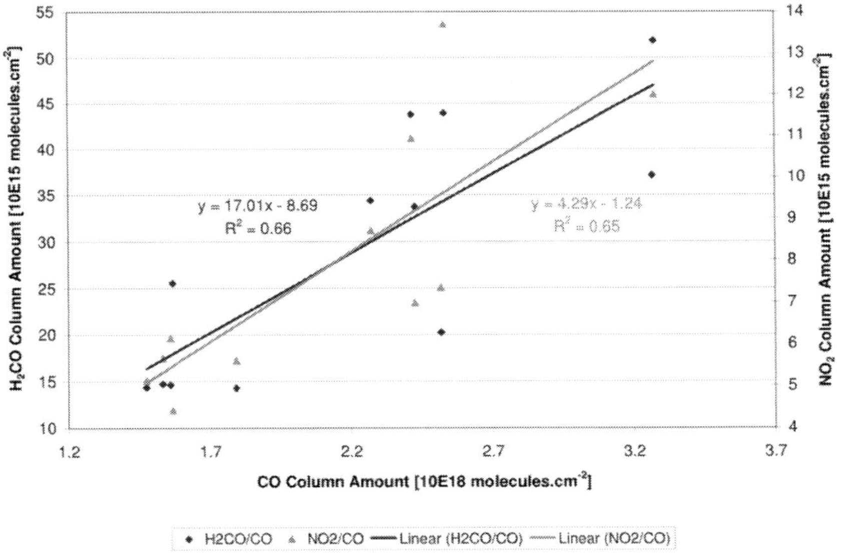

Figure 1: H_2CO and NO_2 column amounts plotted against CO column amounts to derive emission ratios for the active fires on the 8th February 2009.

3. Characterising the Aging of the Smoke Plume Using Changing Emission Ratio Values

OMI UV Aerosol Index was used to track the movement of the smoke plume that was emitted from the first fires and thereby to define the analysis regions for the evolution of the trace gases H_2CO, NO_2 and CO. Figure 2a shows enhanced OMI UV Aerosol Index revealing the location of the smoke plume on the 8th February 2009. Gaps in the data to the edge of each satellite swath artificially break up the plume into three areas, one that is close to and incorporates the active fire region, the main section that extends across the south island of New Zealand and a small region to the east of New Zealand. On the 9th and 10th of February 2009 the smoke plume is located to the northeast of New Zealand's north island. Emission ratios were calculated using OMI H_2CO and NO_2 and AIRS CO total column data over the region of the travelling plume on the 9th and 10th February 2009 and separately for the two largest of the three separate regions identified on the 8th February 2009, the "active fire" grid and the main section or "Area 2".

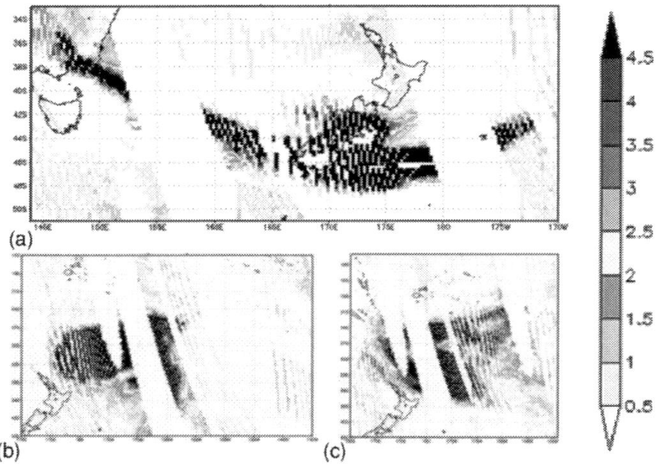

Figure 2: The mapped product of the UV Aerosol Index data on (a) the 8th February 2009 (144E-170W, 33S-51S), (b) the 9th February 2009 (170E-140W, 15S-45S) and (c) the 10th February 2009 (165E-154W, 10S-45S) for the travelling plume. The coloured area of enhanced aerosol in each map represents the area within which data analysis for the trace gases was concentrated on each day.

The total column amounts of H_2CO and NO_2 derived from OMI spectra over the region of the smoke plume showed no significant enhancements after the 10th February 2009. This is because H_2CO and NO_2 are highly reactive with atmospheric lifetimes of the order of hours [21]. In comparison CO has a lifetime in the order of 2 months [22] and so changes in the emission ratio of H_2CO and NO_2 with respect to CO as the smoke ages will be dominated by the rapid changes in the reactive gases. Figure 3 shows the changing emission ratios of H_2CO and NO_2 as the smoke aged from the youngest smoke measured over the active fire area on the 8th February 2009 to the oldest smoke measured to the north of New Zealand on the 10th February 2009. The emission ratios of each gas go to zero at background levels since an emission ratio is the *enhanced* concentration of the trace gas divided by the *enhanced* concentration of a reference gas.

The pseudo time-series of NO_2 emission ratios shown in Figure 3b reveals a rapid decline with age in the Black Saturday smoke plume. NO_2 declines with respect to CO from 0.004 ± 0.001 mol.mol^{-1} on the 8th February over the active fire, to near background levels over Area 2 on the 8th February and over the areas of the 9th and 10th February when considering the levels of uncertainty. This decline is likely to be a symptom of the numerous destructive reactions that NO_2 undergoes in the atmosphere. For example, NO_2 oxidises to nitrates, is photolysed during the day to NO and reacts with a number of compounds including $OH^{\cdot}$, HO_2, ClO and BrO [22].

H_2CO is released as a direct product of biomass burning but it has a lifetime of approximately three hours in daylight [21]. This is largely due to destruction by photolysis or reaction with the hydroxyl radical. To a lesser extent, H_2CO reacts with NO_3 resulting in abstraction of a hydrogen atom, aqueous conversion of H_2CO results in formic acid and the addition of HO_2 destroys H_2CO and forms the hydroxyl methyl peroxy radical [22].

Based on its short lifetime as a consequence of destructive reactions, it could be expected that the concentration of H_2CO would appear highest on the 8th February, particularly over the area of the active fires where the emissions were the freshest. Whilst the emission ratio of H_2CO to CO was high over this area (0.017 ± 0.004 mol.mol^{-1}), further emission ratios revealed that the concentration of H_2CO was highest on the 8th February over Area 2 (the emission ratio was 0.024 ± 0.003 mol.mol^{-1}), where the smoke was slightly older, having travelled further from the site of the fires. The high uncertainties in the derivation of emission ratios mean that the increase may not be statistically significant; however, such an increase could result from reactive sources of H_2CO from within the plume itself. H_2CO is a known product of the oxidation of many hydrocarbons that would be prominent in a smoke plume [21, 22]. In particular, model studies have indicated that non methane volatile organic compounds (NMVOCs) such as acetic acid, methanol and ethene can be a significant secondary source of H_2CO in smoke; these compounds are estimated to account for 43 percent of final H_2CO production [23]. It is also possible that the larger H_2CO to CO ratio calculated for the older part of the smoke plume on the 8th February (Area 2) resulted from a greater original emission ratio from the earlier stages of the active fires. This is plausible because the earliest fires would be expected to be dominated by flaming combustion, and CO is associated more with smouldering phase whilst H_2CO can be associated with both flaming and smouldering combustion [24].

The emission ratios on the 9th and 10th February reveal a subsequent decrease in the amount of H_2CO within the plume as it ages. By these days the composition of H_2CO had decreased to background levels within the respective uncertainties of the estimates, indicating that the excess H_2CO from the fires and possible resultant reactions within the plume had disappeared.

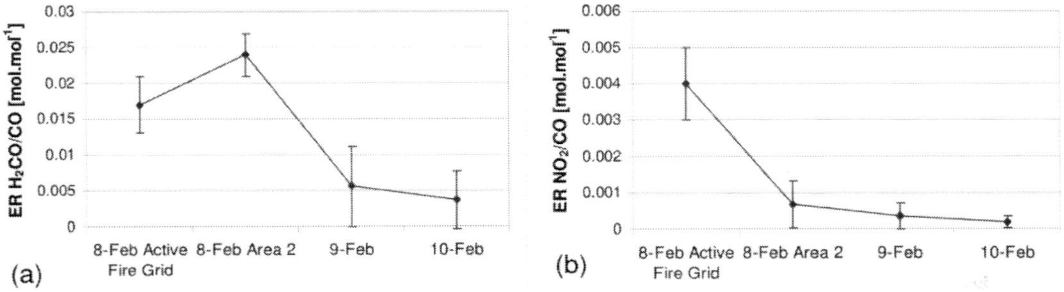

Figure 3: Emission ratios with respect to CO for (a) H_2CO and (b) NO_2, plotted as a "pseudo time-series" as the smoke ages from the youngest smoke measured over the active fire area on the 8th February 2009 to the oldest smoke measured to the north of New Zealand on the 10th February 2009.

4. Conclusions

OMI UV Aerosol Index has been used to track the smoke plume emitted from the Black Saturday bushfires. Comparison of the changing column amounts of H_2CO and NO_2 from OMI with CO columns from AIRS reveals information about the changing composition of the smoke plume as it ages. Large enhancements in all three of these trace gases were observed in the plume. NO_2 column amounts exhibited a rapid decline with time, whilst H_2CO column amounts peaked once the plume had begun to age, then declined and eventually dissipated after the third day. The increase after aging may be attributed to reactive sources of the trace gas from within the plume itself.

Emission ratios with respect to CO from the fires of 0.017 ± 0.004 mol.mol^{-1} for H_2CO and 0.004 ± 0.001 mol.mol^{-1} for NO_2 were determined. This emission ratio for H_2CO is in broad agreement with previous values in the literature. The result for NO_2 is especially significant because it is the first measurement of an emission ratio for this important gas from Australian forest fires.

Acknowledgements

Analyses and visualizations used in this paper were produced with the Giovanni online data system, developed and maintained by the NASA Goddard Earth Sciences (GES) Data and Information Services Center (DISC). The authors acknowledge the mission scientists and Principal Investigators who provided the data used in this research effort. We would also like to thank the reviewers of this paper for their very constructive comments and suggestions and all at the University of Wollongong's Centre for Atmospheric Chemistry for their support.

References

[1] Wang W C, Yung Y L, Lacis A A, Mo T and Hansen J E 1976 Greenhouse effects due to man-made perturbations of trace gases *Science* **194** 685-90

[2] Wofsy S C 1976 Interactions of CH_4 and CO in the Earth's Atmosphere *Annu. Rev. Earth Planet. Sci.* **4** 441-69

[3] Crutzen P J, Heidt L E, Krasnec J P, Pollock W H and Seiler W 1979 Biomass burning as a source of atmospheric gases CO, H_2, N_2O, NO, CH_3Cl and COS *Nature* **282** 253-56

[4] Crutzen P J and Andreae M O 1990 Biomass burning in the tropics: impact on atmospheric chemistry and biogeochemical cycles *Science* **250** 1669

[5] Meyer C P, Luhar A K and Mitchell R M 2008 Biomass burning emissions over northern Australia constrained by aerosol measurements: I--Modelling the distribution of hourly emissions *Atmos. Environ.* **42** 1629-46

[6] Andreae M O and Merlot P 2001 Emission of trace gases and aerosols from biomass burning *Global Biogeochem. Cycles* **15** 955-66

[7] Victoria Police 2009 *Victoria Police Bushfire Update* Accessed on: 4/09/2009 Available from: http://www.police.vic.gov.au/content.asp?Document_ID=19190.

[8] Royal Netherlands Meteorological Institute 2002 *OMI Algorithm Theoretical Basis Document: Volume III- Clouds, Aerosols and Surface UV Irradiance* Ministry of Transport Public Works and Water Management Accessed on: 12/10/2009 Available from: http://www.knmi.nl/omi/research/documents/data/OMI_ATBD_Volume_3_V2.pdf.

[9] NASA 2009 *Giovanni-3 online users manual: Data parameter index* Accessed on: 10/09/2009 Available from: http://disc.sci.gsfc.nasa.gov/giovanni/additional/users-manual/G3_manual_parameter_appendix.shtml

[10] Braak R, de Laat J, Torres O, Veihelmann B, Veefkind J P, Kroon M and Levelt P *OMI UV absorbing aerosols as a tracer of Australian biomass burning aerosols* Royal Netherlands Meteorological Institute Accessed on: 12/10/2009 Available from: http://www.knmi.nl/publications/fulltexts/20070419_vienna_egu_braak_australianwildfires.pdf.

[11] Chance K editor 2002 *OMI Algorithm Theoretical Basis Document: Volume IV- OMI Trace Gas Algorithms* Accessed on: 12/10/2009 Available from: http://macuv.gsfc.nasa.gov/doc/ATBD-OMI-04.pdf

[12] Liu X, Bhartia P K, Chance K, Spurr R J D and Kurosu T P 2010 Ozone profile retrievals from the Ozone Monitoring Instrument *Atmos. Chem. Phys.* **10** 2521-37

[13] NASA 2007 *Ozone Monitoring Instrument (OMI)* Accessed on: 5/09/2009 Available from: http://www.nasa.gov/mission_pages/aura/spacecraft/omi.html

[14] OMI Team 2009 *Ozone Monitoring Instrument (OMI) Data User's Guide* Accessed on: 7/06/10 Available from: http://disc.sci.gsfc.nasa.gov/Aura/additional/documentation/README.OMI_DUG.pdf.

[15] Chahine M T, Aumann H, Goldberg M, McMillin L, Rosenkranz P, Staelin D, Strow L and Susskind J 2001 *AIRS Level 2 Algorithm Theoretical Basis Document Version 2.2: AIRS-Team Retrieval For Core Products And Geophysical Parameters* Accessed on: 23/04/2010 Available from: http://eospso.gsfc.nasa.gov/eos_homepage/for_scientists/atbd/docs/AIRS/atbd-airs-L2.pdf

[16] NASA no date *How AIRS Works* Accessed on: 6/09/2009 Available from: http://airs.jpl.nasa.gov/technology/how_AIRS_works/

[17] NASA Goddard Earth Sciences Data and Information Services Centre 2009 *Giovanni online data system* [database on the Internet] Available from: http://disc.sci.gsfc.nasa.gov.giovanni

[18] Yokelson R J, Urbanski S P, Atlas E L, Toohey D W, Alvarado E C, Crounse J D, Wennberg P O, Fisher M E, Wold C E, Campos T L *et al.* 2007 Emissions from forest fires near Mexico City *Atmos. Chem. Phys.* **7** 5569-84

[19] Paton-Walsh C, Jones N, Wilson S, Meier A, Deutscher N, Griffith D, Mitchell R and Campbell S 2004 Trace gas emissions from biomass burning inferred from aerosol optical depth *Geophys. Res. Lett.* **31** L05116

[20] Paton-Walsh C, Jones N B, Wilson S R, Haverd V, Meier A, Griffith D W T and Rinsland C P 2005 Measurements of trace gas emissions from Australian forest fires and correlations with coincident measurements of aerosol optical depth *J. Geophys. Res., [Atmos.]* **110** D24305

[21] Warneck P 1988 *Chemistry of the Natural Atmosphere* 41 (London:Academic Press Inc.)

[22] Brasseur G P, Orlando J J and Tyndall G S editors 1999 *Atmospheric Chemistry and Global Change* (New York:Oxford University Press)

[23] Stavrakou T, Müller J F, De Smedt I, Van Roozendael M, van der Werf G R, Giglio L and Guenther A 2009 Evaluating the performance of pyrogenic and biogenic emission inventories against one decade of space-based formaldehyde columns *Atmos. Chem. Phys.* **9** 1037-60

[24] Yokelson R J, Griffith D W T and Ward D E 1996 Open-path Fourier transform infrared studies of large-scale laboratory biomass fires *J. Geophys. Res.* **101** 21067-80

IOP Conf. Series: Earth and Environmental Science **11** (2010) 012024 doi:10.1088/1755-1315/11/1/012024

Utilising temperature differences as constraints for estimating parameters in a simple climate model

Roger W Bodman[1, 3], **David J Karoly**[1] **and Ian G Enting**[2]
[1]School of Earth Sciences, The University of Melbourne, Vic 3010, Australia
[2]MASCOS, 139 Barry St, The University of Melbourne, Vic 3010, Australia

E-mail: rwbodman@unimelb.edu.au

Abstract. Simple climate models can be used to estimate the global temperature response to increasing greenhouse gases. Changes in the energy balance of the global climate system are represented by equations that necessitate the use of uncertain parameters. The values of these parameters can be estimated from historical observations, model testing, and tuning to more complex models. Efforts have been made at estimating the possible ranges for these parameters. This study continues this process, but demonstrates two new constraints. Previous studies have shown that land-ocean temperature differences are only weakly correlated with global mean temperature for natural internal climate variations. Hence, these temperature differences provide additional information that can be used to help constrain model parameters. In addition, an ocean heat content ratio can also provide a further constraint. A pulse response technique was used to identify relative parameter sensitivity which confirmed the importance of climate sensitivity and ocean vertical diffusivity, but the land-ocean warming ratio and the land-ocean heat exchange coefficient were also found to be important. Experiments demonstrate the utility of the land-ocean temperature difference and ocean heat content ratio for setting parameter values. This work is based on investigations with MAGICC (Model for the Assessment of Greenhouse-gas Induced Climate Change) as the simple climate model.

1. Introduction

The overall aim of our research program is to develop probabilistic estimates for temperature changes in the 21st Century. Current research is focused on developing approaches for constraining the parameters of a simple climate model, to see if it is possible to use temperature observations, including temperature and ocean heat content differences, for this purpose. The model being used is the latest version of MAGICC (Model for the Assessment of Greenhouse-gas Induced Climate Change), version 6.3 [1], [2], [3]. This is an updated version of the model used throughout the IPCC process, and features new and comprehensive calibration to 19 of the CMIP3 AOGCMs (Third Climate Model Intercomparison Project Atmosphere/Ocean General Circulation Models) [4] and 10 of the C4MIP carbon-cycle models (Coupled Carbon Cycle Climate Model Intercomparison Project) [5].

MAGICC is a simplified Earth system model that represents the movement of energy through the atmosphere, land, mixed ocean layer and deep ocean due to changes in greenhouse gas concentrations and radiative forcing. Four boxes are used to represent the land and ocean in the Northern and Southern Hemispheres, with a single atmosphere. The land is assumed to have zero heat storage capacity, whilst an upwelling diffusion model represents the oceans. Emissions from a given scenario

[3] Author to whom any correspondence should be addressed.

are processed though gas-cycle models to arrive at annual changes in atmospheric concentrations. A set of simplified expressions converts these concentrations into radiative forcings, which are summed to estimate the total change in forcing. A carbon-cycle model is used to track changes in carbon pools within the land, ocean and atmosphere.

The simplified nature of this type of model necessitates the use of a series of parameterised equations. There are 11 parameters that determine the response of the model's climate system to a perturbation. A 'pulse response method' has been used to identify the relative importance of these climate system parameters. Four of the key parameters are climate sensitivity, ocean vertical diffusivity, the warming ratio of land to ocean and the heat exchange coefficient between the land and ocean. Based on this, we have tested a new approach to constraining these four parameters using historical global-mean temperature observations, the land-ocean temperature difference and a ratio based on ocean heat content.

Northern and Southern Hemisphere temperatures, land and ocean temperatures, and ocean heat uptake have been used to constrain parameter values in previous work [6]. However, this information is highly correlated to global-mean temperatures. Previous studies have shown that land-ocean and Northern-Southern Hemisphere temperature differences are only weakly correlated with global mean temperature for natural internal climate variations [7]. Based on this, we have tested an alternative approach using area-average temperature differences between the land and ocean to provide extra independent information. In addition, we have developed a new constraint based on the ratio of ocean heat content in the upper 100m layer to the heat content of the 700m layer, which has been tested and found to have a reduced correlation to global-mean temperature, as discussed in Section 2.2.

We have used three sets of temperature observations, from the Climatic Research Unit at the University of East Anglia [8], the NOAA National Climatic Data Center [9], and the NASA Goddard Institute for Space Studies [10]. Ocean heat content observations are from the recent work of Domingues et al. [11].

2. Parameter Settings

2.1. Ratio of land to ocean warming

We begin by considering the parameter RLO, the ratio of land to ocean warming. The standard model setting is for RLO = 1.3, which has been determined from long-term AOGCM results for equilibrium conditions. However, Sutton et al. [12] determined that this ratio should be around 1.5 for present day non-equilibrium conditions, based on AOGCM results and recent observations. Using the mean temperature anomaly for the last 30 years of observations (1979-2008) from the three sets of global-mean temperature anomalies results in an observed mean RLO of 1.58. The observed land and ocean temperature anomalies are shown in Figure 1.

Testing the goodness-of-fit between the model results and observations for the land minus ocean temperature differences we obtained, for the mean case, RLO between 1.5 and 1.6, as shown in Figure 2 and Figure 3, which agrees well with Sutton et al. and our updated calculation.

2.2. Ocean vertical diffusivity, K

Now considering the ocean vertical diffusivity parameter K, the new version of MAGICC includes calculation of the ocean heat content at different ocean depths, which can be compared to observations, such as the recent data of Domingues et al. [11]. Figure 4 shows the timeseries for ocean heat content (OHC) for the upper ocean down to depths of 100, 300 and 700m for the observations and the model results (with standard parameter settings). Note that the model appears to be diffusing too much heat away from the surface, a problem that is also seen in the CMIP3 model results [13].

We have checked the correlation between global-mean temperatures and ocean heat content. As expected, these are highly correlated. However, we also tested two other potential measures: the ocean heat content difference between the top 100m layer and the ocean heat content for the 700m deep layer; and also the ratio of these two. The difference has a correlation of nearly 0.7, but the ratio has a very small correlation value. The 100/700m ratio correlation between the model surface temperatures and OHC is about 0.6, rather higher than for the observations (perhaps as a consequence of model limitations such as missing horizontal advection processes and natural variability). These results are included in Table 1.

Running the model with the standard parameter settings, except for varying the K parameter, and testing the fit between the model results and the observed ocean heat content, we obtain the result shown in Figure 5. The best fit occurs at K = 0.5 (the smallest value tested), much lower than the standard setting of 2.3. Although a K value of 0.5 is probably too low, it addresses the problem of the model diffusing too much heat downwards as noted above (other ocean parameters including the upwelling rate need to be adjusted, something that will be addressed in further research). We have proceeded with a value of K = 1.0 to ascertain the impact this has on the temperature projections discussed in Section 3.

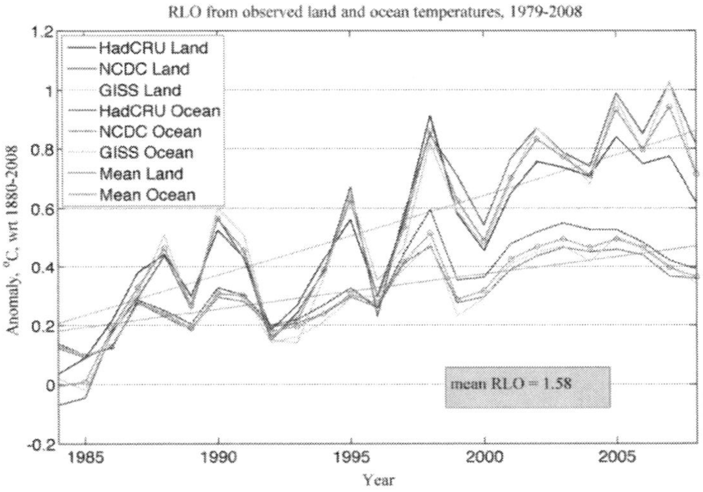

Figure 1: Observed land and ocean temperature anomalies wrt 1880-2008 for 30 years, 1979-2008.

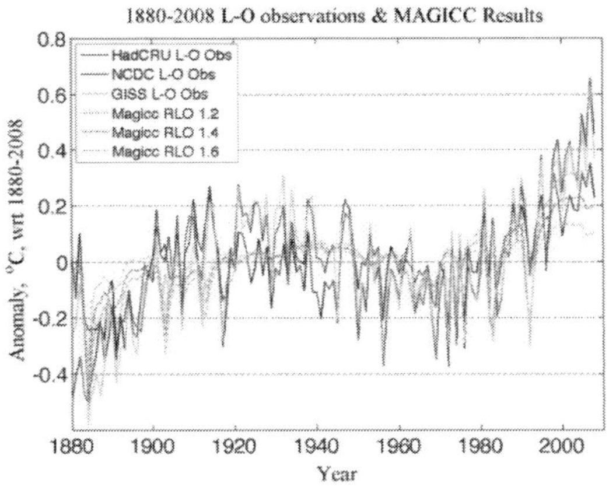

Figure 2: Observed and modelled land minus ocean temperature differences, with variable RLO (Standard parameters dT2x=3.0, K=2.3, *xklo*=1.0).

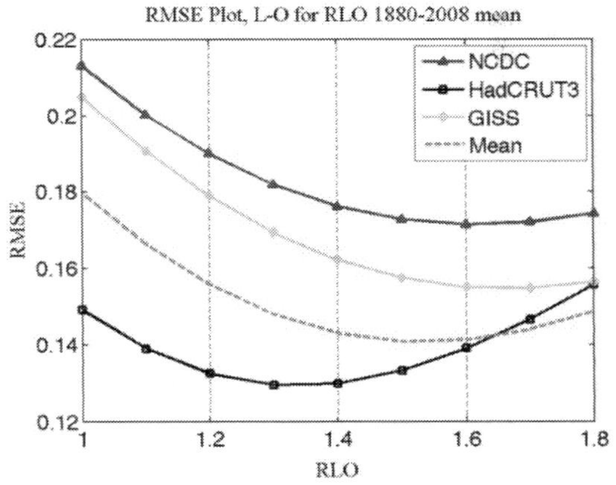

Figure 3: RMSE result for Figure 2, indicating a minimum RMSE error for RLO = 1.5 to 1.6 in the mean case.

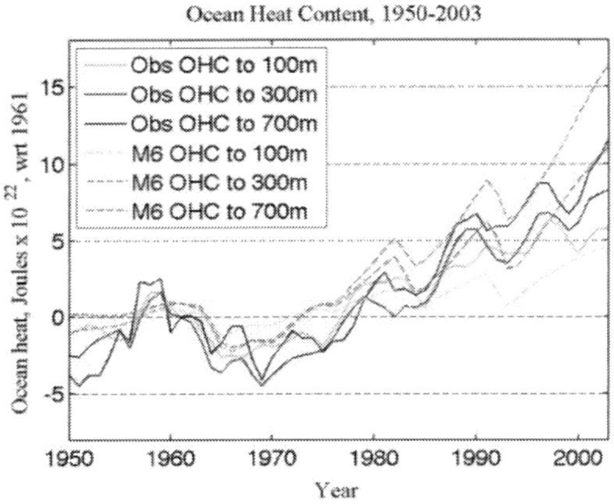

Figure 4: Ocean heat content from observations [11] and MAGICC for 1950-2003.

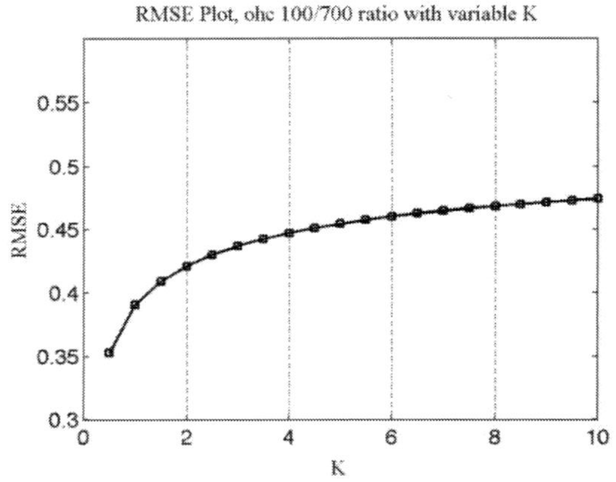

Figure 5: RMSE plot for the modelled and observed OHC 100/700 ratio for variable K.

Table 1: Correlation for global-mean temperature anomaly (Tgav) to ocean heat content (OHC) over 1984-2003for historical observations and model results.

Correlation between: and OHC at:	Mean Obs and OHC Obs	MAGICC Tgav and MAGICC OHC
100m	0.83	0.93
300m	0.88	0.89
700m	0.84	0.88
100 – 700 difference	0.66	0.84
300 – 700 difference	0.40	0.79
100/300 ratio	-0.45	0.52
100/700 ratio	-0.15	0.63

2.3. Land-ocean heat exchange coefficient, xklo

Now, we consider *xklo*, the land-ocean heat exchange coefficient, and test for the goodness-of-fit between the observed land-ocean temperature difference and the model results, for revised parameters values of RLO=1.6 and K=1.0, but keep the climate sensitivity (dT2x) at 3.0. We obtain a RMSE result pointing towards *xklo* in the range 1.5 to 3.5, with almost no variation over this range; see Figure 6. The standard value is 1.0 but we now set *xklo* = 1.5.

2.4. Climate sensitivity, dT2x

Then, taking the new parameter settings for RLO=1.6, K=1.0 and *xklo* = 1.5, we can test for climate sensitivity using the global-mean temperature. We arrive at a climate sensitivity of between 1.5 and 3.0°C for the best fit between the observed temperature record and the model results as illustrated in Figure 7. The standard value is 3.0; we proceed with dT2x set to 2.5°C.

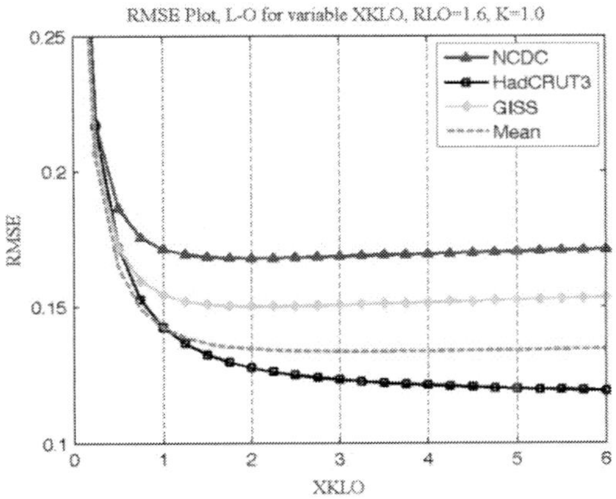

Figure 6: L-O RMSE plot for *xklo*, with RLO=1.6, K=1.0 and dT2x =3.0.

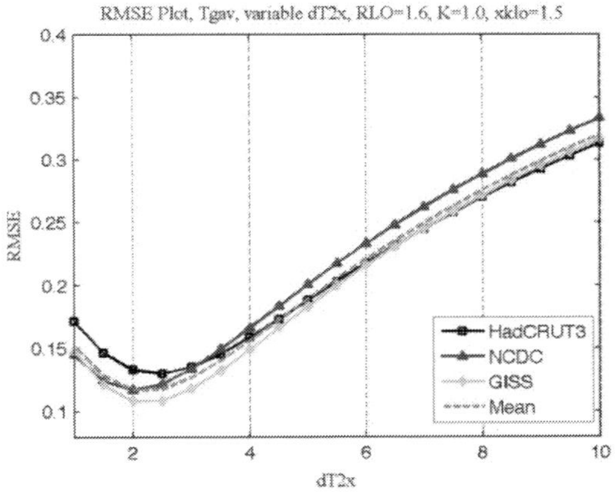

Figure 7: Tgav RMSE plot for variable dT2x, with RLO=1.6, K=1.0 and *xklo*=1.5.

3. Projected warming

Finally, we test the new parameter settings, in this case using two scenarios, to gauge the impact of these revised settings on global-mean temperatures into the 21st Century. The scenarios used here are the SRES (Special Report on Emissions Scenarios [14]) A1FI high growth path and a 450ppm CO_2-equivalent stabilisation scenario. The results are shown in Figure 8. For the A1FI scenario, the change in 2100 is 0.4°C greater warming than for the standard parameter values, and for the 450ppm stabilisation scenario the difference is negligible. In the A1FI scenario, the reduced ocean diffusivity leads to greater mixed layer warming over time, resulting in greater surface warming which is amplified by the larger RLO value, despite a lower climate sensitivity setting.

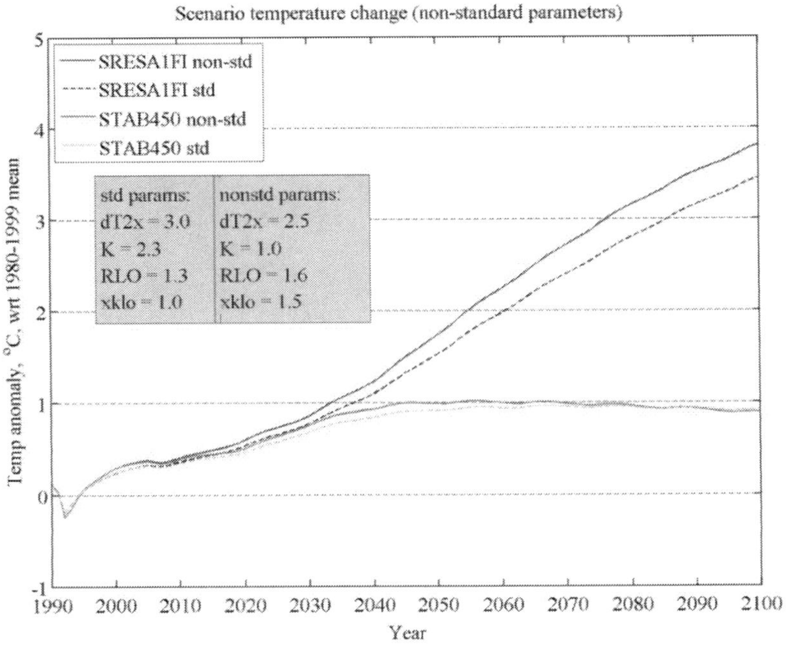

Figure 8: Comparison of two scenarios, SRES A1FI and 450ppm CO_2-eqivalent stabilisation, with standard and revised parameter settings.

4. Conclusions

We have demonstrated that land-ocean temperature differences and the ratio of ocean heat content for different ocean depths can provide two additional observational constraints for setting parameter values in a simple climate model. Using these observational constraints, projections with revised model parameters give higher global warming estimates than with standard parameters for the A1FI emissions scenario. However, uncertainty in observations, aerosol forcing and the carbon cycle, as well as the climate model, requires a more sophisticated analysis that takes into account this uncertainty, the interaction of the model parameters and the cross-correlation of the constraints. Further work is required to evaluate the model's ocean temperature profile and heat uptake characteristics to align more closely with observations.

Acknowledgments
We wish to thank Tom Wigley and Malte Meinshausen for making MAGICC available for use in this study. The research was supported by the Australian Research Council through the Discovery Projects funding scheme (project number FF0668679).

References

[1] Wigley TML and Raper SCB 2001 Interpretation of high projections for global-mean warming *Science* **293** 451-4
[2] Raper SCB, Gregory JM and Osborn TJ 2001 Use of an upwelling-diffusion energy balance climate model to simulate and diagnose A/OGCM results *Climate Dynamics* **17** 601-13
[3] Meinshausen M, Raper SCB and Wigley TML 2008 Emulating IPCC AR4 atmosphere-ocean and carbon cycle models for projecting global-mean, hemispheric and land/ocean temperatures: MAGICC 6.0 *Atmospheric Chemistry and Physics Discussions* **8** 6153-272
[4] Meehl G, Covey C, Delworth T, Latif M, McAvaney B, Mitchell JFB, et al. 2007 The WCRP CMIP3 multi-model dataset: A new era in climate change research *Bulletin of the American Meteorological Society* **88** 1383-94
[5] Friedlingstein P, Cox P, Betts R, Bopp L, von Bloh W, Brovkin V, et al. 2006 Climate-carbon cycle feedback analysis: results from the C4MIP model intercomparison *Journal of Climate* **19** 3337-53
[6] Meinshausen M, Meinshausen N, Hare W, Raper SCB, Frieler K, Knutti R, et al. 2009 Greenhouse-gas emission targets for limiting global warming to 2degC *Nature* **458** 1158-62
[7] Karoly DJ and Braganza K 2001 Identifying global climate change using simple indices *Geophysical Research Letters* **28** 2205-8
[8] Climatic Research Unit. Temperature. [cited 26 Feb 2010]; Available from: http://www.cru.uea.ac.uk/cru/data/temperature/
[9] National Climatic Data Center. Global surface temperature anomalies. [cited 26 Feb 2010]; Available from: http://www.ncdc.noaa.gov/oa/climate/research/anomalies/index.html
[10] Goddard Institute for Space Studies. GISS surface temperature analysis. [cited 26 Feb 2010]; Available from: http://data.giss.nasa.gov/gistemp/
[11] Domingues CM, Church JA, White NJ, Gleckler PJ, Wijffels SE, Barker PM, et al. 2008 Improved estimates of upper-ocean warming and multi-decadal sea-level rise *Nature* **453** 1090-3
[12] Sutton RW, Dong B and Gregory JM 2007 Land/sea warming ratio in response to climate change: IPCC AR4 model results and comparison with observations *Geophysical Research Letters* **34** L02701
[13] Wijffels SE 2009 New estimates of ocean warming rates and how well they are simulated by AR4 Greenhouse2009 Perth
[14] Nakicenovic N and Swart R. IPCC Special Report on Emissions Scenarios. Cambridge: Cambridge University Press 2000.

IOP Conf. Series: Earth and Environmental Science **11** (2010) 012025 doi:10.1088/1755-1315/11/1/012025

A case study of consensus modelling for tracking oil spills

Brian King[1], Ben Brushett[1,2] and Charles Lemckert[2]

[1]Asia-Pacific ASA, PO Box 1679 Surfers Paradise, QLD, 4817 Australia
[2]School of Engineering, Griffith University, Gold Coast, QLD, 4222 Australia

Email: bking@apasa.com.au, bbrushett@apasa.com.au, c.lemckert@griffith.edu.au

Abstract. Metocean forecast datasets are essential for the timely response to marine incidents and pollutant spill mitigation at sea. To effectively model the likely drift pattern and the area of impact for a marine spill, both wind and ocean current forecast datasets are required. There are two ocean current forecast models and two wind forecast models currently used operationally in the Australia and Asia Pacific region. The availability of several different forecast models provides a unique opportunity to compare the outcome of a particular modelling exercise with the outcome of another using a different model and determining whether there is consensus in the results. Two recent modelling exercises, the oil spill resulting from the damaged Pacific Adventurer (in Queensland) and the oil spill from the Montara well blowout (in Western Australia) are presented as case studies to examine consensus modelling.

1. Introduction

The two ocean current forecast models currently being used operationally in the Australian and greater Asia Pacific region are the US Navy Coastal Ocean Model (NCOM) and the Australian BLUElink model. Both of these models were developed for large to mesoscale ocean circulation, and as such neither model includes the effects of tidal currents. This lack of tidal current forcing limits the effectiveness of the models in shallow near coastal waters, where tidal currents are important and can be the dominant driving force in water circulation. Asia-Pacific ASA has developed an aggregation tool which combines the effects of both coastal tidal currents and large scale oceanic currents. The tool functions by spatially and temporally interpolating the large scale currents on to the higher resolution, tidal currents using vector addition, thus producing a very effective current forecast dataset able to be used in both open ocean and coastal waters alike.

The two wind forecast models used in this study were the US Global Forecast System (GFS) and the US Navy Operational Global Atmospheric Prediction System (NOGAPS). Wind forcing is a very important factor when modelling the drift trajectories of marine pollutant spills. This is due to the oil slick moving with the water body as a result of the currents, as well as the oil slipping across the water surface due to the effects of the wind.

The availability of several different forecast models provides an additional opportunity to compare the outcomes of a particular modelling exercise. This is done by varying the input data from one forecast model to another and comparing the results. If the outcomes are similar, then there is consensus between the datasets, and the modeller can be confident that the forecast is more reliable. If there is a discrepancy between the forecasts, then there is no consensus, which suggests that the forecast may not be as reliable. In such a situation it is necessary for the modeller to further interogate the input data to ascertain which may be the most reliable forecast.

Two recent emergency responses involving pollutant forecast modelling were investigated using input data from the Environmental Data Server (EDS) in OILMAP v6, to forecast the fates and trajectory of the spilled pollutants. The first was the Pacific Adventurer oil spill which occurred off Moreton Island, Queensland; and the second was the Montara well blow out which occurred in the Timor Sea, Western Australia.

2. Metocean forecast models and ASA modelling software

2.1. BLUElink

There are several components to the BLUElink system, including operational forecasts, reanalysis and data assimilation. The operational forecasts from BLUElink were derived from the Ocean Model Analysis and Prediction System (OMAPS-fc). This system uses the Ocean Forecasting Australia Model (OFAM) [1]. The 3D model has a horizontal resolution of 1/10° in the Australian region (90°E – 180°E, 75°S – 16°N). The vertical resolution is a 47 layer z-coordinate system, with the topmost layer being 10m thick. Forecast length is 144 hours, at intervals of 24 hours [2]. Data assimilation is controlled by the BLUElink Ocean Data Assimilation System (BODAS) which is an ensemble optimal interpolation (EnOI) scheme that assimilates Sea Surface Temperature (SST), Sea Surface Height (SSH) and temperature and salinity profiles. Atmospheric fluxes are provided by the Global Atmospheric Prediction System (GASP) [3].

2.2. NCOM

The Navy Coastal Ocean Model is based on the Princeton Ocean Model (POM) and has global coverage with a horizontal resolution of 1/8°. Vertical resolution is a σ-z coordinate system with 19 σ-coordinate layers in the upper 137m (topmost surface layer thickness of 1m) and 21 z-coordinate layers from 137m to 5500m. Forecast length is 72 hours, at time intervals of 6 hours. Data assimilation is controlled by the Modular Ocean Data Assimilation System (MODAS) which assimilates temperature, salinity and SSH. Atmospheric forcing is provided by the Navy Operational Global Atmospheric Prediction System (NOGAPS) atmospheric fluxes [4].

2.3. GFS

The Global Forecasting System (GFS) is a global spectral numerical model operationally run by the US National Oceanic and Atmospheric Administration (NOAA). The T254 version (used in this study) provides global coverage with a horizontal resolution of 1/2° with 64 unequally spaced vertical layers. The forecast length is 180 hours, and the temporal resolution is 6 hours [5].

2.4. NOGAPS

The Navy Operational Global Atmospheric Prediction System (NOGAPS) is the principal source of atmospheric forcing for the US Navy ocean models (e.g. NCOM) and short term numerical weather prediction (NWP). NOGAPS has global coverage, with horizontal resolution ~ 1/2°. The forecast length of the NOGAPS product is 144 hours with temporal resolution of twelve hours (at 00 and 12 UTC) and updates at 06 and 18UTC to enable background forecasts. Outputs from the model include momentum flux, both latent and sensible heat fluxes, precipitation, solar and long wave radiation and surface pressure, as well as 10 metre U and V wind velocities [6,7].

2.5. OILMAP v6

OILMAP is a modelling system designed for predicting the physical and chemical fates and effects of hydrocarbon spills in the marine environment. It contains a trajectory model which uses a large number of Lagrangian particles to simulate the trajectory, as well as a number of algorithms which account for hydrocarbon spreading, evaporation, emulsification, entrainment, and shoreline interactions. Time and space varying environmental data (speed and direction of winds and water currents) provide forcing to simulate the drift of hydrocarbon slicks. To account for variability in the inputs (such as wind gusts) uncertainty particles are included in the model runs. These uncertainty particles are subjected to winds and water currents that have been varied by up to ±30% of their strength and ±30 degrees in direction.

Dispersion of the hydrocarbon slick is controlled by the horizontal dispersion parameter. A typical value for this during strong winds and rough seas is 35 m^2s^{-1}, as the strong winds and large waves tend to rapidly break the slick up and disperse it. During calmer conditions with light winds and calm seas typical values can be as low as 0.5 m^2s^{-1}. Other parameters include wind factor and wind angle, these values control the amount the slick is influenced by the effects of the wind by defining the relative movement of the slick across the water surface in relation to the strength and direction of the wind.

Predictions of the OILMAP model have been validated worldwide [8] and in Australia [9] by field observations and by hind-casting past hydrocarbon spills. For consensus forecasting purposes in this case study, all parameters were kept consistent between each model run, only the input winds and currents were changed; this ensured only the input data would have an effect on the modelling forecasts.

2.6. HYDROMAP

The two-dimensional tidal currents (which were aggregated with ocean currents from NCOM and BLUElink) were generated using an advanced ocean/coastal model, HYDROMAP. The HYDROMAP model has been tested and verified through field measurements throughout the world over past twenty-five years [10]. HYDROMAP employs a sophisticated multi resolution gridding system, which enables six levels of spatial resolution. The sub-gridding allows for higher resolution of currents within areas of greater bathymetric and coastline complexity. The input data for this hydrodynamic model was Geoscience Australia 250m resolution gridded bathymetry, and open boundaries loaded with the eight dominant tidal constituents from the TPXO7.1 TOPEX-Poseidon global inverse tidal model [11].

2.7. COASTMAP Environmental Data Server (EDS)

The EDS automatically downloads, stores, catalogues and disseminates metocean forecast and hindcast data, such as winds and oceanic currents for use within OILMAP. The format of the data collected is often disparate, however the EDS is able to serve the data in a commonly accessible form. Currently available in the Australasian region are NCOM and BLUElink forecast ocean currents, HYDROMAP tidal currents and GFS and NOGAPS forecast winds. Also included in the EDS is an aggregation tool which takes the oceanic currents and aggregates them with tidal currents. This aggregation process spatially interpolates the ocean currents from their coarse resolution grids onto a finer variable resolution grid, at the resolution of the tidal current model. The ocean forecast currents and winds are also temporally interpolated from 24 or 6 hourly intervals to one hourly intervals.

3. Consensus modelling case studies

3.1. Case Study 1 – Pacific Adventurer spill

In the early hours of the morning on the 11th of March 2009 the Pacific Adventurer encountered severe weather conditions off the Coast of Moreton Island. Due to this severe weather, 31 shipping containers were lost overboard. In the process several of the containers ruptured the ships fuel tanks which resulted in the loss of 270 tonnes of heavy fuel oil to the marine environment at the water surface [12]. Asia-Pacific ASA provided modelling support to determine the likely fates and possible shoreline strikes of this heavy fuel oil.

Figure 1 shows the various model runs completed using OILMAP, and environmental forecast data from the COASTMAP EDS. NCOM and BLUElink forecast currents aggregated with tidal currents provided the current forcing, whilst GFS and NOGAPS wind forecast models provided wind forcing. As shown, there is a general consensus between the model forecasts. All four model forecasts show that the shorelines on the northern end of Moreton Island and the beaches near Kawana will be impacted, with the possibility of shoreline impacts to the beaches both north and south of the Kawana Beach region. The best correlation between the model predicted shoreline impacts and observed shoreline impacts was attained by using NCOM predicted currents aggregated with tidal currents, and the GFS forecast winds.

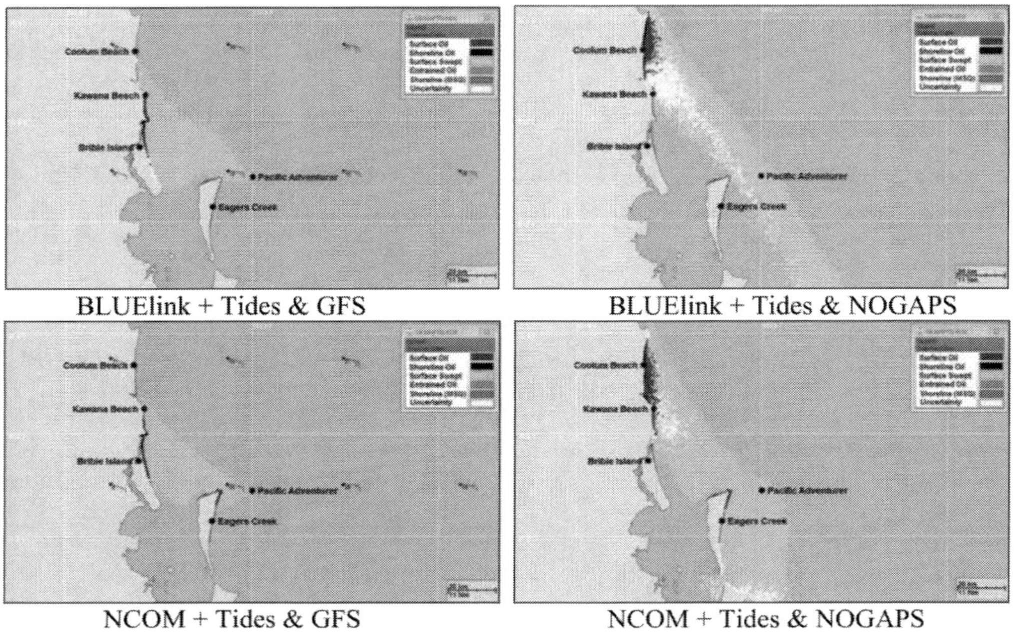

Figure 1. The four different model runs completed when forecasting the Pacific Adventurer spill. The dark grey dots represent the surface oil, the light grey dots represent the water surface swept by the oil over the 3 days modelled, the yellow represents the "uncertainty particles" used by the model, the black represents the oil stranded on the shorelines, as predicted by the model, and the red indicates the full extent of the shoreline oil stranding as reported by Maritime Safety Queensland (MSQ). Note that the model shoreline stranded particles can be difficult to see under the red observed shoreline stranding of oil, however the swept area gives an indication of the shorelines impacted.

3.2. Case Study 2 – Montara Well head blowout

At approximately 4 am WST on the 21st of August 2009, a leak formed at the Montara well head, located approximately 680 km west of Darwin off the Kimberley coast in Western Australia. It was estimated that around 400 barrels per day of crude oil was being discharged into the sea from the well head platform above the water surface. The leak continued for 74 days until the well was successfully "killed" on the 3rd November 2009. This equates to a total of almost 30,000 barrels of crude oil being released into the marine environment [13]. Asia-Pacific ASA was involved in the provision of modelling support throughout this incident. The forecasting of the spill proved to be very difficult, especially in the initial stages of the incident. At the beginning there was no consensus between the forecast models, with a different direction of travel predicted for the NCOM plus tidal currents and the BLUElink plus tidal current forecast data (see figure 2). As shown, the NCOM tidal aggregation predictions show the oil plume moving towards the West, whereas the BLUElink tidal aggregation shows the oil plume moving in a Northerly direction. Two surface drifters were deployed which showed the currents moving in an easterly direction (different to both NCOM and BLUElink). They also showed that the currents were tidally influenced (as indicated by the tidal oscillations in the buoy trajectories). This shows that for predicting drift patterns in this region, the addition of the tidal component to the surface currents is important. As the incident continued, the forecast datasets proved to better resolve the surface currents in the region when compared to drifter tracks, location of predicted and observed surface oil, and when directly comparing the NCOM and BLUElink currents with hindcast currents. Of the 13 weeks that the oil was tracked, approximately 10 weeks returned good current forecast data.

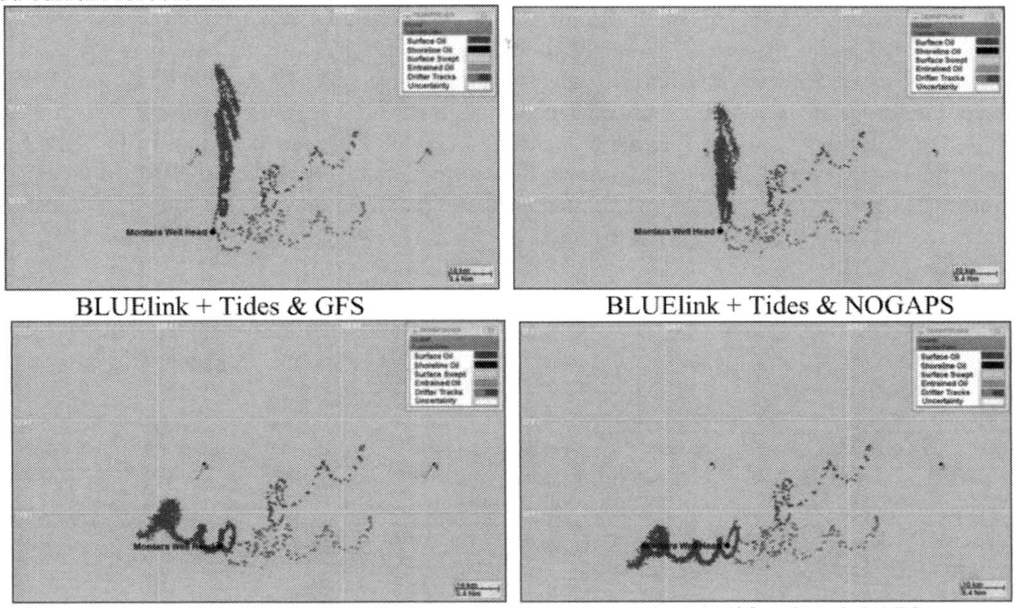

BLUElink + Tides & GFS BLUElink + Tides & NOGAPS

NCOM + Tides & GFS NCOM + Tides & NOGAPS

Figure 2. The four different model runs completed when forecasting the Montara well blowout. The light grey area represents surface area swept by the oil over the first three days of the modelled spill, the dark grey represents surface oil particles and the red and maroon points show the tracks of the surface drifters which were deployed to ground truth the currents.

4. Discussion

Different forcing was dominant for each of the incidents studied, due to the nature of the environmental conditions at the time of the spills. Wind forcing was dominant over surface currents for the Pacific Adventurer spill due to the incident occurring at the end of a large storm, which produced model winds at an average of 20 knots, up to a maximum of 28 knots (over the 3 days modelled), whilst surface currents had less of an effect as they averaged around 1 knot with maximum of 2.5 knots. This is in contrast to the Montara incident which occurred in comparatively calm weather with model winds that only averaged around 5 knots, up to a maximum of 10 knots and where tidally governed surface currents averaged around 0.7 knots with a maximum of up to 1.4 knots.

When consensus between forecast models is achieved, the outcome is potentially more reliable, however as shown in the two case studies above, consensus is not always reached. One reason for this is the positioning of mesoscale eddies, which have spatial extents in the order of tens of kilometers. As the two aforementioned global current forecast models (NCOM and BLUElink) have spatial resolutions of approximately 10 km, their ability to effectively model mesoscale eddies can be limited. To adequately resolve mesoscale eddies, a much finer resolution is required. Problems arise with ocean forecast models when eddies are misplaced. This is particularly evident as the eastern side of a clockwise rotating eddy will spin an object to the south, whereas if the eddy is moved such that the object is now placed on the western side, the particle will be pushed in the complete opposite direction, towards the north.

One possible reason that NCOM performed slightly better than BLUElink in the Pacific Adventurer incident is that the incident occurred quite close to shore in shallow waters. NCOM has a grid which can resolve to a minimum of 5m depth, where the BLUElink model is currently only able to resolve to a minimum of 20m water depth, thus making it less effective in shallow waters.

5. Conclusions

As shown in the Pacific Adventurer case study, all four metocean forecast datasets can be used to effectively model the movement of oil in the marine environment, however the Montara case study shows that this is not always the case, and a sound understanding of the inputs and their strengths and limitations is required to effectively forecast the drift of oil. If consensus is reached between the models, then the forecaster has a higher degree of confidence of a more reliable outcome; if consensus is not reached, then there is a lower degree of confidence. It was also evident that in both cases the tidal circulation component of the surface currents was important. This was due to both incidents occurring in relatively shallow coastal environments (less than 200m deep). Without the inclusion of tides, the predictions may not have been as effective.

Acknowledgements

This research was supported under the Australian Research Council's Linkage Projects funding scheme LP0991159.

References

[1] Andreu-Burello I, Brassington G, Oke P and Beggs H 2010 Including a new data stream in the BLUElink Ocean Data Assimilation System *Aust. Met. and Oceanogr. J.* **59** 77-86

[2] Australian Bureau of Meteorology. BLUElink> Ocean Model Analysis and Prediction System version 1.0 (OceanMAPSv1.0) Technical Specification [document on the Internet]. Canberra [updated July 2007; cited 2010 March 5]. Available from: http://www.bom.gov.au/oceanography/forecasts/technical_specification.pdf

[3] Brassington G B, Pugh T, Oke P R, Freeman J, Andreau-Burrel I, Huang X and Warren G 2009 *Operational Ocean Data Assimilation for the BLUElink Ocean Forecasting System*, Fifth WMO Symposium on the Assimilation of Observations for Meteorology, Oceanography and Hydrology, Melbourne, 5-9 October 2009

[4] Barron C N, Birol Kara A, Rhodes R C, Rowley C and Smedstad L F 2007 *Validation Test Report for the 1/8° Global Navy Coastal Ocean Model Nowcast/Forecast System.* Naval Research Laboratory, Stennis Space Centre

[5] Environmental Modelling Centre. The GFS Atmospheric Model [document on the Internet] [Updated August 28 2003; cited 2010 March 5]. Available from: http://www.emc.ncep.noaa.gov/gmb/moorthi/gam.html

[6] Rosmond T E, Tiexiera J, Peng M and Hogan T 2002 Navy Operational Global Atmospheric Prediction System (NOGAPS): Forcing for Ocean Models *Oceanogr. Mag.* **15** 99-108

[7] Rosmond T E 1992 The design and testing of the Navy Operational Global Atmospheric Prediction System *Weather. Forecast.* **7** 262-272.

[8] Spaulding M, Kolluru V, Anderson E and Howlett E 1994 Application of three-dimensional oil spill model WOSM/OILMAP to hindcast the Braer Spill *Spill. Sci. Technol. B.* **1** 23

[9] King B, McAllister F and Hubbert G 1999 Data requirements for calibration and validation of the spill model OILMAP in Marine Meteorology and Related Oceanographic Activities. WMO Report No. 44. *MARPOLSER 98 Proceedings* **1** WMO/TD – No. 959 35-52

[10] Zigic S, Zapata M, Isaji T, King B and Lemckert C 2003 *Modelling of Moreton Bay using an ocean/coastal circulation model* Coast and Ports Australasian Conference 9 -12 September Auckland, New Zealand

[11] Egbert G D and Erofeeva S Y 2002 Efficient inverse modeling of barotropic ocean tides *J. Atmos. Ocean. Technol.* **19** 183-204

[12] Asia-Pacific ASA 2009 *Independent Assessment of the Shoreline Cleanup Operations For the Pacific Adventurer Oil Spill* Report to Maritime Saftety Queensland by APASA. Gold Coast

[13] PTTEP Australasia. Frequently Asked Questions Montara Incident [document on the Internet]. West Perth. [cited 2010 January 2] Available from: http://www.au.pttep.com/faq.asp#Q3

Improved regional climate modelling through dynamical downscaling

Stuart Corney[1], Jack Katzfey[2], John McGregor[2], Michael Grose[1], Greg Holz[1], Chris White[1], James Bennett[1,3], Suzie Gaynor[1] and Nathan Bindoff[1,2]

[1]Climate Futures for Tasmania, Antarctic Climate and Ecosystems CRC, University of Tasmania, Private Bag 80, Hobart, Tasmania 7001, Australia
[2]Centre for Australian Weather and Climate Research, CSIRO Marine and Atmospheric Research, PMB 1, 107 Station St, Aspendale, Victoria 3195, Australia
[3]Hydro Tasmania Consulting, 89 Cambridge Park Drive, Cambridge, Tasmania, 7170, Australia

Email: Stuart.Corney@acecrc.org.au

Abstract. Coupled Ocean-Atmosphere General Circulation Models (GCMs) provide the best estimates for assessing potential changes to our climate on a global scale out to the end of this century. Because coupled GCMs have a fairly coarse resolution they do not provide a detailed picture of climate (and climate change) at the local scale. Tasmania, due to its diverse geography and range of climate over a small area is a particularly difficult region for drawing conclusions regarding climate change when relying solely on GCMs. The foundation of the Climate Futures for Tasmania project is to take the output produced by multiple GCMs, using multiple climate change scenarios, and use this output as input into the Conformal Cubic Atmospheric Model (CCAM) to downscale the GCM output. CCAM is a full atmospheric global general circulation model, formulated using a conformal-cubic grid that covers the globe but can be stretched to provide higher resolution in the area of interest (Tasmania). By modelling the atmosphere at a much finer scale than is possible using a coupled GCM we can more accurately capture the processes that drive Tasmania's weather/climate, and thus can more clearly answer the question of how Tasmania's climate will change in the future. We present results that show the improvements in capturing the local-scale climate and climate drivers that can be achieved through downscaling, when compared to a gridded observational data set. The underlying assumption of this work is that a better simulated current climatology will also produce a more credible climate change signal.

1. Introduction

The Climate Futures for Tasmania (CFT) project uses a dynamical downscaling approach to deliver the high quality projections of Tasmania's climate out to 2100. Dynamical downscaling global climate models (GCMs) for Tasmania is a way of incorporating Tasmania's complex topography and maritime-influenced climate to provide a clearer picture of regional variations and impacts of projected climate change. Climate models can reproduce many aspects of the weather system to describe the overall climate and, as such, they are our best tool for assessing potential changes in the future climate.

The regional downscaling undertaken by CFT improves on previous work relying on only GCM model output in three ways:

1. Mean values of the relevant variables in the modelling output is closer to observed data;
2. Spatial patterns of these variables in the downscaled models are more like the observed data than the patterns produced by the GCMs;
3. Downscaled models have a more realistic seasonal variability than GCMs.

2. Modelling Program

Dynamical downscaling uses either a nested limited-area model or a stretched-grid global model. In both cases forcing data are taken from a host GCM in order to downscale climate information from the coarse spatial scales of GCMs to a finer scale suitable for regional studies. The use of dynamical downscaling allows us to demonstrate changes in the local climate of Tasmania like seasonality, while maintaining the relationships between different climate variables. The Climate Futures for Tasmania project has made the technical and financial commitment to undertake an extensive dynamical downscaling process involving multiple GCMs and IPCC SRES emissions scenarios [1]. The GCMs chosen for downscaling were CSIRO Mk3.5, GFDLCM 2.0, GFDLCM 2.1, MIROC 3.2 (medium resolution), UKMO HadCM 3 and MPI/ECHAM 5. These models were chosen based on an assessment of their ability to reproduce present-day climate variability in the Australian region [2]. The two emissions scenarios [1] were chosen to give a range of likely projected futures, from a high (A2) to a low (B1) increase in atmospheric content of CO_2.

CFT undertook the dynamical downscaling program using the CSIRO Conformal Cubic Atmospheric Model (CCAM). CCAM has been the mainstay of the dynamical downscaling program undertaken at CSIRO for more than a decade [3,4]. CCAM is a full atmospheric global climate model, based on using a conformal-cubic grid. To allow for downscaling experiments, CCAM can be configured to use a stretched grid by utilising the Schmidt [5] transformation of the coordinates and dynamical equations. A stretched grid allows for higher resolution in the region of interest.

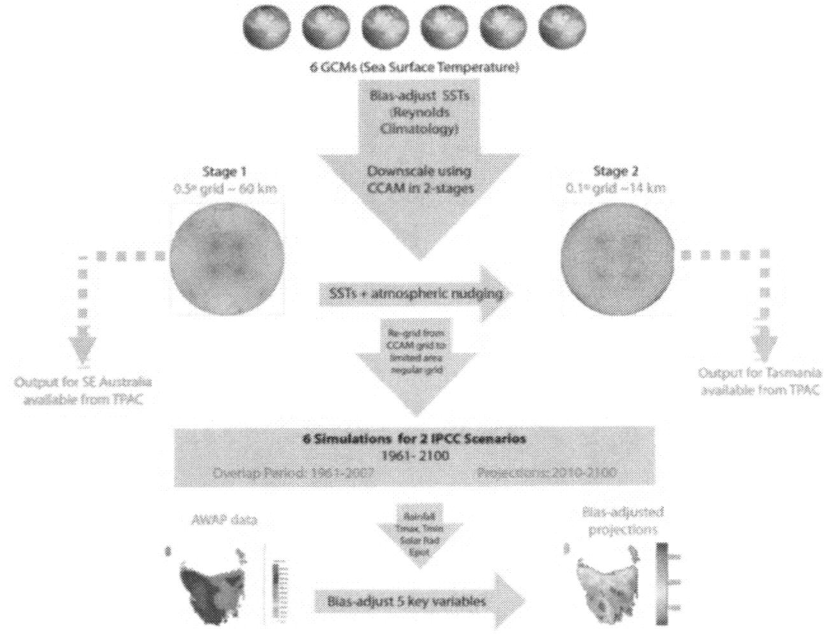

Figure 1. Schematic of the downscaling process used for the Climate Futures for Tasmania project.

The downscaling process employed by the project is outlined in figure 1. In order to achieve a desired final resolution of 0.1°, a two stage downscaling process was undertaken. The two-stage process is employed as it has been shown using CCAM that the downscaling process can produce less accurate advection of synoptic systems if the ratio of grid size between the parent and downscaled model is too large. For CFT, the first stage (intermediate model) involved downscaling from the host GCM to a grid with the high resolution face of the cubic conformal grid covering all of Australia at a resolution of approximately 0.5°. The second stage placed the high-resolution face over Tasmania and the Bass Strait islands at an approximate resolution of 0.1°.

CCAM is a global, stretched-grid model, and as such can be configured to require only bottom boundary conditions. The intermediate model runs in this configuration, taking only monthly interpolated sea surface temperature (SST) and ice cover from the host GCM. It does not use any forcing of the atmospheric components between the GCM and CCAM (it does however use the same radiative forcing and atmospheric composition as the host GCM). The monthly SSTs are bias-corrected against Reynolds SST climatology [6] to correct for biases in the GCM SSTs.

The high resolution (0.1°) model is forced using output from the corresponding intermediate (0.5°) model. Like the intermediate model, the high-resolution model uses bias-corrected SST from the host GCM as a bottom boundary condition. However unlike the intermediate model the atmosphere of the high-resolution model is forced using several variables from the intermediate model. We have implemented a scale-selective filter [7] to spectrally nudge the surface pressure, wind fields, temperature and atmospheric moisture above 850 hPa. Spectral nudging allows the model to evolve freely at small length-scales, with the regional atmosphere at large length-scales specified by the intermediate model.

The downscaling process from the GCM through the 0.5° model, to the high resolution 0.1° model was identical for each of the 12 downscaling simulations undertaken by CFT. Each process resulted in modelling output featuring 6-hourly data for 140 variables over the period 1961-2100.

3. Downscaling Results

For regions of the length-scale of a few hundred kilometres, dynamical downscaling adds value to the results from global climate models by modelling changes at finer spatial and temporal scales. Downscaled models often also have an improved mean value of many observables due to improved topographic resolution. Local topography strongly affects, for example, the mean temperature and total rainfall.

The improved spatial pattern of the dynamically downscaled modelling output can be clearly seen in the annual mean temperature at the three different grid sizes in figure 2. The top left panel of figure 2 shows the observed pattern of annual mean temperature available from the Bureau of Meteorology (http://www.bom.gov.au/jsp/awap). The effects of topographic features such as the central plateau, the ridge of mountains down the west and the Ben Lomond mountain range in the north east are clearly visible, as are the warmer temperatures around the coast and in the Tamar Valley in the mid-north. The corresponding annual mean temperature from a GCM (GFDLCM 2.1 is shown, but all GCMs give similar results) is shown in the top right panel. This has very little spatial variability as Tasmania is represented by only a few cells (in fact Tasmania is represented by a single cell in GFDLCM 2.1). The bottom two panels of figure 2 show the annual mean temperature from the corresponding downscaled 0.5° (left) and 0.1° (right) models. The 0.5° output starts to show some differentiation of temperature across the state. In the 0.1° model most of the features (for example the cooler central plateau, warmer coastal regions etc) present in the observed map are also represented.

The improved spatial variability is quantified through spatial correlation (R^2). Table 1 displays the correlation values for the mean temperature and total rainfall of the model output compared against gridded observations from the Australian Water Availability Project (AWAP) [8]. The correlation clearly improves as the resolution increases.

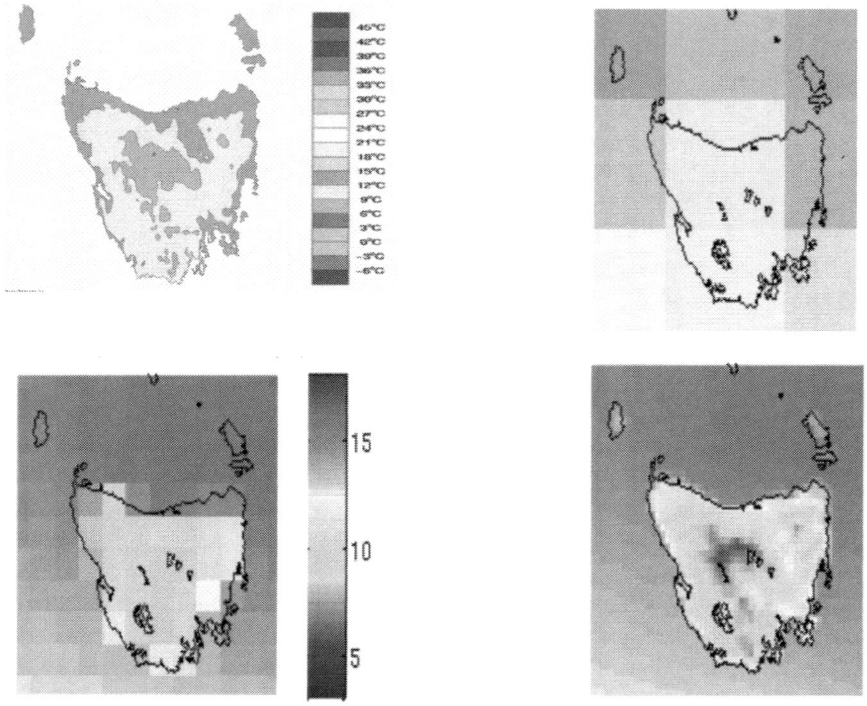

Figure 2. Annual mean temperature of Tasmania from the Bureau of Meteorology (top left) and as model output from an example host GCM GFDLCM2.1 at top right, and the corresponding downscaled 0.5°model (bottom left) and 0.1° model (bottom right). The period considered is 1961-1990 for all figures. All GCMS give similar results.

Table 1: Spatial correlation (R^2) over Tasmania between the AWAP gridded observations and GFDLCM2.1 for the raw GCM and downscaled to 0.5° and 0.1°. Both temperature and rainfall are calculated over the period 1961-1990. This improvement is similar across the six GCMs

Model resolution	Annual mean temperature	Annual total rainfall
GCM	0.22	0.01
0.5°	0.72	0.50
0.1°	0.99	0.69

A second measure of the improvement offered by dynamical downscaling is the difference between the statewide mean output by the models compared to the observed (AWAP) value. The GCM shown in figure 2 has only one land cell. This cell has a mean temperature of 12.5° Celsius. If we consider only the land cells of both the 0.5° and 0.1° resolution models the mean temperature of Tasmania is very close to the mean AWAP temperature of 10.4° C.

Seasonality of rainfall is crucial for climate adaptation work. Different regions of the state have different seasonal patterns of rainfall. The top images in figure 3 show the mean annual rainfall as represented in the AWAP dataset (left) and the model mean output (right). The bottom graphs in figure 3 show how the seasonal rainfall changes for three regions of the state (the west, the east and the north) as well as statewide. The 0.1° CFT model output correctly models the different seasonal patterns of the various regions of Tasmania, from the strongly seasonal west coast to the almost uniform distribution on the east coast. As GCMs represent the state by a single cell (or at most a few cells), the output cannot display the regional differences in seasonality that are observed.

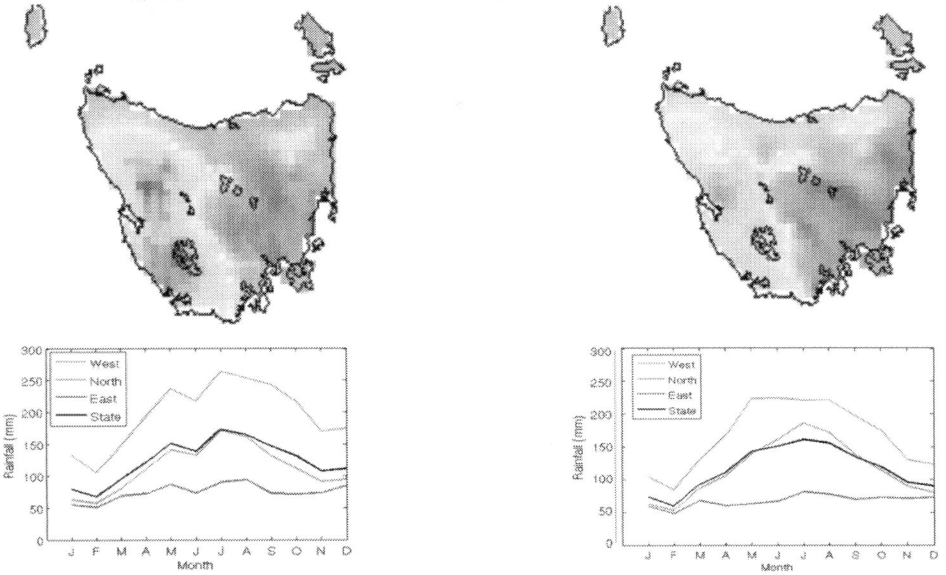

Figure 3. AWAP mean annual rain (top left) and the model mean annual rain (top right) for the period 1961-1990. The bottom graphs show the seasonal variation in rainfall for three regions of Tasmania and the statewide mean (AWAP on left, model mean on right).

4. Bias-adjustment

Despite the high skill of the CFT modelling output in reproducing the mean and spatial and temporal variance of the observed climate, some applications require modelling output to be calibrated closely to observed data. Applications such as biophysical and hydrological models have strongly non-linear responses. Similarly, capturing the correct spatial pattern of variables such as precipitation is important when attempting to project future changes to extreme events, as magnitudes can vary greatly over relatively short distances.

In order to allow the CFT modelling output to be used in such applied models, or for analyses of extremes, we have bias adjusted the modelling output of five key variables – daily maximum temperature (tmax), daily minimum temperature (tmin), daily rainfall, daily potential evaporation and daily solar radiation. We have developed a bias-adjustment process based on the percentile binning method of Panofsky and Brier [9].

The percentile binning method has an advantage over other bias-adjustment methods as it ensures the current model climate matches the observations over the entire probability distribution, and then applies this adjustment to the probability distribution into the future period. It does not assume that the distribution of the variable being adjusted is normally distributed about a mean.

The bias-adjustment process was applied on a daily, cell-by-cell basis for each of the land grid cells. The overlap period between the modelling output and the AWAP observed dataset, from 01/01/1961 to 31/12/2007 formed the training period, where we calculated the adjustments made to each percentile bin. These adjustments were then applied to the entire modelling output (01/01/1961 to 31/12/2100). In order to remove the effect of climate change when calculating the percentile bin of each daily value, the modelling output was detrended using a 30-year running mean. However in order to preserve the climate change signal, the adjustment was applied to the original (non-detrended) time series

The ability of the bias-adjustment process to correct the spatial pattern of the model to more closely match the AWAP data for rainfall is demonstrated in figure 4. The bias-adjustment process removed the observed seasonal bias in rainfall. The spatial correlation between the model and AWAP increases from between 0.8 - 0.87 for raw model data to 0.99 when using bias-adjusted model data.

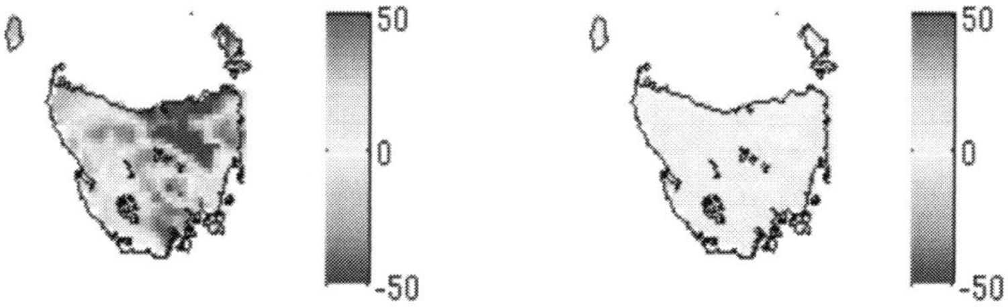

Figure 4. Results of the percentile-binning bias-adjustment process. The left panel shows the percentage difference in mean annual in rainfall (1961-1990) between AWAP and the raw model mean. The right panel is the difference between AWAP and the bias adjusted model mean rainfall.

5. Conclusions

Climate Futures for Tasmania has undertaken an extensive modelling program in order to deliver high-quality projections of Tasmania's climate to 2100. We have demonstrated the high level of skill the downscaled models have in reproducing the recent climate of Tasmania. Compared to GCM model output the dynamical downscaling process gives improved climate information by giving mean patterns closer to observed values, by improving the spatial pattern of variables across the state and by allowing us to simulate the seasonality of variables in different regions of Tasmania.

Acknowledgements

This work was supported by the Australian Government's Cooperative Research Centre Program through the Antarctic Climate and Ecosystems Cooperative Research Centre (ACE CRC). Climate Futures for Tasmania is possible with support through funding and research of a consortium of state and national partners.

References

[1] Nakicenovic N & Swart R 2000 Special Report on Emissions Scenarios A Special Report of Working Group III of the Intergovernmental Panel on Climate Change Cambridge United Kingdom

[2] Smith I & Chandler E 2009 Refining rainfall projections for the Murray Darling Basin of south-east Australia-the effect of sampling model results based on performance *Climatic Change* in press

[3] McGregor JL, 2005 *CCAM: geometric aspects and dynamical formulation* CSIRO Atmospheric Research Melbourne Victoria

[4] McGregor JL & Dix MR 2008 An updated description of the Conformal-Cubic Atmospheric Model in K Hamilton & W Ohfuchi (eds) *High Resolution Simulation of the Atmosphere and Ocean*, Springer 51-76

[5] Schmidt F 1977 Variable fine mesh in spectral global model *Beitr. Phys. Atmos.* **50** 7

[6] Reynolds RW 1988 A real-time global sea surface temperature analysis *Journal of Climate* **1** 12

[7] Thatcher M & McGregor J 2009 Using a scale-selective filter for dynamical downscaling with the conformal cubic atmospheric model *Monthly Weather Review* **137** no 6 11

[8] Raupach MR, Briggs PR, Haverd V, King EA, Paget M & Trudinger, CM 2008, *Australian Water Availability Project (AWAP), final report for Phase 3* CSIRO Marine and Atmspheric Research, Canberra, Australia

[9] Panofsky HA & Brier GW 1968 *Some applications of statistics to meteorology* The Penn. State University Press University Park Pa Reichle & Koster

Interannual modes of variability of Southern Hemisphere atmospheric circulation in CMIP3 models

S Grainger[1], C S Frederiksen[1] and X Zheng[2]

[1] Centre for Australian Weather and Climate Research, Bureau of Meteorology, Melbourne, Australia

[2] College of Global Change and Earth System Science, Beijing Normal University, Beijing, China

E-mail: S.Grainger@bom.gov.au

Abstract. The atmospheric circulation acts as a bridge between large-scale sources of climate variability, and climate variability on regional scales. Here a statistical method is applied to monthly mean Southern Hemisphere 500hPa geopotential height to separate the interannual variability of the seasonal mean into intraseasonal and slowly varying (time scales of a season or longer) components. Intraseasonal and slow modes of variability are estimated from realisations of models from the Coupled Model Intercomparison Project Phase 3 (CMIP3) twentieth century coupled climate simulation (20c3m) and are evaluated against those estimated from reanalysis data. The intraseasonal modes of variability are generally well reproduced across all CMIP3 20c3m models for both Southern Hemisphere summer and winter. The slow modes are in general less well reproduced than the intraseasonal modes, and there are larger differences between realisations than for the intraseasonal modes. New diagnostics are proposed to evaluate model variability. It is found that differences between realisations from each model are generally less than inter-model differences. Differences between model-mean diagnostics are found. The results obtained are applicable to assessing the reliability of changes in atmospheric circulation variability in CMIP3 models and for their suitability for further studies of regional climate variability.

1. Introduction

Identifying and evaluating modes of atmospheric circulation variability in climate models has a role in understanding climate change on both hemispheric and regional scales. The variability of Southern Hemisphere (SH) extratropical circulation has been examined in a number of studies using observational and reanalysis data [e.g. 1]. Using Coupled General Circulation Models (CGCMs) from the Coupled Model Intercomparison Project Phase 3 (CMIP3) dataset [2], individual modes of variability such as the Southern Annular Mode (SAM) and El-Niño Southern Oscillation (ENSO) have also been examined [e.g. 3, 4, 5].

Recently, Frederiksen and Zheng [6], using reanalysis data, examined interannual modes of seasonal mean variability of the extratropical SH atmospheric circulation. In particular, they identified interannual modes of variability related to (internal) intraseasonal variability, and slowly varying (time scales of a season or longer) internal dynamics and external forcings. Here, we estimate the interannual modes of variability from CMIP3 CGCMs and compare them with those found by Frederiksen and Zheng [6].

2. Data

The SH atmospheric circulation is represented by monthly mean 500hPa geopotential height in summer (December-January-February, DJF) and winter (June-July-August, JJA). The reference data is from the National Centers for Environmental Prediction (NCEP) National Center for Atmospheric Research reanalysis [7]. Data for 1951-2000 were obtained on a 2.5° × 2.5° longitude/latitude grid.

Data from the last 50 years of the CMIP3 twentieth century coupled climate simulation (20c3m) are taken to represent the second half of the 20th century. 70 realisations from 23 models have been obtained. Information about the modelling groups and model documentation can be found at http://www-pcmdi.llnl.gov/ipcc/model_documentation/ipcc_model_documentation.php. Model data were mapped from their original grid to the NCEP reanalysis grid so that the model interannual modes of variability can be directly compared with the NCEP reanalysis modes [4, 8].

For all data, the 2.5° × 2.5° grid is sub-sampled to 5° × 5° and thinned towards the poles as described by Frederiksen and Zheng [6] so that the data are approximately weighted by area. Monthly anomalies were obtained by subtracting the climatological monthly mean at each grid point.

3. Methods

3.1. Statistical model

The principles of the statistical model used are described in detail in Frederiksen and Zheng [9]. The monthly mean anomaly of a climate variable, x, after the removal of the annual cycle, from a single realisation can be decomposed as

$$x_{ym} = \mu_y + \varepsilon_{ym} \quad (1)$$

where m (m = 1, 2, 3) is the month index in the season, and y (y = 1, ..., Y) is the year index in a sample of Y years [10]. In (1) μ_y is the slowly varying component, independent of month. For an atmospheric variable from a CGCM realisation, such as geopotential height, μ_y is related to external radiative forcings (e.g. greenhouse gases, ozone), slowly varying SST and sea-ice boundary forcing and slowly varying internal atmospheric dynamics. ε_{ym} is the intraseasonal component or internal residual monthly departure of x_{ym} from μ_y. ε_{ym} is assumed to comprise a stationary stochastic process that is statistically independent and identically distributed with respect to y [10], with zero mean if the sample of Y years is large enough.

In practice, the intraseasonal and slow components in (1) cannot be directly calculated. However, Zheng and Frederiksen [10] showed that the total seasonal mean and intraseasonal covariances can be estimated from monthly mean data. The slow covariance is then estimated as a residual, i.e.

$$\hat{\mathbf{V}}[\mu_y] = \hat{\mathbf{V}}[x_{yo}] - \hat{\mathbf{V}}[\varepsilon_{yo}] \quad (2)$$

where $\hat{\mathbf{V}}[\ ]$ denotes the estimated covariance matrix between all grid points, and o denotes the average over an index (i.e. m).

3.2. Comparison between NCEP reanalysis and CMIP3 model realisations

The intraseasonal and slow modes of variability were estimated using eigenvalue decomposition of the intraseasonal and slow covariance matrices to obtain Empirical Orthogonal Functions (EOFs). For each CMIP3 realisation, the "best match" to each NCEP reanalysis mode was determined. Generally, this will be the CMIP3 realisation EOF with the highest pattern correlation to the NCEP reanalysis EOF, although the EOFs were inspected to check the agreement between areas of maximum loading. For the slow modes of variability, correlations between the projected time series of the EOF and sea surface temperature (see Frederiksen and Zheng [6] for examples) were also inspected.

How well the "best match" reproduces the NCEP reanalysis can be quantified using a metric of the form proposed by Taylor [11]:

$$S = \frac{4(1+R)^2}{(\hat{\sigma}+1/\hat{\sigma})^2(1+R_0)^2} \qquad (3)$$

where $\hat{\sigma}$ is the estimated standard deviation of the CMIP3 realisation mode normalised by the standard deviation of the NCEP reanalysis mode, R is the pattern correlation between the two modes, R_0 is the maximum attainable pattern correlation, and S is the score, with values between 0 and 1. Here R_0 is defined as the maximum pattern correlation found in the CMIP3 realisations for that mode. S is therefore a relative score, with a value of 1.0 indicating the best representation of the NCEP reanalysis mode across the CMIP3 realisations.

4. Results

Frederiksen and Zheng [6] examined in detail the intraseasonal and slow modes of variability estimated from the NCEP reanalysis. The modes that they obtained were found to have similar structures to modes found in earlier observational and dynamical studies. Figure 1 shows Taylor diagrams [11] of the "best match" in the CMIP3 20c3m realisations to the leading four intraseasonal and slow modes of variability in the NCEP reanalysis for DJF and JJA. In each plot, the standard deviation of the CMIP3 realisation mode has been normalised by the standard deviation of the corresponding NCEP reanalysis mode.

For the DJF intraseasonal modes (figure 1a) intraseasonal EOF-1 (IEOF-1), the intraseasonal component of SAM, has the best reproduced structure, with a median pattern correlation of about 0.91. The structures of IEOFs -2, -3 and -4, mid-latitude planetary wave-4 modes, are less well reproduced than for SAM, with median pattern correlations between 0.5 and 0.6, and maximum pattern correlations between 0.8 and 0.9. The median normalised standard deviations of all four leading DJF intraseasonal modes are between 0.75 and 1.0, indicating that the estimated variance in the CMIP3 realisations is generally less than that estimated by the NCEP reanalysis. DJF slow EOF-1 (SEOF-1), the slow component of SAM, has the best reproduced structure of the DJF slow modes (figure 1b), with a median pattern correlation of about 0.89. However, there is a larger spread of estimated variances than for the intraseasonal component of SAM. The structures of the other slow modes are less well reproduced than the slow component of SAM. SEOF-2 and SEOF-3, two modes of atmospheric response to ENSO, have median pattern correlations of about 0.5, while SEOF-4, the South Pacific Wave, has a median pattern correlation of about 0.3. The median normalised standard deviation of all four slow modes is less than 1.0, but for SEOF-3 the 90th percentile is almost 1.5.

The match to the JJA intraseasonal modes is shown in figure 1c. The structure and normalised standard deviation in the CMIP3 realisations of IEOF-1, the intraseasonal component of SAM, is little changed from DJF. However, JJA IEOF-2 (mid-latitude planetary wave-3) has generally higher pattern correlations than for DJF IEOF-2, usually in the range 0.5-0.85. IEOF-4 appears to be preferred to IEOF-3 (both mid-latitude planetary wave-3 modes) in the models, with nearly 90% of the estimated normalised standard deviations of IEOF-4 greater than 1.0. In contrast, the median normalised standard deviations of the first three IEOFs are all about 1.0. The JJA slow modes of variability appear to be least well reproduced in the CMIP3 realisations (figure 1d). The pattern correlation of the slow component of SAM (SEOF-1) in particular is much lower than for DJF. The median normalised standard deviation of all four slow modes (SEOFs -2, -3 and -4 are two modes of atmospheric response to ENSO and the South Pacific Wave respectively) is about 0.75.

In order to quantify the results shown in figure 1, (3) has been used to obtain relative scores for the CMIP3 20c3m realisations for each mode. Figure 2 shows the relative scores for DJF SEOF-1 and SEOF-2 with realisations grouped according to model. For DJF SEOF-1 (figure 2a) it is clear that most models reproduce the slow component of SAM relatively well. However some models, notably CSIRO-Mk3.5, FGOALS-g1.0 and INGV-SXG, have much lower estimated variances than the NCEP reanalysis, and consequently lower relative scores. There is a wider range of relative scores for DJF SEOF-2 (figure 2b) than for DJF SEOF-1. While some models reproduce the atmospheric response to ENSO variability relatively well, e.g. CSIRO-Mk3.5, ECHAM5/MPI-OM, GFDL-CM2.0, GFDL-

CM2.1 and MRI-CGCM2.3.2, others have relatively poor responses, e.g. BCCR-BCM2.0 and the three GISS models. These results are consistent with the findings of van Oldenborgh et al. [5] for ENSO in CMIP3 model oceans.

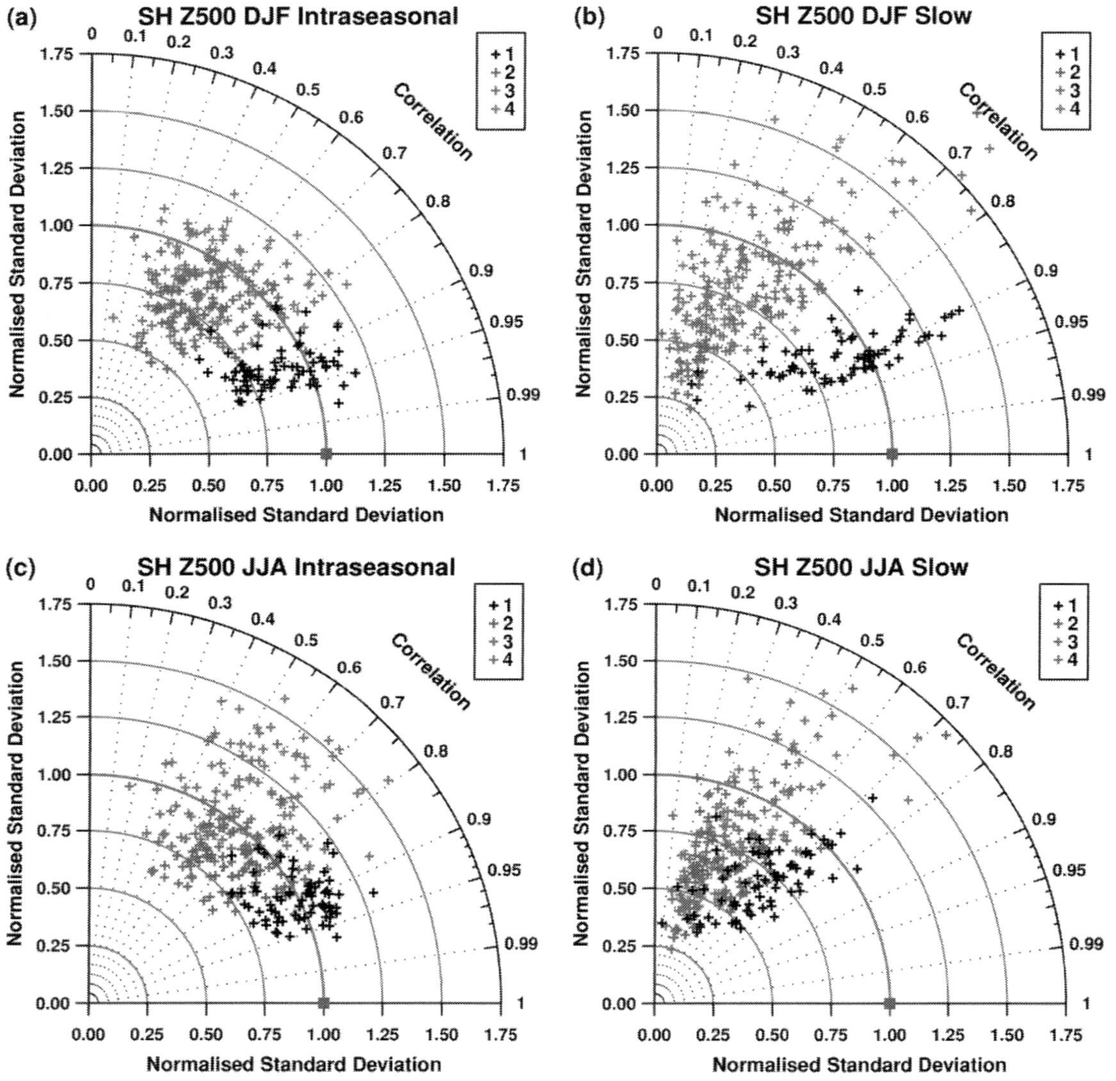

Figure 1. Taylor diagrams of the best match in the CMIP3 20c3m realisations to the leading four NCEP reanalysis SH 500hPa geopotential height EOFs of the (a) DJF intraseasonal, (b) DJF slow, (c) JJA intraseasonal, and (d) JJA slow components of variability.

A feature of the results in figure 2 is that the range of relative scores from realisations within the same model is generally less than inter-model differences. This is also seen in the relative scores for the other slow and intraseasonal modes of variability (not shown), although there is a greater spread of results in winter than in summer. Further, when the intraseasonal and slow eigenvalues are summed (i.e. a measure of the amount of intraseasonal and slow covariance in each CMIP3 realisation) the results (not shown) also suggest that the inter-model differences are generally larger than differences between realisations from the same model. This suggests that how well each CMIP3 model reproduces

20th century atmospheric circulation variability can be measured by diagnostics averaged over all realisations from each model.

Figure 3 shows three diagnostics averaged over all realisations from each CMIP3 model for summer and winter. Figure 3a shows the ratio of the intraseasonal covariance sums in the CMIP3 models to the NCEP reanalysis. The models generally have lower intraseasonal covariance than the NCEP reanalysis in DJF, but in JJA the ratios in the models are centred on 1.0. The slow covariance ratio (figure 3b) is generally lower than 1.0, particularly in JJA. There is also a greater spread of results within a season than for the intraseasonal covariance, with a factor of about 4 between the highest and lowest slow ratios, compared with about 2.5 for the intraseasonal ratio. Figure 3c shows the model mean overall relative score, the relative score averaged over the leading four intraseasonal and slow modes in a season, weighted by the estimated standard deviation of the NCEP reanalysis mode. Clear differences between the model overall relative scores can be seen, particularly in DJF. The majority of the model overall relative scores are lower in JJA than in DJF, mainly due to the lower estimated variances of the model slow modes.

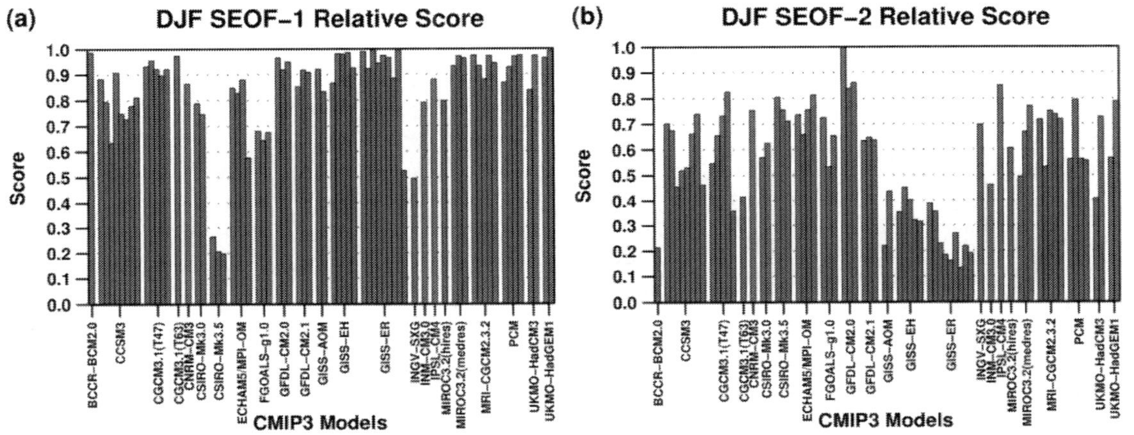

Figure 2. Relative scores of the best match in CMIP3 20c3m realisations to NCEP reanalysis SH 500hPa geopotential height for (a) DJF SEOF-1, and (b) DJF SEOF-2.

5. Conclusions

In this paper, intraseasonal and slow modes of variability of SH 500hPa geopotential height for DJF and JJA have been estimated from CMIP3 model 20c3m realisations. These have been evaluated against those estimated from NCEP reanalysis data for 1951-2000. It was found that the intraseasonal and slow components of SAM had the best reproduced structure in the models. In general, the intraseasonal modes of variability are better reproduced in the models than the slow modes, particularly in JJA. Diagnostics were proposed to quantify how well the models reproduce the NCEP reanalysis modes of variability. It was found that differences between realisations from the same model are generally less than inter-model differences. When these diagnostics were averaged over all realisations from each model, clear differences between model diagnostics can be seen. This provides guidance in assessing the reliability of changes in atmospheric circulation variability in CMIP3 climate change scenarios. The suitability of models for further study of regional climate variability may also be gauged from such diagnostics.

Acknowledgments

SG is supported by the Australian Climate Change Science Program of the Australian Department of Climate Change. XZ is supported by Projects 40975062/40875062 NSFC and the SRF for ROCS, SEM China.

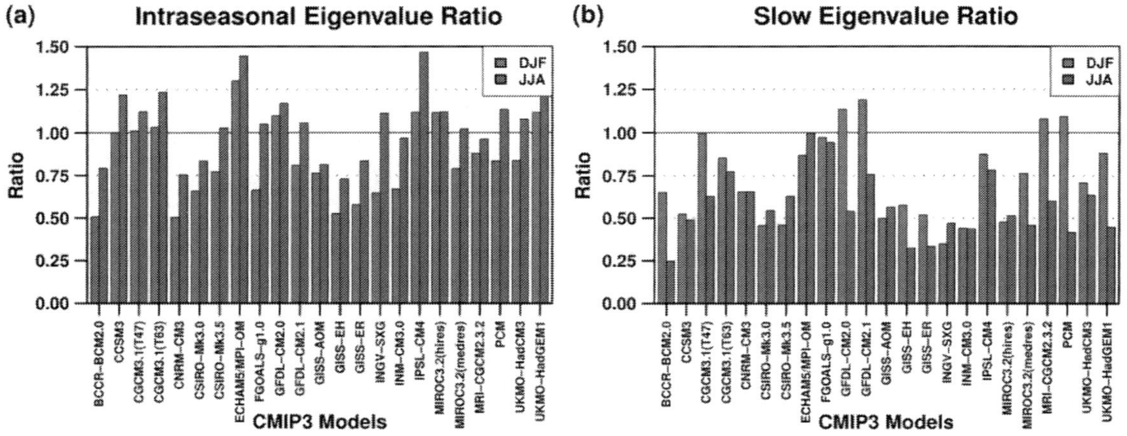

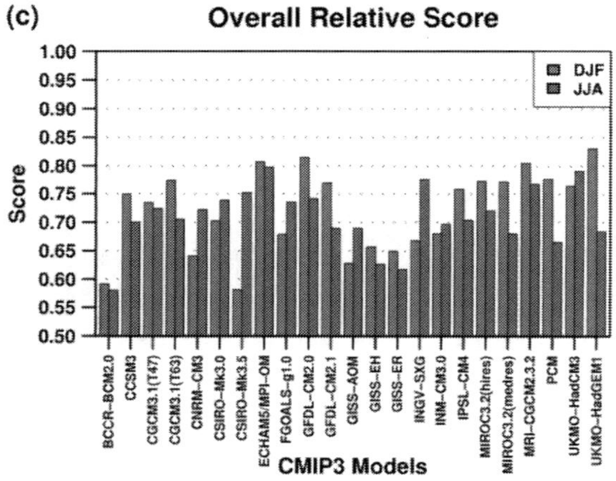

Figure 3. Averages over all 20c3m realisations in each CMIP3 model for DJF and JJA for (a) intraseasonal eigenvalue ratio, (b) slow eigenvalue ratio, and (c) overall relative score.

References

[1] Kiladis G N and Mo K C 1998 *Meteorology of the Southern Hemisphere* no 49, ed D J Karoly and D G Vincent (Boston: American Meteorological Society) pp 307-36
Frederiksen J S and Frederiksen C S 1993 *J. Atmos. Sci.* **50** 3148-63
L'Heureux M L and Thompson D W J 2006 *J. Clim.* **19** 276-87
[2] Meehl G A, Covey C, Delworth T, Latif M, McAvaney B, Mitchell J F B, Stouffer R J and Taylor K E 2007 *Bull. Am. Meteorol. Soc.* **88** 1383-94
[3] Randall D A *et al.* 2007 *Climate Change 2007: The Physical Science Basis. Contribution of Working Group I to the Fourth Assessment Report of the Intergovernmental Panel on Climate Change*, ed S Solomon, D Qin, M Manning, Z Chen, M Marquis, K B Averyt, M Tignor and H L Miller (Cambridge: Cambridge University Press) pp 589-662
[4] Miller R L, Schmidt G A and Shindell D T 2006 *J. Geophys. Res.* **111** D18101 doi:10.1029/2005JD006323
[5] van Oldenborgh G J, Philip S Y and Collins M 2005 *Ocean Sci.* **1** 81-95
[6] Frederiksen C S and Zheng X 2007 *Clim. Dyn.* **28** 849-66
[7] Kalnay E *et al.* 1996 *Bull. Am. Meteorol. Soc.* **77** 437-71
[8] Grainger S, Frederiksen C S and Zheng X 2008 *ANZIAM J.* **50** C399-412
[9] Frederiksen C S and Zheng X 2007 *Frontiers in Turbulence and Coherent Structures* vol 6, ed J Denier and J S Frederiksen (Singapore: World Scientific Publications) pp 87-120
[10] Zheng X and Frederiksen C S 2004 *Clim. Dyn.* **23** 177-91
[11] Taylor K E 2001 *J. Geophys. Res.* **106** 7183-92

Can we define climate using information theory?

J Walter Larson

Mathematics and Computer Science Division, Argonne National Laboratory, Argonne, IL, USA
Computation Institute, University of Chicago, Chicago, IL, USA
School of Computer Science, The Australian National University, Canberra, ACT, Australia

E-mail: larson@mcs.anl.gov

Abstract. The standard definition of climate is, by convention, based on a thirty-year sample. But why? One way to define the sampling period for constructing climatologies is to ask: What is a sufficient sample to construct probability density functions (PDF) for key meteorological variables? I propose an information-theoretic framework for evaluating climatic sampling periods based on level of detail and associated uncertainties in the estimated PDF, the Shannon entropy growth curve and its discrete derivative, and the Kullback-Leibler divergence. I compute these quantities for 235 years of daily data from the Central England Temperature record and use these statistics to compare popular sampling periods and discuss the feasibility of determining an optimal sampling period.

1. Introduction

It is standard practice to compute climatic quantities by using a thirty-year sample of meteorological data—a convention defined by the World Meteorological Organisation. The thirty-year sampling period was chosen at the 1935 International Meteorological Organisation conference [1, 2, 3]. The central role of the sampling period in "defining" climate motivates tests for assessing the relative informativeness of this choice.

Trewin [2] summarises potential criteria for assessing climatic averaging periods. The climatic mean as a descriptor has limits, and numerous applications require information about the underlying distribution [4, 5]. Thus, criteria for sampling periods for probability density function (PDF) construction—which will provide more information about quantiles and extrema—are desirable. The question then becomes: How long a sampling period is required to derive PDFs that are variable by variable sufficiently complete to describe the whole record?

I propose a method for assessing climate sampling intervals through the following questions:

Q1 Given sample of W years for a variable, how much detail is present in, and how confident can one be of an estimate of, the PDF?

Q2 What is the information content of a sample of W years?

Q3 How well does a sample of W years reflect the whole available data record of $Y > W$ years?

Q4 What is the incremental change in a sample's information content with respect to W?

Q5 How well does a sample of W years represent a sample of $W + 1$ years?

These questions will be answered using an optimal PDF construction technique described in section 4, coupled with information-theoretic concepts presented in section 2. The goal is to identify—if possible—an "optimal" sampling period W^*. I apply this framework to daily average temperature data from the Central England Temperature record (CET).

2. Information theory

Information theory [6] is a means for quantifying overall structure in probability mass functions (PMFs) and PDFs for discrete and continuous variables, respectively. Its antecedents lie in Boltzmann's statistical mechanical definition of thermodynamic entropy. Previous applications of information theory to climate include predictability ([7] and references therein), model evaluation [8, 9], variability [9], and causality [10].

For a discrete variable $X \in \{x_1, \ldots, x_N\}$, the PMF $\vec{\pi}$ is defined such that π_i is the probability that $X = x_i$; the probabilities π_i satisfy the conditions $0 \leq \pi_i \leq 1$, for $i \in \{1, \ldots, N\}$, and $\sum_{i=1}^{N} \pi_i = 1$. The Shannon entropy (SE) $H(X)$ is

$$H(X) = -\sum_{i=1}^{N} \pi_i \log [\pi_i]. \tag{1}$$

The base of the logarithm in (1)—and in (2) below—defines the units for information; for bases 2 and e, $H(X)$ is measured in *bits* and *nats*, respectively. The normalisation properties of the probabilities π_i imply $H(X) \geq 0$. Heuristically, $H(X)$ quantifies the "surprise" in X. The uniform distribution $\pi_1 = \cdots = \pi_N = 1/N$ maximises $H(X)$; for a system with N states, $0 \leq H(X) \leq \log N$.

Consider two distinct PMFs for X, $\vec{\zeta}$ and $\vec{\pi}$. The additional information—or gain—required to describe $\vec{\pi}$ given $\vec{\zeta}$ is the Kullback-Leibler divergence (KLD)

$$D_{KL}(\vec{\pi} \parallel \vec{\zeta}) = \sum_{i=1}^{N} \pi_i \log \left[\frac{\pi_i}{\zeta_i}\right]. \tag{2}$$

Formulations of the SE and KLD exist for a continuous variable x; one replaces the PMFs $\vec{\pi}$ and $\vec{\zeta}$ with the PDFs $p(x)$ and $q(x)$, respectively, and the summations over discrete states in (1) and (2) with integrals with respect to x [6]. Numerical computation of the SE is sensitive to the discretisation $dx \rightarrow \Delta x$. Thus, one must take care in discretising or binning continuous data to form a PMF or PDF, respectively. This issue is discussed in detail in section 4.

Consider a climate variable for which a sample of Y years is available. Suppose we draw from this sample a contiguous subset spanning a period of $W < Y$ years. Estimating the sample's PDF answers **Q1**. If we compute the SE for this sample, $H(W)$ answers **Q2**. The KLD $D_{KL}(P \parallel Q)$ for the PDFs $Q(W)$ and $P(Y)$ estimated from the windows spanning W and Y years, respectively, answers **Q3**. The *discrete derivative* [11] $\Delta H(W) \equiv H(W) - H(W-1), W \geq 1$ answers **Q4**. The KLD $D_{KL}(P(W + \Delta W) \parallel Q(W))$ answers **Q5**.

3. The Central England temperature record

The CET is longest observational record for surface air temperature. Manley's original CET [12] comprises monthly averages beginning in 1659. The daily CET [13] (figure 1(a)) spans the period 1772–present. I used daily averages for the period 1772–2006, obtained from the British Atmospheric Data Centre [14]. Parker et al. [13] describe the data selection and processing methods used to create the daily CET. Archived temperatures are rounded to the nearest 0.1°C. The distribution of sample means versus sampling period W is shown in the box-whisker plot in figure 1(b). There is evidence the CET exhibits low-frequency oscillations beyond the century timescale [15] and a secular warming trend (figure 1(c)) [16].

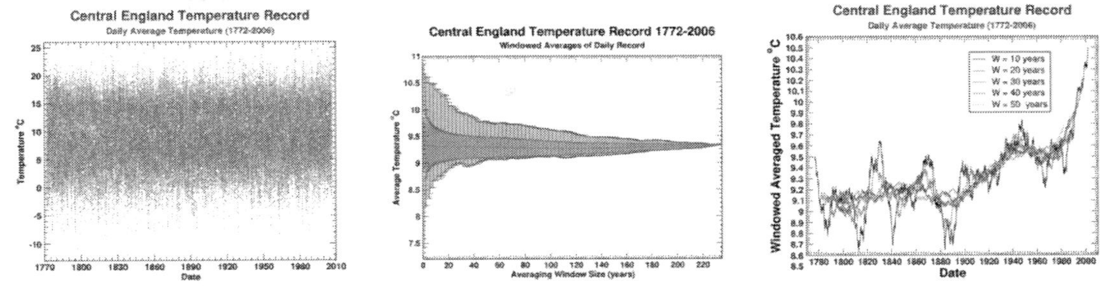

(a) (b) (c)

Figure 1. Daily CET 1772–2006: (a) timeseries, (b) distributions of windowed time averages, and (c) running averages versus time for various sampling window sizes W. In (b), box of width twice the sample standard deviation is centered on the mean, whiskers indicate minima/maxima.

4. Numerical method for PDF and entropy estimation

Computing the SE (1) for a continuously valued quantity requires a robust scheme for PDF estimation. If a binning-based method is used, the main perils in discretisation are choosing too few (*oversmoothing*) or too many (*overfitting*) bins. Discretisation must be performed carefully because the SE is sensitive to the number of bins used to compute it from a sample of a continuous variable. Scott [17] provides an excellent overview of PDF estimation.

I employed Knuth's [18] optimal binning scheme, because it is compatible with the coarse-graining of continuous data for information estimation purposes. The scheme is derived from Bayesian principles using the following assumptions: a discrete uniform prior distribution for the number of bins within a feasible range, a Dirichlet prior distribution for the bin probabilities, and a piecewise constant PDF with uniform bins. Prior knowledge I of the data sample's range V (and implied bin boundary locations) and size N is also included to derive the posterior probability $p(M|\vec{d}, I)$ that a PDF with M bins describes the data sample $\vec{d}$ [18]

$$p(M|\vec{d}, I) \propto \left(\frac{M}{V}\right)^N \frac{\Gamma(\frac{M}{2})}{\Gamma(\frac{1}{2})^M} \frac{\prod_{k=1}^{M} \Gamma(n_k + \frac{1}{2})}{\Gamma(N + \frac{M}{2})}. \quad (3)$$

In 3, $\Gamma(\cdot)$ is the Gamma function and n_k is the number of counts in each bin. The value of M that maximises $p(M|\vec{d}, I)$ yields the most probable piecewise constant, uniform-bin-width PDF for $\vec{d}$. Maximising the logarithm of the RHS of (3) is computationally easier and equivalent to maximising the RHS of (3). The mean bin probability density μ_i and its variance σ_i^2 are [18]

$$\mu_i = \left(\frac{M}{V}\right)\left(n_i + \frac{1}{2}\right)\left(N + \frac{M}{2}\right)^{-1} \quad (4)$$

$$\sigma_i^2 = \left(\frac{M}{V}\right)^2\left(n_i + \frac{1}{2}\right)\left(N - n_i + \frac{M-1}{2}\right)\left(N + \frac{M}{2} + 1\right)^{-1}\left(N + \frac{M}{2}\right)^{-2}. \quad (5)$$

Note that even when data is absent from a bin, $\mu_i \neq 0$, and the μ_i are normalised by construction. The average signal-to-noise ratio is $SNR = \sum_{i=1}^{M} \mu_i/\sigma_i$. The optimal value of M from (3), the resulting μ_i and σ_i^2 from (4) and (5), respectively, and the associated SNR answer **Q1**.

The rounding of the CET data presents a problem for the optimal binning scheme; this truncation adds fine structure to the PDF's broad-scale shape. This fine structure spawns numerous local maxima in the RHS of (3), and growth at large M (cf. figure 5D in [18]). In fact, the optimal binning scheme is an excellent detector of severely truncated data [19]. Truncation effects can be removed by adding a random uniform deviate that brackets the rounded value by half the truncation value; this will not replace lost information [19] but does allow estimation of the PDF's large-scale structure. I have smoothed the CET data by adding to each observation a uniform random deviate $\delta_i \in [-0.05, 0.05)$; this smoothing allows PDF estimation using (3) and, when truncated to the nearest 0.1°C, yields the original timeseries. Multiple random smoothings of the CET have been performed, and the resulting quantities computed from these datasets are consistent with the results presented here.

SE and KLD values were calculated by first estimating the PDF through maximisation of the RHS of (3). SE and KLD uncertainty estimates were computed through windowed resampling of the data combined with Monte Carlo simulation of a large number (10,000) of "neighboring" PDFs determined from (4) and (5) for each sample. Random numbers are generated by using standard techniques [20]. Sampling was implemented by sliding a window of W years one year at a time across the total dataset of Y years. This process results in $Y - W + 1$ samples for a window spanning W years; the CET has $Y = 235$ years, and $W = 1$ and $W = 10$ years result in 235 and 226 distinct samples (or 2.35×10^6 and 2.26×10^6 PDFs), respectively. For each sample, SE and KLD quantities were computed, yielding one ensemble of values for each statistic for each value of W. For each ensemble, the sample mean, standard deviation, minimum, maximum, and key percentiles were computed. The end result is a set of distributions—$H(W)$, $\Delta H(W)$, et cetera—that allow us to gauge the relative informativeness and representativeness of the sampling period W, and to intercompare sampling periods.

5. Information-theoretic evaluation of climatic sampling periods

Figure 2 shows PDF estimates for the daily CET. The optimal value of M grows steadily with increasing W but with variation for fixed W (figure 2(a)). The SNR grows steadily with W (figure 2(b)), with a narrowing in the range of values of SNR. Growth in M and SNR with W implies more detailed PDFs and greater confidence in their structure (figures 2(c)–2(f)).

Figure 3 shows percentiles of H for the CET. The median value of $H(W)$ climbs rapidly for $1 \leq W \leq 10$ years, continues to climb steadily for $10 < W \leq 60$ years, climbs less rapidly for $60 < W \leq 100$ years, and converges to a near-constant value $H \approx 4.55$ bits beyond the century timescale (figure 3(a)). The range of values of $\Delta H(W)$ narrows rapidly for $1 \leq W \leq 10$ years, and then drops off steadily with increasing sample size W (figure 3(b)). The interdecile ranges of ΔH at $W = 30$ and $W = 50$ years correspond roughly to the interquartile ranges at $W = 10$ and $W = 30$ years, respectively.

Figure 4 shows percentiles for the total and incremental information gains $D_{KL}(P(Y) \parallel Q(W))$, and $D_{KL}(P(W) \parallel Q(W-1))$, respectively. The representativeness of a subsample of W years increases rapidly for $1 \leq W \leq 10$ years, with the median of $D_{KL}(P(Y) \parallel Q(W))$ dropping by a factor of 10 over this range and continuing to decrease steadily out to $W = 100$ years (figure 4(a)). The incremental information gain $D_{KL}(P(W) \parallel Q(W-1))$ also drops in a similar fashion as adding one year to a sample becomes less significant (figure 4(b)). The 90th percentiles at $W = 50$ and $W = 30$ years correspond roughly to the 30th and 10th percentiles at $W = 30$ and $W = 10$ years, respectively.

The information statistics produced here do not identify an immediate candidate for an optimal sampling period W^*. Picking such a value would move beyond viewing distributions of SE and KLD statistics versus W to implicitly assuming that all realisations for fixed W are equally representative. It is clear, however, that for sampling periods less than 30 years, much is to be gained from expanding the sample size by one year. It is also likely that expanding a sample of thirty years by one or more decades is informative.

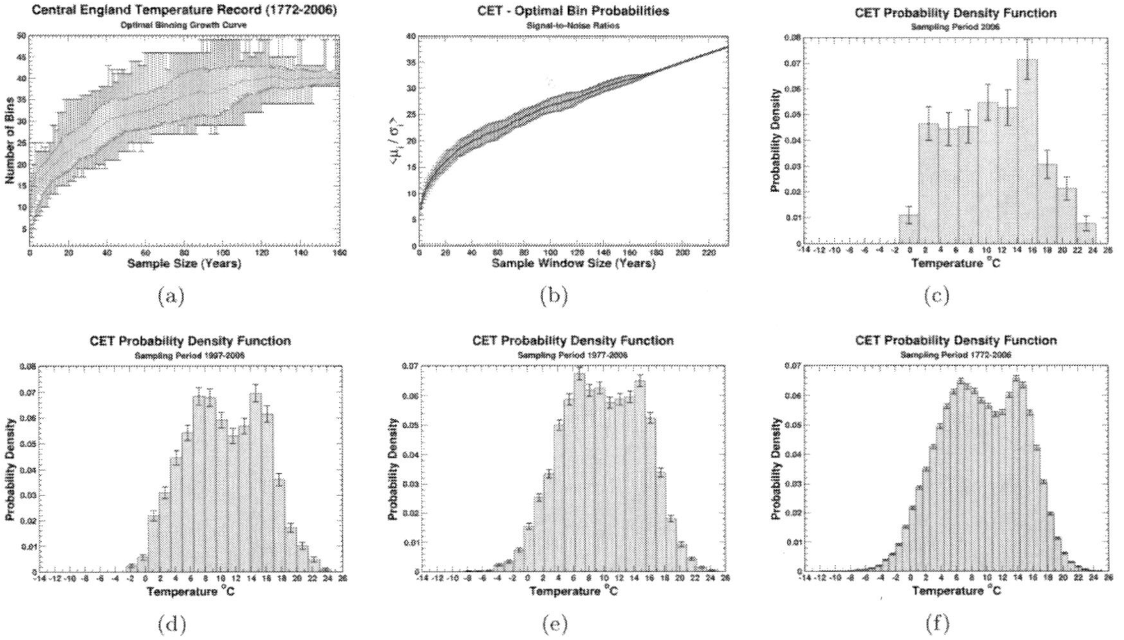

Figure 2. Optimal binning of the CET data: (a) growth curve for number of bins, (b) SNR and estimated PDFs for windowed samples covering periods (c) the year 2006, (d) 1997–2006, (e) 1977–2006, and (f) 1772–2006. Boxes and whiskers in (a-b) are defined as for figure 1(b).

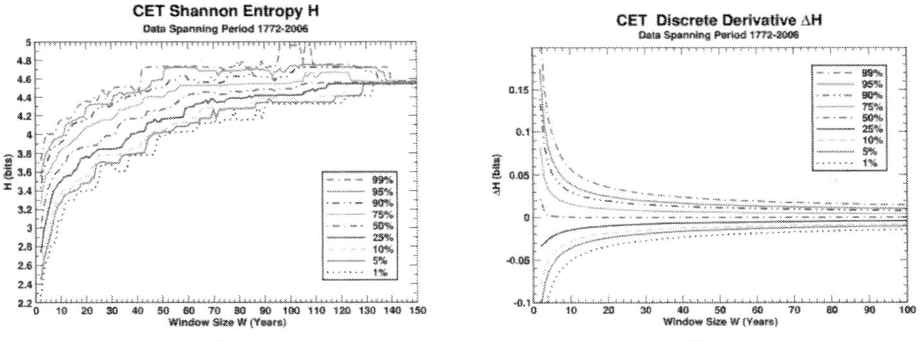

Figure 3. Shannon entropy estimation for CET: (a) SE as a function of sampling window size and (b) marginal entropy increase ΔH.

6. Conclusions and future work

An information-theoretic framework for evaluating the definition of climate based on sampling periods has been presented and applied to the CET. The framework allows direct intercomparison between sampling periods. The question posed by this paper's title may be answerable using these techniques, but the results presented here are preliminary. The presence of a secular trend in the CET complicates the decision-making process regarding an optimal sampling period W^*; techniques analogous to those used to evaluate normals in the presence of secular trends [2] must be developed achieve this goal. The overall utility of this framework

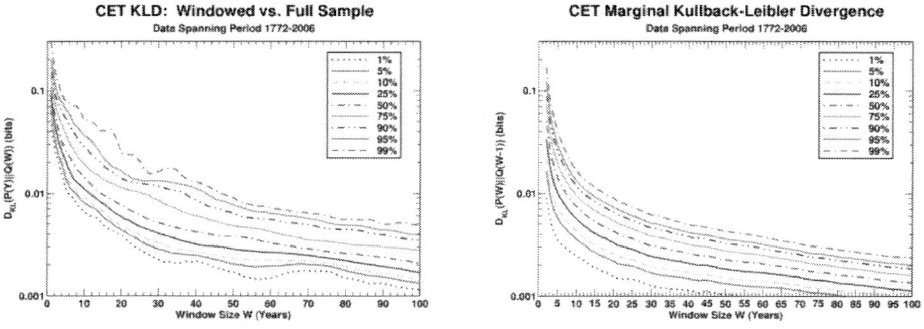

(a) (b)

Figure 4. Kullback-Leibler divergences for the CET: (a) KLD for sample of W years versus the whole dataset and (b) Marginal KLD $D_{KL}(P(Y) \parallel Q(W))$.

must also be evaluated through application to a wider variety of variables for both observational data streams and reanalyses. These endeavours constitute areas of future investigation.

Acknowledgments

This work was supported by the Office of Advanced Scientific Computing Research, Office of Science, U.S. Department of Energy (DOE), under Contract DE-AC02-06CH11357.

References

[1] International Meteorological Organization 1937 *Proceedings of the Meetings in Danzig and Warsaw 29–31 August and 12 September 1935* (Leyden: Secretariat of the IMO)
[2] Trewin B 2007 The role of climatological normals in a changing climate Tech. Rep. WDMCP-No. 61 World Meteorological Organization Geneva
[3] Hulme M, Dessai S, Lorenzoni I and Nelson D R 2009 *Geoforum* **40** 197–206
[4] Guttman N B 1989 *Bulletin of the American Meteorological Society* **70** 602–607
[5] Kunkel K E and Court A 1990 *Bulletin of the American Meteorological Society* **71** 201–203
[6] Cover T M and Thomas J A 2006 *Elements of Information Theory* 2nd ed (New York: Wiley-Interscience)
[7] DelSole T and Tippett M 2006 *Reviews of Geophysics* **45** RG4002
[8] Shukla J, DelSole T, Fennessy M, Kinter J and Paolino D 2006 *Geophysical Research Letters* **33** L07702
[9] Larson J W 2009 *Proceedings of the 18th World IMACS Congress and MODSIM09 International Congress on Modelling and Simulation* ed Anderssen R S, Braddock R D and Newham L T H (Modelling and Simulation Society of Australia and New Zealand and International Association for Mathematics and Computers in Simulation) pp 2639–2646
[10] Knuth K H, Gotera A, Curry C T, Huyser K A, Wheeler K R and Rossow W B 2005 *Proceedings of the Earth-Sun System Technology Conference (ESTO 2005)* (Adelphi, MD)
[11] Crutchfield J P and Feldman D P 2003 *Chaos: An Interdisciplinary Journal of Nonlinear Science* **13** 25–54
[12] Manley G 1974 *Quarterly Journal of the Royal Meteorological Society* **100** 389–405
[13] Parker D E, Legg T P and Folland C K 1992 *International Journal of Climatology* **12** 317–342
[14] British Atmospheric Data Centre, Hadley Centre, UK Meteorological Office 2009 Historical Central England Temperature (CET) Data `http://badc.nerc.ac.uk/data/cet/`
[15] Benner T C 1999 *International Journal of Climatology* **19** 391–403
[16] Jones P D and Hulme M 1997 *Climates of the British Isles, Present, Past, and Future* ed Hulme M and Barrow E (London: Routledge) pp 173–196
[17] Scott D W 1992 *Multivariate Density Estimation: theory, practice, and visualization* (New York: Wiley)
[18] Knuth K H 2006 Optimal data-based binning for histograms `http://arxiv.org/abs/physics/0605197`
[19] Knuth K H, Castle J P and Wheeler K R 2006 *Proceedings of the 17th meeting of the International Association for Statistical Computing—European Regional Section: Computational Statistics (COMPSTAT 2006)* ed Rizzi A and Maurizio V (Springer)
[20] Press W H, Teukolsky S A, Vetterling W T and Flannery B P 1992 *Numerical Recipies in Fortran 77: The Art of Scientific Computing* Second ed (New York: Cambridge University Press)

17th National Conference of the Australian Meteorological and Oceanographic Society IOP Publishing
IOP Conf. Series: Earth and Environmental Science **11** (2010) 012029 doi:10.1088/1755-1315/11/1/012029

Towards the development of a climate model evaluation system in the Centre for Australian Weather and Climate Research

Lawrie Rikus, Ben Hu, Martin Dix, Ian Watterson and Tracey Elliott

ACCESS Model Evaluation Team, The Centre for Australian Weather and Climate Research, GPO Box 1289K, Melbourne, Victoria 3001

E-mail: l.rikus@bom.gov.au

Abstract. The paper describes the philosophy of the climate model evaluation scheme being developed within CAWCR as well as the database of observational data-sets which under-pins it. It argues that model evaluation should measure 'fitness-of-purpose', that it should be objective and that it should be based on the largest possible number of observational data-sets. Time series plots of smoothed observational data and the relevance of the disparity between data from different sources are discussed. The paper calls for active participation of the Australian climate research community in the project.

1. Introduction

The ACCESS model evaluation team is developing a climate model evaluation scheme (based on one developed at the United Kingdom's MetOffice [1,2]) for use by the Centre for Australian Weather and Climate Research (CAWCR) climate research community. Although we also calculate a number of standard model evaluation scores such as the Watterson M-score [3], the Climate Prediction Index (CPI) [4] , the Reichler and Kim performance index [5] and the Model Climate Performance Index (MCPI) [6] the CAWCR scheme is designed to specifically evaluate the suitability of a given model for a specific task.

At the heart of the scheme is the concept of metrics which objectively measure the performance of the models against some specified criteria, with sets of metrics defined to signify the model's suitability for specific purposes. Metrics are generally scalar measures and are evaluated using observational data-sets [6]. Comparisons of data-sets for a number of parameters clearly shows that it would be inadvisable to rely on any single observational data-set in this context; they can vary substantially even in their global mean monthly climatology. Hence we need to assemble a comprehensive and wide-ranging collection of observational data-sets to ensure that we have a good estimate of the observational uncertainty and subsequently the range of validity for each metric. These data-sets need to be converted to conform to common, well defined data-set and variable conventions to facilitate their ready use in model evaluation.

This paper describes the philosophy of the CAWCR model evaluation scheme and the collection of observational data-sets that is being assembled to underpin it.

2. The Philosophy of the CAWCR Scheme

The main idea of the scheme follows Banks et al [1]; evaluation of models should measure fitness of purpose. This requires knowledge of the relevant climate processes as well as the needs of the users who want to apply the model to that purpose. The criteria for a satisfactory model should be coded as an objective set of metrics which characterize the performance of the facets of the model's behavior that are crucial for the specified purpose. The model error metrics must lie within defined uncertainty limits for the model to be defined as a 'good' model for that purpose. If a model does not satisfy the criteria it is deemed to be unsuitable for that specific purpose. Note that this principle implies that the relative performance between models is not important; each model satisfies the criteria or it does not. Note that experience [6] shows that different models show different strengths and it is unlikely that one climate model will be satisfactory for all purposes.

2.1. Climate model metrics

While some metrics will measure the conformance of the model to basic physical principles (e.g. conservation of mass and energy), most will be based on a comparison of two dimensional model fields with gridded observations or on errors in scalar quantities such as climate or blocking indices. As it is essential that the models be assessed in the context of the range of observational uncertainty[2,6] each observational data-set should have quantified uncertainty limits but this is often not readily available. As shown in the section on observations below, the spread between data-sets for at least some observational fields shows that the observational uncertainty can be quite large. Using a collection of data from different sources facilitates at least an estimate of the uncertainty to be made. The task of combining observational data from different sources is a complicated statistical problem but we would argue that it must done objectively with no subjective ranking of the different data sources allowed; any ranking should have an objective basis.

2.2. Metrics for Australian climate research

Part of the task of setting up the model evaluation scheme is that of defining sets of metrics for purposes relevant to the climate modeling and analysis community in Australia. This requires input from that community so that a comprehensive and well-founded set of metrics can be defined, implemented and used productively. For the specific purpose of evaluating a model's Australian climate simulation there are a number of possible metrics which should be considered. These can come from time series analysis of mean Australian region fields and include measures based on spectral decomposition, the period, phase and strength of the mean seasonal signal and trend analysis. For Australian wide rainfall evaluation the metrics could include the Southern Oscillation Index (SOI), the Indian Ocean Dipole (IOD), the eastern Australian blocking index and indices for the strength and position of the southern hemisphere storm track and the Australian monsoon.

Given the demand for regional climate change projections, metrics for the large scale climate drivers known to be required for down-scaling techniques should also be included if the model is to be used for down-scaling. For climate change purposes the definition of metrics should include the quality of the simulation of large scale climate drivers but also needs to reflect the performance of the physics of the model and its response to known forcing mechanisms.

3. The CAWCR climate observational dataset repository

To under-pin the climate model evaluation scheme we are collecting observational data-sets from a large number of sources. These inherently have different time and spatial resolutions, different variable names, units and conventions and span different time periods. Before these data-sets can be used effectively they need to be converted to a standard convention for file, variable names and standard units and undergo quality control. The convention chosen for the CAWCR archive was to follow the Coupled Model Intercomparison Project Phase 5 (CMIP5) requirements [7]. Most of the data sets have already been converted and are available on the National Computational Infrastructure (NCI) machines in Canberra with the possibility that they will be incorporated into the Earth Systems Grid in the near future. They are stored as a time series of monthly means over the period of availability as well as monthly mean climatologies. Researchers requiring access to the data should contact one of the authors. All of the data has been quality assessed but it is likely that more problems remain to be discovered; a process which will be facilitated by having a number of users for the data.

Included in the data is a database of metric scores defined for each variable for a set of models against each observational data-set for each month of the time period for which the model and data overlap. There are a set of scripts to add new models as results become available. There are plans to link in a comprehensive set of meta-data for each observational data-set including all the original meta-data as well as a comprehensive record of the processing performed to ensure compliance to the data conventions.

4. The treatment of observational datasets

As part of the quality control process in assembling the data-set collection the time series of various area means are plotted together for each variable. Here we present some examples which demonstrate the wide variance in data-sets. Note that all plots show seasonally filtered and smoothed time series. Figure 1 shows the time series of global and southern mid-latitude means for precipitation over the ocean for all the data sources collected and processed up to the beginning of February 2010.

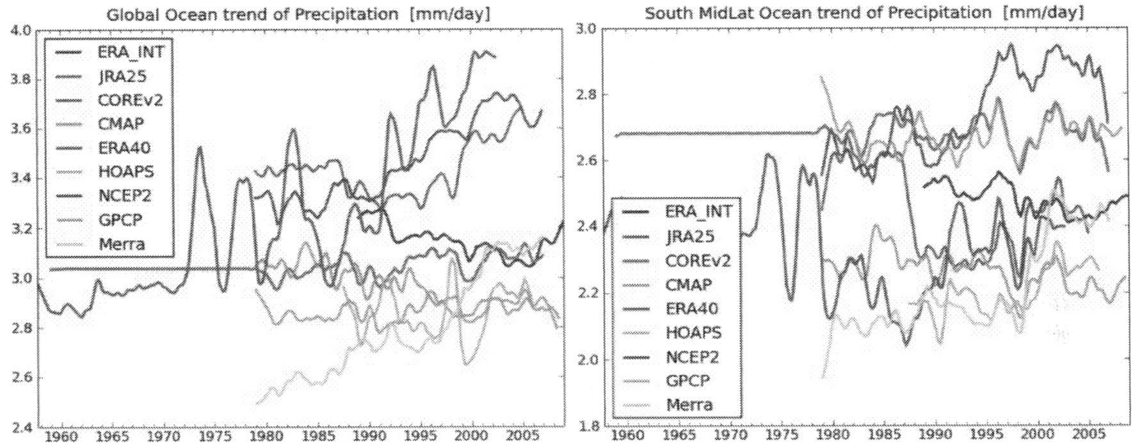

Figure 1. The time series for global mean oceanic precipitation in units of mm/day (left) and the southern mid-latitude oceanic mean (right) from the data-sets in the archive.

Note the anomalous behavior of some of the re-analysis derived data-sets (ERA-40, JRA25, NCEP2) in the global mean which, at least for ERA-40, stems from a well known problem with its satellite retrieval methods over the tropics [8] and causes an apparent overestimation relative to the rest of the data-sets for the last half of the period and large inter-annual variability after about 1970. The problems are not as obvious for ERA-40 and JRA25 in the southern mid-latitude plot although it is possible that there is a problem with the NCEP2 data. The flat line structure in the early part of the COREv2 data is due to the fact that it is an 'artificial' data-set developed to drive ocean models and uses annual mean values for years prior to 1980.

Figures 2 and 3 show the corresponding time series for net downwards surface longwave and shortwave radiation respectively. Again there is a substantial spread in values and the data series fall into two obvious groups. In both cases the groups for the global mean case are different than for the southern mid-latitude case. The ISCCP and HOAPS data, both satellite derived, show significant inter-annual variability relative to the rest.

In the absence of objective measures it is not always obvious which data-set (if any) is preferable to any of the others. Precipitation on the scales considered here is derived from satellite data using inversion algorithms with links to rain gauge records [9] or from re-analyses which derive the precipitation from a synthesis of substantial meteorological data and the numerical weather prediction model physics and which now seem to provide data of comparable quality [10]. The case for the surface radiation fluxes is similar but with even fewer genuine surface measurements to constrain the satellite algorithms. On the basis that all data-sets contain some useful information the best way to treat the disparity is to use all of them but the best way to do that is not obvious; a simple mean approach ignores relative quality and the strong possibility that the different data-sets are not independent. This is clearly a fertile area for research.

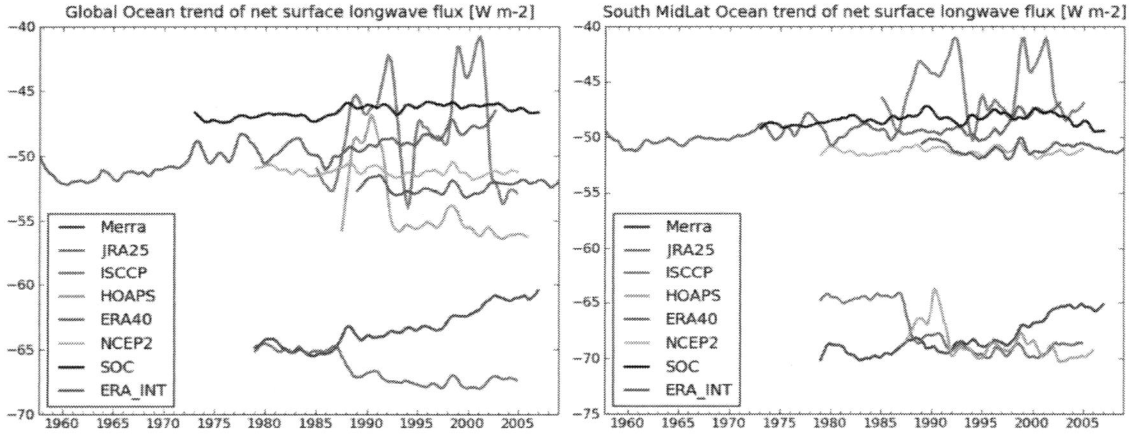

Figure 2. The time series for global mean net surface longwave flux in units of Wm^{-2} (left) and the southern mid-latitude oceanic mean (right) from the data-sets in the archive.

5. Conclusions

This paper has argued that model evaluation should be application dependent, data driven, use all available observational data-sets objectively and be accessible to the whole climate research community.

Before the model evaluation scheme introduced in this paper becomes fully functional there are a number of tasks to complete and a number of problems to be solved. One of the primary tasks is to develop a full suite of climate metric sets for use in CAWCR as well as the general Australian climate research community. This requires community consultation and the participation of experts in the areas where the models are to be assessed. The developing observational data-set collection needs to be expanded, quality controlled and converted to the standard conventions and the actual storage set up to allow community access. This will require us to develop and entrain expertise in assessing data sets over a number of different specialist areas; again participation of outside experts is required. Finally more study is required to find the optimal way to use the data from disparate data-sets in a fully objective manner allowing an accurate assessment of the uncertainty in the climate data used to evaluate the models.

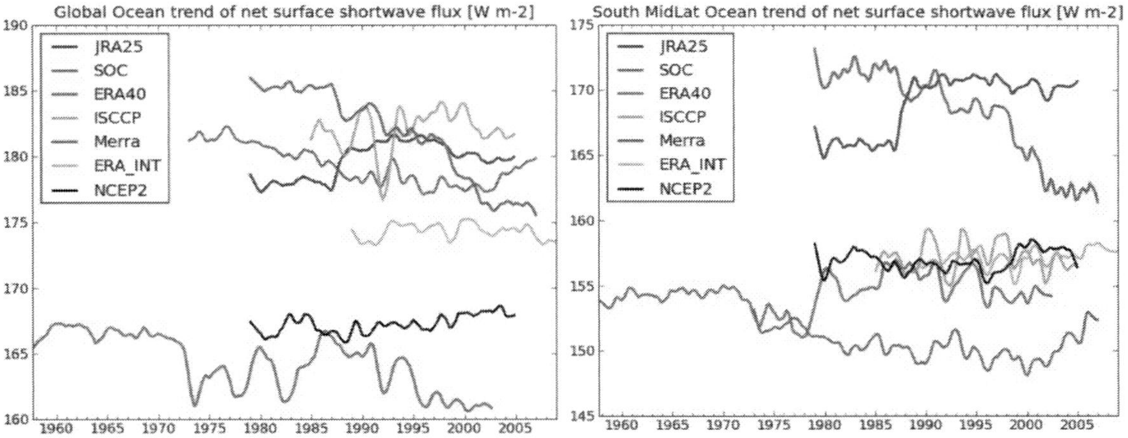

Figure 3. The time series for global mean net surface shortwave flux in units of Wm^{-2} (left) and the southern mid-latitude oceanic mean (right) from the data-sets in the archive.

References

[1] Banks H *et al* 2008 Evaluation of HadGEM3-AO *Integrated Climate Programme deliverable* Q2 (Internal report available on request from authors at UKMO)

[2] Williams K D and Webb M J 2009 A quantitative performance assessment of cloud regimes in climate models *Clim. Dyn.* **33** 141-157

[3] Watterson I G 1996 Nondimensional measures of climate model performance *Intl. J. Climatol.* **16** 379-391

[4] Murphy J M, Sexton D M H, Barnett D N, Jones G S, Webb M J, Collins M and Stainforth D A 2004 Quantification of modelling uncertainties in a large ensemble of climate change simulations *Nature* **430** 768-772

[5] Reichler T and Kim J 2008 How well do coupled models simulate today's climate? *Bull. Am. Meteorol. Soc.* **89** 303-311

[6] Gleckler P J, Taylor K E and Doutriaux C 2008 Performance metrics for climate models *J. Geophys. Res.* **113** doi:10.1029/2007JD008972

[7] CMIP5 Coupled Model Intercomparison Project Data Description 2010 [cited 2010 Jan 20]. Available from: http://cmip-pcmdi.llnl.gov/cmip5//data_description.html

[8] Troccoli A and Kallberg P 2004 Precipitation correction in the ERA-40 reanalysis *ERA-40 Project Report Series* **13**. Available from: http://www.ecmwf.int/publications

[9] Huffman G J, Adler R F, Arkin P, Chang A, Ferraro R, Gruber A, Janowiak J, Mcnab A, Rudolf B and Schneider U 1997 The Global Precipitation Climatology Project (GPCP) combined precipitation data set *Bull. Am. Meteorol. Soc.* **78** 5-20

[10] Simmons A J, Willet K M, Jones P D, Thorne P W and Dee D 2010 Low-frequency variations in surface atmospheric humidity, temperature and precipitation: inferences from reanalyses and monthly gridded observational datasets *J. Geophys. Res.* **115** D01110, doi:10.1029/2009JD012442

IOP Conf. Series: Earth and Environmental Science **11** (2010) 012030 doi:10.1088/1755-1315/11/1/012030

NSW-IMOS: An Integrated Marine Observing System for Southeastern Australia.

Moninya Roughan[1,2], Bradley D Morris[1,2], Iain M Suthers[3,2]

[1] Coastal and Regional Oceanography Laboratory, School of Mathematics and Statistics, University of New South Wales, Sydney NSW, 2052, Australia.
[2] Sydney Institute of Marine Science, Mosman, NSW, 2088, Australia.
[3] School of Biological Earth and Environmental Sciences, University of New South Wales, Sydney NSW, 2052, Australia.

E-mail: mroughan@unsw.edu.au

Abstract.
The East Australian Current (EAC) and its eddy field dominate the shelf circulation along the coast of SE Australia. It impacts the climate, weather and fisheries along the most populated part of the Australian coastline. Yet there is much we don't know about the EAC and its role in the transport of heat, generation of eddies, its role in upwelling and the associated biological processes. In this paper we introduce the New South Wales node of the Integrated Marine Observing System (NSW-IMOS). The observing system is designed to complement extant observations in the region and capture key continental shelf processes such as slope water intrusions and encroachments of the EAC, the generation of eddies and the biological response to such events.

1. Introduction

The East Australian Current (EAC) flows poleward from the Coral Sea to the Tasman Sea. It impacts the coastal ocean along its path, particularly along the coast of southeastern Australia where the EAC and its eddy field dominate the shelf circulation [1]. It is in this region that the ocean has warmed more than 2°C/100yr in the past century [2]. The effect of increased ocean temperatures will impact on our climate, weather and marine ecosystems, where some species will prosper and others will decline [3]. This region is also where that the majority of the Australian population live (`http://www.censusdata.abs.gov.au`). In 2007 the Australian government commenced funding implementation of the Australian Integrated Marine Observing System (IMOS, `http://imos.org.au/`, [4]).

The goal of NSW-IMOS is to devise an observing system which will comprehensively examine the physical and ecological interactions of the East Australian Current and its eddy field with coastal waters and to assess the impacts of climate change. More specifically:

- To investigate the EAC, its separation and resultant eddy field along the coast of SE Australia: Determine seasonal and spatial variability in the separation of the EAC from central NSW; Determine the formation, frequency, and evolution of EAC eddies; Understand air sea interactions, particularly to determine the ontogeny of East Coast Lows and severe winter storms in relation to warm core eddies.

- To quantify key continental shelf processes along the coast of SE Australia: Quantify the impact of key processes such as onshore encroachment of the EAC, slopewater intrusions, upwelling, downwelling and internal waves; Quantify the seasonal and decadal variation in the EAC and its southward extension; Examine the coastal wind and wave climate in driving nearshore currents and the northward sediment transport.
- To determine the biological response to oceanographic and climate effects.

We will achieve these goals through an integrated monitoring program along the NSW continental shelf which includes: a network of oceanographic mooring arrays, high frequency coastal radar to enhance spatial coverage, monthly biogeochemical sampling, regular coastal and bluewater surveys by autonomous ocean gliders and an underwater vehicle, and deployment of acoustic receivers for the tracking of tagged fish. The full NSW-IMOS plan can be obtained from `http://imos.org.au/nswimos.html` [5].

The *in situ* data when combined with satellite observations, enable the modelling required to explain the role of the oceans in seasonal prediction and climate change. Sustaining the project will allow identification and management of climate change in the coastal marine environment. It will also provide the observations necessary to better understand and predict the fundamental connections between coastal biological processes and regional/oceanic phenomena that influence biodiversity. A review of the current literature on the East Australian Current is outlined in [6] which sets the framework under which this ocean observing system was designed. The observing system presented here represents a new way forward for integrated observations of the waters of southeastern Australia and goes some way towards filling some of the knowledge gaps that exist in our understanding of the EAC.

2. NSW-IMOS Instrumentation Plan

2.1. Sydney National Reference Transect

To build on our long history (since 1942) of hydrographic observations at the 50 m and 100 m Port Hacking stations, we devised a National Reference Transect (NRT) off Sydney. This transect consists of three moorings shore normal off Bondi, a fourth mooring and four hydrographic sampling locations off Port Hacking (Table 1). This area is typically downstream of the EAC separation and is often influenced by EAC eddies.

2.1.1. Moored Observations The Ocean Reference Station (ORS) is the only site of sustained moored observations along the coast of NSW, operating since Nov 1989 [7]. The ORS is located 3 km off Bondi Beach, in 65 m of water (Table 1). Seven wave-rider buoys have been maintained (for up to 30 years) by the Manly Hydraulics Laboratory (MHL, NSW Department of Commerce) [8]. The buoys are typically in 70 m water depth and between 6−12 km from the coast (Figure 1).

Presently the SYD100 and SYD140 moorings (Table 1) consist of a bottom mounted TRDI 300 kHz acoustic Doppler current profiler (ADCP) housed in a rigid frame with gimbal mount, which measures velocity through the water column in 8 m bins with a 5 min sampling interval. Temperature is recorded at 5 min intervals using a series of Aquatech 520T, and 520PT Aqualoggers at 8 m intervals throughout the water column (co-inciding with the middle of each ADCP bin) to within 20 m of the surface. Additional pressure/temperature loggers are placed at 4 depths (bottom, sub-surface and 2 mid-water column) so as to estimate mooring lay-over. These moorings were initially deployed in June 2008 and apart from some data losses due to instrument failure a comprehensive data set has been returned. In time the PH100 site will be instrumented with two Wetlabs water quality meters (WQM) near bottom and sub-surface, that consists of a SeaBird CTD, as well as measurements of fluorescence, dissolved oxygen and turbidity. It is our intention that as the program progresses these data will be telemetred back to land and made available in near real-time.

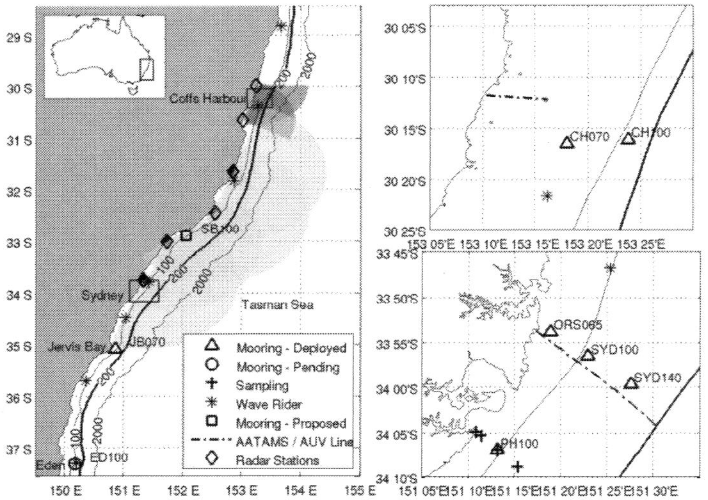

Figure 1. Map showing locations of the present (deployed), pending and proposed moorings as well as the hydrographic sampling locations and the wave-rider buoys. Subpanels (right) show expansions of the Coffs Harbour (top) and Sydney (bottom) regions. Dashed lines show the listening post transects. Diamonds show the locations of the HF Radar stations. Dark shading shows the coverage of the pending WERA HF Radar array at Coffs Harbour and light shading shows the coverage of the proposed CODAR array in Stockton Bight.

2.1.2. Biogeochemical Sampling In addition to the historic biogeochemical sampling, monthly sampling along the NSW-IMOS NRT (Table 1) includes hydrographic measurements, Conductivity, Temperature and Depth (CTD) profiles, water quality, chlorophyll sampling at PH050 and PH100 plus a plankton net sample and a water sample for genetic analysis at PH100. Monthly sampling along the NSW-IMOS NRT off Port Hacking also serves to calibrate fluorometric observations obtained from the instrumented moorings and ocean colour estimates of chlorophyll obtained from satellites (e.g. MODIS), chlorophyll, total suspended solids and coloured dissolved organic material (CDOM).

Table 1. Details of NSW-IMOS mooring sites and biogeochemical sampling sites (PH). * CSIRO funded sampling for temperature and nutrients. @ IMOS funded sampling for chlorophyll-a, pigments (HPLC) CDOM, PAR, phytoplankton and zooplankton. #The exact location is still to be finalised.

Platform Code	Latitude °S	Longitude °E	Depth (m)	Dist (km)	Principal Axis (°T)	Local Isobath (°)	Maintained Since
CH070	30°17.00′	153°17.91′	70	15			August 2009
CH100	30°16.07′	153°23.80′	100	22.2			August 2009
ORS065	33°53.88′	151°18.9′	65	2.1	13.4	16	1989
SYD100	33°56.63′	151°23.03′	100	9.9	18.3	26	June 2008
SYD140	34°00.08′	151°27.92′	140	19	25.0	28	June 2008
PH100	34°06.98′	151°13.14′	100	6.1	-	-	December 2009
JB070	35°05′	150°51′	70	1.85	-	-	July–Oct 2009
ED100#	37°18.96′	150°19.50′	100	25	-	-	N/A
PH025	34°04.94′	151°10.79′	25	0.9	-	-	April 1997
PH050*	34°05.35′	151°11.35′	50	2.1	-	-	December 1942
PH100*@	34°06.98′	151°13.14′	100	6.1	-	-	May 1953
PH125	34°08.88′	151°15.37′	125	11.1	-	-	April 1997

2.2. Additional moorings along the NSW Coast

Additional moorings will be deployed along the NSW coastline to complement the arrays beyond the NSW borders.

Coffs Harbour, 30° ***S:*** Oceanographically this area is typically upstream of the separation point of the EAC, (downstream of the Great Barrier Reef (GBR)) and biologically this area has already been recognised nationally with the designation of the Solitary Island Marine Park. Coral species found there are highly diverse with unique associations of tropical species near their southern latitudinal range as well as subtropical species that are absent or rare in the GBR. This region (and further north to Cape Byron) is exemplified by strong warm currents and a complex wave climate opposing the mean flow. Two moorings similar in configuration to the Sydney moorings have been deployed at the 70 m (CH070) and the 100 m (CH100) isobaths (Table 1) since late 2009.

Stockton Bight, 33° ***S:*** The Stockton Bight-Myall Shore region is known for its biological productivity. It is typically downstream of the EAC separation area and is on the receiving end of a seemingly persistent nutrient supply associated with the separation of the EAC. We plan to deploy a mooring equipped with CTD, ADCP and thermistor string (Figure 1), pending further funding.

Eden, 37° ***S:*** The mooring will again be similar to the SYD140 mooring consisting of a bottom mounted ADCP and a thermistor string throughout the water column (Figure 1), to be deployed in 2010. This will give a record of the predicted extension of the EAC into southern waters.

2.3. HF Radar

A medium range WERA HF Radar system will be deployed off Coffs Harbour in mid 2010 (Figure 1). The system provides spatial maps of surface currents ($\sim$ 1 m depth), as well as wave height, period and direction, and possibly wind direction at hourly intervals. We anticipate coverage out to $\sim$ 70 km offshore with a resolution of $1 - 1.5$ km. As the shelf is very narrow this range should see well out beyond the 2000 m isobath and generally cover the width of the narrow EAC jet. This will provide information on shelf edge processes, flow topography interactions and the generation of EAC instabilities. We intend to deploy a long range CODAR array in the Stockton Bight region when further funding becomes available (Figure 1). This will enable monitoring of the EAC separation and eddy field.

2.4. Autonomous Ocean Gliders

Ocean gliders will be deployed on a regular basis (4 per year, since late 2008) to map the 3-D structure of the EAC and its eddy field. To date five successful Slocum glider deployments have been undertaken in the EAC eddy field. The continuous sub-surface optical measurements obtained by the glider represent a significant advance on the occasional vertical light profiles from ship-based CTDs [9].

3. Preliminary Results from the Sydney Mooring Array: June 2008–June 2009

The 5 minute velocity data from the 3 Sydney moorings (Table 1, Figure 1) have been subsampled to hourly and lowpass filtered with a 36 h cut off frequency and a cosine taper applied to the last 5%. The data were rotated to alongshore (v) and across-shore (u) as per the calculation of the principal axis after consideration of the local isobath direction (Table 2). At all three sites across the shelf, mean alongshore velocities ($-0.1, -0.16, -0.1$ ms^{-1}, at ORS, SYD100, SYD140 respectively) are typically an order of magnitude larger than across-shore velocities. Mean magnitudes ($0.18, 0.23, 0.17$ ms^{-1}) are directed poleward ($184°, 175°, 179°$ respectively). Inshore (ORS) velocities are typically variable with a range of $v = -1.19$ to 0.78 ms^{-1}, $u = -0.19$ to 0.19 ms^{-1} where negative indicates poleward (onshore) transport. Midshelf (SYD100) velocities

Table 2. Mooring data statistics calculated from the hourly, depth-averaged, lowpass filtered data for the mooring array off Sydney (Figure 1, Table 1). Note: due to instrument failure at SYD140 maximum velocities recorded were anomalously lower than at SYD100.

	u (ms^{-1})			v (ms^{-1})			T (°C)		
	ORS	SYD100	SYD140	ORS	SYD100	SYD140	ORS	SYD100	SYD140
Mean	-0.01	0.01	0.00	-0.09	-0.19	-0.10	17.5	17.0	17.0
Std	0.03	0.05	0.06	0.22	0.26	0.17	1.87	1.93	2.25
Skewness	-0.35	-0.10	0.02	-0.34	-0.84	-0.42	0.64	0.78	0.61
Max	0.19	0.25	0.33	0.78	0.89	0.50	23.8	24.3	24.7
Min	-0.19	-0.24	-0.34	-1.19	-1.39	-1.29	13.8	13.3	12.6

are marginally stronger ranging $v = -1.39$ to 0.89 $ms^{-1}, u = -0.24$ to 0.25 ms^{-1}. At the shelf break velocities are typically greater, however due to instrument failure during an extreme event this is not reflected in the statistics (range $v = -1.29$ to 0.5 $ms^{-1}, u = -0.34$ to 0.33 ms^{-1}). Standard deviation in alongshore velocity is greatest midshelf (0.26) with a skewness of -0.84 indicating the predominance of poleward velocities along the shelf. Maximum poleward velocities off Sydney typically occur with EAC eddy intrusion events where the eddy moves across the continental shelf. Currents are generally baroclinic, however at times during strong equatorward flow barotropic currents exist (Figure 2). Vertical velocity shear is greatest during strong poleward current events, and at the inner shelf.

As with velocity, maximum temperatures co-incide with EAC eddy intrusions where mooring knockdowns of up to 30 m vertically have been recorded. At all three sites maximum water temperatures occurred during May 2009 (end of Autumn) of $23.8 - 24.7$°C across the shelf. Temperature exhibits strong stratification throughout the water column. The vertical temperature range is similar across all three sites ranging $\sim 12 - 24$°C. Standard deviation in temperature ranges $1.8 - 2.2$°C across the shelf with positive skewness. The onshore encroachment of eddies significantly alters the temperature field and mixed layer depth.

4. Discussion: National Context and Future Directions

The IMOS array described here gives for the first time simultaneous measurements of the EAC along the NSW coastline. It also provides timeseries of parameters not previously recorded (such as fluorescence, dissolved oxygen and pH). The data sheds new light on EAC interactions with the continental shelf and cross shelf processes including frequency, intensity and duration of events. With time the biological response to such events will also be quantified. The spatial and temporal coverage allowed for by the array will provide context for the more patchy historical data. It will also contribute to improved model parameterisation and prediction capability

An advantage of IMOS when compared to some international counterparts is that there is a level of federal co-ordination, particularly with regards to the distribution of new funding for the extension of IMOS through to 2013. The national backbone that integrates and links all the observations in a national context consists of the data management facility (`http://imos.aodn.org.au`), a satellite remote sensing facility and a network of nine national reference stations where identical measurements of physical and biogeochemical parameters are undertaken on a regular (nominally monthly) basis.

The development of a complete system from observations to hindcasts, nowcasts and forecasts will likely take decades. What we present here is the initial implementation over the first seven years and the vision for the future. While the initial focus has been on the physical environment

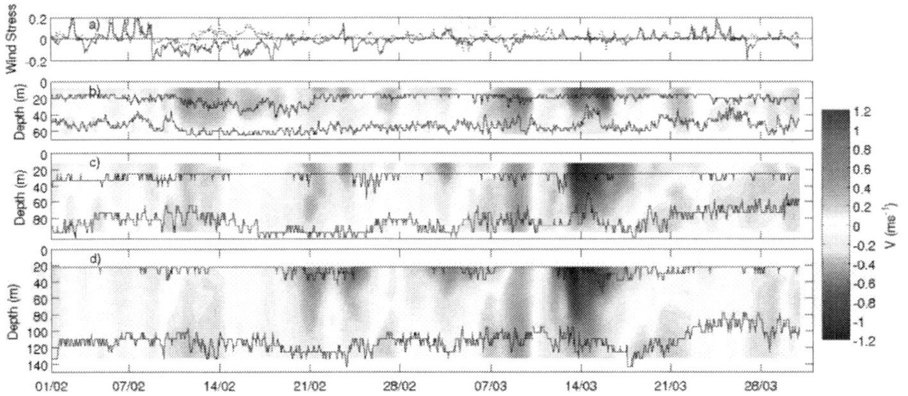

Figure 2. a) Surface wind stress (Pa), alongshore (-), across-shore (...), b–d) Alongshore velocity (v) at b) ORS065, c) SYD100, d) SYD140 for 1 Feb – 31 Mar 2009. Wind stress was calculated from the u and v components of wind velocity measured at Sydney Airport. Solid black lines represent the depth of the surface and bottom mixed layers.

this is by no means setting limitations on future directions. Already the observing system includes a series of shore normal arrays of listening posts for tagged fish, deployed off Sydney and Coffs Harbour to coincide with the locations of our moorings (Figure 1). This will be extended to the south coast in 2011. The data will give a better understanding of relationship between oceanographic conditions and the movement of many species of tagged fish and sharks. Furthermore as more reliable and cost effective biogeochemical sensors become available the system will be augmented to include more parameters.

Acknowledgments

We are grateful for the support of our partners NSW DECCW, Oceanographic Field Services, and Manly Hydraulics Laboratory. Integrated Marine Observing System (IMOS) is supported by the Australian Government through the National Collaborative Research Infrastructure Strategy and the Super Science Initiative. Data from the ocean reference station (ORS065) were provided by Sydney Water Corporation. This is Sydney Institute of Marine Science contribution number 0035.

References

[1] Hamon B V and Tranter D J 1971 *Aust. Nat. Hist.* **17** 129–133

[2] Ridgway K R, Coleman R C, Bailey R J and Sutton P 2008 *J. Geophys. Res.* **113, C08039**

[3] Hobday A J, Okey T A, Poloczanska E S, Kunz T J and Richardson A J Impacts of climate change on Australian marine life Tech. rep. Australian Greenhouse Office, Canberra

[4] Roughan M, Suthers I M and Meyers G *Climate Change Monitoring Strategy* ed You J and Hendersen-Sellars A (Sydney University Press) ISBN: 978-1-920899-41-7

[5] Roughan M Suthers I M, Morris B and Others 2009 NSW-IMOS Node Science and Implementation Plan July 2009–2013. Tech. rep. Sydney Institute of Marine Science Mosman, Sydney NSW, 2088, Australia

[6] Suthers I M, Roughan M, Baird M E and Young J Submitted *Deep Sea Res. 30 May 2010*

[7] Tate P M, Wyllie S and Lawson N 1994 *Proceedings 57th Water Industry Conference*

[8] Kulmar M, Lord D and Sanderson B 2005 *Proceedings Coasts and Ports Australasian Conference*

[9] Baird M E, Suthers I M, Griffin D A, Hollings B, Pattiaratchi C, Everett J D, Roughan M, Oubelkheir K and Doblin M In Press *Deep Sea Research 2010*

Teleconnections associated with the intensification of the Australian monsoon during El Niño Modoki events

A S Taschetto[1], R J Haarsma[2], A Sen Gupta[1], C C Ummenhofer[1] and M H England[1]

[1] Climate Change Research Centre, University of New South Wales, Sydney, NSW 2052 Australia

[2] Royal Netherlands Meteorological Institute, De Bilt, The Netherlands

E-mail: a.taschetto@unsw.edu.au

Abstract. In this study we investigate the teleconnection between the central-western Pacific sea surface temperature (SST) warming, characteristic of El Niño Modoki events, and Australian rainfall using observations and atmospheric general circulation model experiments. During Modoki events, wet conditions are generally observed over northwestern Australia at the peak of the monsoon season (i.e. January and February) while dry conditions occur in the shoulder-months (i.e. December and March). This results in a shorter but more intense monsoon season over northwestern Australia relative to the climatology. We show that, apart from the well-known displacement of the Walker circulation, the anomalous warming in the central-western equatorial Pacific also induces a westward-propagating disturbance associated with a Gill-type mechanism. This in turn generates an anomalous cyclonic circulation over northwestern Australia that reinforces the climatological mean conditions during the peak of the monsoon season. The anomalous circulation leads to convergence of moisture and increased precipitation over northern Australia. This response, however, only occurs persistently during austral summer when the South Pacific Convergence Zone is climatologically strengthened, phase-locking the Gill-type response to the seasonal cycle. The interaction between the interannual SST variability during El Niño Modoki events and the evolution of the seasonal cycle intensifies deep convection in the central-west Pacific, driving a Gill-type response to diabatic heating. The intensified monsoonal rainfall occurs strongly in February due to the climatological wind conditions that are normally cyclonic over northwestern Australia.

1. Introduction

The sea surface temperature (SST) variability of the tropical Pacific has important implications for climate worldwide, primarily associated with El Niño-Southern Oscillation (ENSO) events. The eastern Pacific warm water associated with El Niño episodes induces an anomalous Walker cell that leads to subsiding air and relative dry conditions over the maritime continent. When the maximum SST warming is located in the central Pacific instead, an anomalous double Walker cell is observed in the troposphere with the joint ascending branch over the warm anomaly and descending branches over the eastern Pacific and the Indonesian region [1]. This "flavour" of El Niño, termed El Niño Modoki, has been associated with climate impacts distinct from conventional El Niño events (e.g., [1, 2, 3, 4]). The Modoki phenomenon has become more frequent than traditional El Niños since the late 1970s [1].

Over Australia, dry conditions are observed in December and March during El Niño Modoki events [5,6]. In addition, unusually wet conditions occur in northern Australia during the peak of the monsoon season, particularly in February [6]. As the positive rainfall anomaly in February counteracts the negative one in December, analysis of the averaged austral summer season therefore gives a false impression that Modoki does not affect the Australian monsoon. This behaviour is the opposite to that observed during traditional El Niño events and seems to be a robust feature across different datasets (figure 1).

Over northwestern Australia, this bi-modal rainfall anomaly behaviour in austral summer indicates that Modoki events tend to delay the normal monsoon onset and advance the monsoon retreat over the region. Modoki does not necessarily reduce the total amount of rainfall received in the year, but it leads to a rearrangement of the annual precipitation cycle in northwestern Australia, producing a shorter and more intense monsoon season.

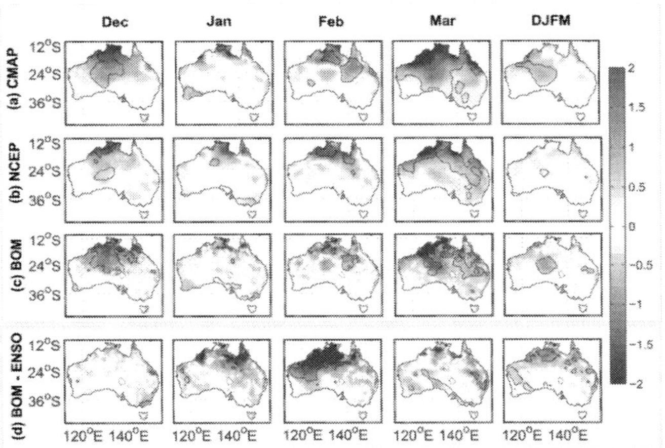

Figure 1. Composites of rainfall anomalies during Modoki (a-c) and conventional El Niño (d) events from December to March and for the December to March (DJFM) average using the (a) CMAP, (b) NCEP/NCAR Reanalysis and (c-d) BOM datasets. Modoki years based on Ashok et al. (2007): 1979/80, 1986/87, 1990/91, 1991/92, 1992/93, 1994/95, 2002/03, 2004/05. El Niño years: 1957/58, 1965/66, 1969/70, 1972/73, 1982/83, 1987/88, 1997/98. Units in mm day^{-1}. Areas within the thin black line are statistically significant at the 90% level according to a two-tailed t-test.

The reason why Australia experiences dry conditions with a warm anomaly in the equatorial Pacific is already known to be due to Walker circulation changes and the anomalous subsiding air that inhibits convection, the formation of clouds and thus reduces rainfall. An explanation for the opposite rainfall response during the peak of the monsoon season has been proposed by [6]. Using observations and numerical experiments, the authors have shown that northwestern Australian rainfall is enhanced in February due to an increase in moisture convergence caused by an anomalous cyclonic circulation over the region. However, the mechanisms responsible for generating the anomalous cyclonic circulation were not established. More recently, [7] suggested that this circulation anomaly is a remote response to anomalous heating in the tropical Pacific as developed in the Gill-Matsuno theory.

In this study we address the physical link between the warm Pacific SST anomalies associated with Modoki events and the anomalous cyclonic circulation that causes enhanced rainfall in northwestern Australia during February. This study complements [7] by considering the mean wind state and by examining the initialization of the air-sea process in the AGCM that leads to a Gill-Matsuno response.

2. Data and methods

2.1. Data sets and climatologies

In this study we use the rainfall datasets from the Australian Bureau of Meteorology (BOM; [8]), the CPC Merged Analysis of Precipitation (CMAP; [9]), and the National Center for Environmental Prediction (NCEP) / National Center for Atmospheric Research (NCAR) Reanalysis [10]. In addition, the global SST analysis from the Hadley Centre (HadISST1; [11]), the outgoing longwave radiation

(OLR) from the NOAA Interpolated OLR [12] and horizontal winds from the NCEP/NCAR reanalysis are also used here. The use of different datasets corroborates the analysis of [7] that is based on the NCEP/NCAR Reanalysis only.

We focus our analysis on the period for which more reliable global data exists, and on the period when Modoki events are more frequent, i.e. 1979 to 2006. The El Niño Modoki years used here are the same as those selected by [1]: 1979/80, 1986/87, 1990/91, 1991/92, 1992/93, 1994/95, 2002/03 and 2004/05. The Modoki year starts in June and ends in May of the following year.

2.2. Numerical experiments

A series of four experiments generated with the NCAR Community Atmospheric Model (CAM3) was performed by [6] to test the sensitivity of the Australian rainfall to the location of the SST warming along the equatorial Pacific. Each experiment consisted of 50 members integrated for 12 months. A 1°C anomaly was superimposed onto the SST seasonal cycle around 10°S-10°N over four locations: eastern, central-eastern, central-western and western Pacific. The authors concluded that Australian rainfall is more sensitive to the warm SST anomaly in the central-western Pacific, suggesting that the dateline is a key region for the Modoki forcing. Here we make use of this experiment, hereinafter referred to as the central-western Pacific experiment.

3. Results

3.1. Seasonality of the atmospheric response: Observations

The fact that the northern Australian rainfall anomalies during Modoki events change sign during the peak of the monsoon for the same tropical Pacific SST signature suggests that this response is phase-locked to the mean seasonal cycle. Figure 2(a-f) shows the long-term monthly mean rainfall, the regions with strong convective activity (via low OLR values) and the location of the maximum SST in the southern latitudes. The intensification of the precipitation rate and the broadening of the deep convection in the southern latitudes are most pronounced in January and February, indicating the strengthening of the South Pacific Convergence Zone (SPCZ).

To assess the seasonality of the atmospheric response during Modoki years, we show in figure 2(g-i) the composite seasonal cycle for Modoki years and long-term mean seasonal climatology for the southern tropical Pacific region around the dateline. This region is chosen as it contains the warm SST anomalies in the key central-western Pacific region as suggested by [6], and also because it exhibits strong convective and rainfall responses from the seasonally strengthened SPCZ. Although Modoki events are associated with a statistically significant warming of the SST compared to the mean state, the responses in precipitation and OLR are only significantly different from the climatology from November to February. This demonstrates the seasonality in the interaction between the interannual anomalies from Modoki events and the mean state of the atmosphere.

3.2. The Gill-Matsuno mechanism: Modelled response in February

We assess the simulated atmospheric response to the SST warming in the central-western Pacific for February in figure 3. The positive SST anomaly changes the local surface flux balance via enhanced evaporation. A zone of low pressure forms over the underlying warm waters, driving a convergence of low-level winds over the central-western Pacific (figure 3a).

At the same time the anomalous SST forcing drives additional diabatic heating to the troposphere via enhanced air-sea heat fluxes. In the equatorial regions, any available potential energy generated by diabatic heating is immediately converted to kinetic energy. Therefore, rising motion is also intensified by the convergence of low level winds (via continuity) and by enhanced diabatic heating (via thermodynamics). The anomalous positive vertical velocity transports moisture from the convective mixing layer to the mid troposphere, favouring the development of deep convective clouds.

Deep convection eventually leads to disturbances in the high troposphere generating a Gill-Matsuno-type response [13, 14] in the central-west Pacific and a remote response in the atmosphere via excitation of equatorial waves. The remote response to the SST forcing is seen over Australia with a center of low pressure and anomalous cyclonic circulation (figure 3a). The cyclonic circulation anomaly is also seen in the vertically integrated moisture flux from the surface to 500hPa (figure 3b), indicating anomalous convergence of moisture over northern Australia. The anomalous convergence of moisture and cyclonic circulation enhance the normal atmospheric conditions during the monsoon season over Australia, i.e. a cyclonic circulation in February (figure 2m) and low pressure over the continent, generating intensified convection and increased monsoonal rainfall over the northwest region. Note that only during January and February is the mean circulation cyclonic over northwestern Australia (figure 2l-m). The anomalous cyclonic circulation then reinforces the background state, generating enhanced monsoonal rainfall during the peak of the Australian monsoon.

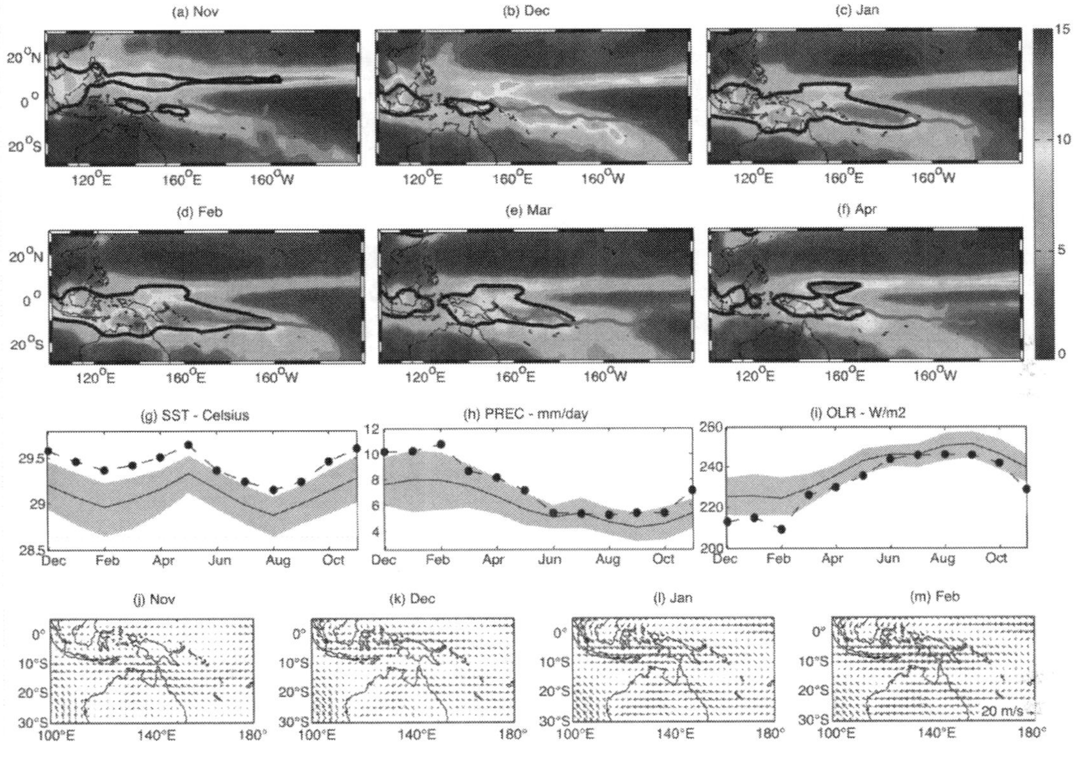

Figure 2. (a-f) Long term monthly mean precipitation (shaded, mm day^{-1}), OLR<220W/m^2 (black line) and maximum SST location (red line, °C) from November to April. (g-i) Annual cycles of SST, precipitation (from CMAP) and OLR based on 1979-2006 mean (solid line) and Modoki years (dashed-circle line) for the averaged southern tropical Pacific region around the dateline, i.e. 0°-10°S, 160°E-160°W. The gray areas represent the 95% confidence level based on a Monte Carlo test. (j-m) Long term monthly mean winds at 850hPa from November to February.

The remote response, based on the Gill-Matsuno theory, is most clearly seen in the simulated streamfunction anomaly fields (figure 3d-e). A horizontal quadrupole vortex structure, consistent with [13], appears at the upper levels of the atmosphere with negative anomaly over Australia. An opposing positive anomaly occurs at low levels over Australia, consistent with the anomalous cyclonic circulation and low sea level pressure anomaly. This indicates a simulated baroclinic response in the tropics during Modoki events. In addition to the return flow west of the heat forcing, the model simulates an anomalous Walker circulation with upward motion in the central-western Pacific in response to the warm equatorial SST (figure 3f). Mass conservation requires a compensating downward motion, which occurs over the Indian Ocean. Interestingly, Australia lies in between the ascending and descending Walker cell anomalies (figure 3f), which allows the Gill return flow to freely influence the rainfall over the continent.

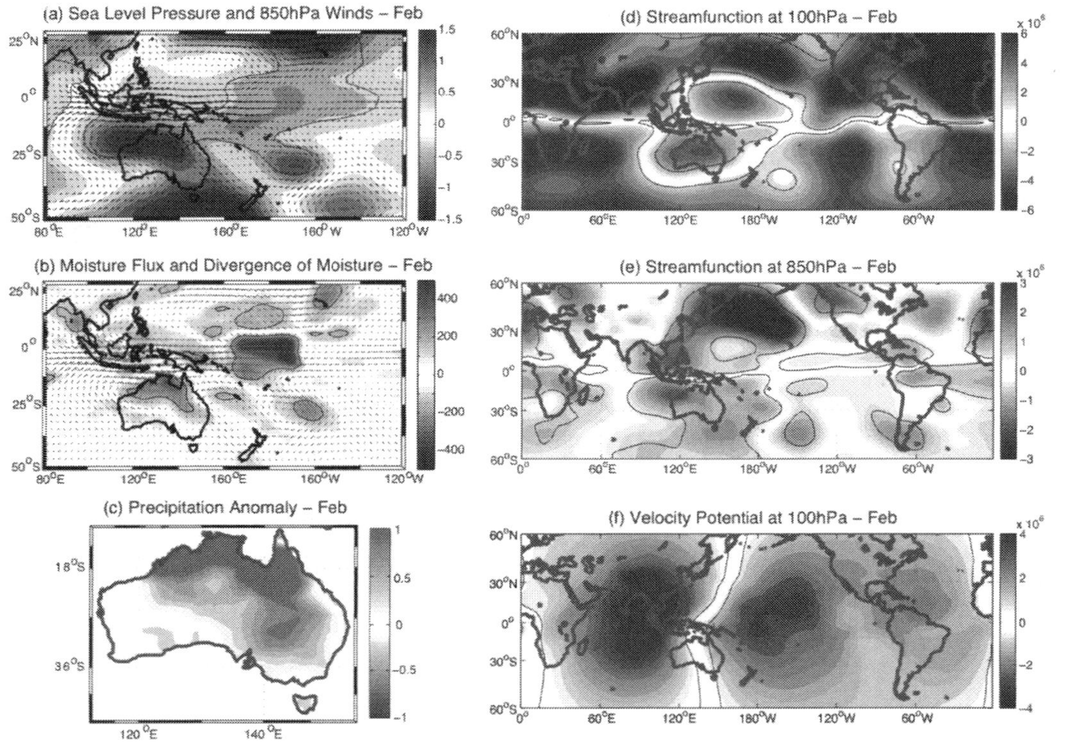

Figure 3. Simulated anomalies in February for the idealized experiment forced with 1°C SST warming in central-western equatorial Pacific (10°S-10°N, 160°E-160°W). (a) Sea level pressure (hPa) and wind anomalies at 850hPa (m s^{-1}). (b) Vertically integrated moisture flux (kg $m^{-1}s^{-1}$) anomaly from the surface to 500hPa and the associated divergence (kg s^{-1}). (c) Precipitation anomaly (mm day^{-1}). (d) Streamfunction anomaly at 100hPa (m^2s^{-1}). (e) Streamfunction anomaly at 850hPa (m^2s^{-1}). (f) Velocity potential anomaly at 100hPa (m^2s^{-1}). Adapted from [7].

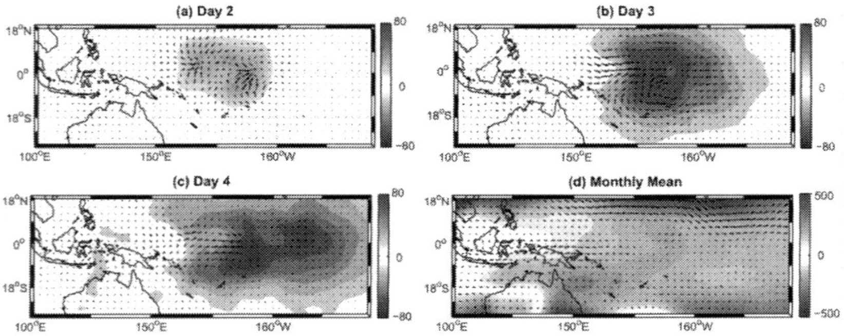

Figure 4. Evolution of the anomalies of sea level pressure and winds at 850hPa on (a) day 2, (b) day 3, (c) day 4, and (d) over 30-60 days average representing the February anomaly.

A preliminary analysis of the daily response to the simulated SST warming in the equatorial Pacific confirms the initialisation of the mechanism described here. Figure 4 shows the sea level pressure and low-level wind anomalies for the first three days of the simulation, after the Modoki-like SST perturbation is superimposed on the surface SST forcing on January 1st. The anomaly plotted in figure 4 is based on a single simulation and as such should be viewed with certain caution. However the response for the first few days is not expected to vary considerably given the predictability of the atmosphere to an initial SST forcing. The 30-60 days mean (representative of February) exhibits a similar solution as that presented in figure 3a based on the 50-member ensemble monthly mean.

4. Conclusions

Northern Australia generally experiences dry conditions during both canonical El Niño and El Niño Modoki events. This results from changes in the Walker circulation that drives anomalous subsidence, weakened convection and reduced rainfall over the continent. However, during El Niño Modoki events, when the location of the maximum SST anomaly is centered on the dateline, at the peak of the monsoon season, an increase in rainfall over northwestern Australia generally occurs. The teleconnection between the warm SST anomaly over the central-western Pacific and the intensified monsoonal rainfall over northwestern Australia can be explained by the Gill-Matsuno theory and the interaction between this interannual response and the climatological mean circulation state.

The short-lived rainfall increase only occurs in January-February phase-locked to the strengthening of the South Pacific Convergence Zone (SPCZ). The intensification of the SPCZ in the southern summer months is conducive to the development of increased deep convection and heavy rainfall in the central-west Pacific. During Modoki years, the warmer than normal SST in the central-west Pacific decreases the local pressure, causing anomalous wind convergence at low levels and a consequent rising motion. In addition, the warm SST anomaly also enhances the supply of latent heat flux to the moist mixing layer. This anomalous configuration during Modoki events, conspires to enhance the convection that occurs during the normal SPCZ intensification during austral summer. Enhanced vertical transport of moisture from the convective mixing layer to the mid troposphere drives intense deep convection over the southern latitudes of the central-western tropical Pacific. Moist convection and enhanced rainfall generates anomalous diabatic heating that eventually leads to disturbances in the high troposphere akin to a Gill-Matsuno-type response. The Gill-Matsuno response is associated with a remote cyclonic circulation over northwestern Australia. The anomalous circulation reinforces the mean cyclonic winds over the region during January and February, thus enhancing the monsoon. As a consequence, there is increased convergence of moisture and enhanced rainfall over the region.

References

[1] Ashok K, Behera S K, Rao S A, Weng H and Yamagata T 2007 *J. Geophys. Res.* **112** doi:10.1029/2006JC003798
[2] Kumar K K, Rajagopalan B, Hoerling M, Bates G and Cane M 2006 *Science* **314** 115–9
[3] Wang G and Hendon H H 2007 *J. Clim.* **20** 4211–26
[4] Weng H, Ashok K, Behera S K, Rao S A and Yamagata T 2007 *Clim. Dyn.* **29** 113–29
[5] Taschetto A S and England M H 2009 *J. Clim.* **22** 3167–74
[6] Taschetto A S, Ummenhofer C C, Sen Gupta A and England M H 2009 *Geophys. Res. Lett.* **36** L12704, doi:10.1029/2009GL038416
[7] Taschetto A S, Haarsma R D, Sen Gupta A, Ummenhofer C C, Hill K J and England M H 2010 *J. Clim., in press*
[8] Jones D and Weymouth G 1997 *Bureau of Meteorology Technical Report 70* 19 pp
[9] Xie P and Arkin P A 1997 *Bull. Amer. Meteorol. Soc.* **78** 2539–58
[10] Kalnay E et al 1996 *Bull. Am. Meteorol. Soc.* **77** 437–71
[11] Rayner N A, Parker D E, Horton E B, Folland C K, Alexander L V, Rowell D P, Kent E C and Kaplan, A 2003 *J. Geophys. Res.* **108** doi:10.1029/2002JD002670
[12] Liebmann B and Smith C A 1996 *Bull. Am. Meteorol. Soc.* **77** 1275–7
[13] Gill A E 1980 *Q. J. R. Meteorol. Soc.* **106** 447–62
[14] Matsuno T 1966 *J. Meteorol. Soc. Japan* **44** 25–43

IOP Conf. Series: Earth and Environmental Science 11 (2010) 012032 doi:10.1088/1755-1315/11/1/012032

Uncertainties in Bathymetric Retrievals

Barbara A Burns, John R Taylor and Harvinder Sidhu

School of Physical, Environmental and Mathematical Sciences, UNSW@ADFA, Canberra ACT Australia

E-mail: b.burns@adfa.edu.au

Abstract. The sensitivity of the remotely-sensed water reflectance to variations in water depth and bottom albedo is investigated using a semianalytical model and the radiative transfer numerical model HYDROLIGHT. Properties of the water column are taken from published observations made on 14 Feb 2001 in the coastal waters of Moreton Bay (southeast Queensland). The sensitivities of reflectance to depth and albedo are computed using partial derivatives derived from the semianalytical model, and with finite differences using the numerical model. These are compared for various depths and bottom conditions, and also to the instrumental resolution of the Hyperion satellite remote sensing system. Sensitivity to albedo is found to be greater than sensitivity to depth under some situations making the depth signal in reflectance ambiguous. These sensitivities are then applied to bathymetry algorithms to produce error bars for depth retrievals under variable bottom albedo conditions. This is done for both the standard algorithm that uses reflectances at 550 and 650 nm corrected for optically deep water reflectance, and a blue/green ratio algorithm intended for use over low albedo areas. In the standard algorithm variations of 10% in albedo produce errors in depth estimates of approximately 0.05m over sandy bottoms which is small compared with the 2m depth resolution required to reach the bottom in the Moreton Bay experiment. The blue/green ratio algorithm showed no skill for bathymetric mapping with this dataset.

1. Introduction

The retrieval of water depth (bathymetry) is complicated by the sensitivity of surface reflectance to the spectral shape and spatial variability of bottom albedo and the turbidity of the water as well as to depth. Bathymetry algorithms address this problem through a range of methods, from simple ratioing of two channels of multispectral satellite imagery [1], to spectral library matching [2], to optimal inversion of analytic models [3,4] sometimes coupled to a numerical radiative transfer model [5]. Based on model-data comparisons, [6] has shown that the precision of bathymetric retrievals is a function of the contribution of the bottom reflectance (dependent on composition) to the remote sensing signal. In this paper we investigate this sensitivity to bottom albedo from first principles and use a semianalytical model to provide error bars on two simple depth retrieval algorithms.

2. Models and Dataset

Two models have been used in this study. The first is the semianalytical model of Lee et al. [3] that relates remote sensing reflectance to water depth and bottom albedo with vertically-averaged attenuation coefficients. This model has been used as the basis for CSIRO's Semi-Analytical Model for Bathymetry, Un-mixing, and Concentration Assessment (SAMBUCA) [5]. It is based on the following analytical expression from [7] for the irradiance reflectance just "below the surface of a homogeneous ocean bounded below by a reflecting bottom at depth *H*":

$$R(0,H) = R_\infty - R_\infty \exp(-(K_d + \kappa_C)H) + A\exp(-(K_d + \kappa_B)H) \quad (1)$$

where

R_∞ = reflectance from optically deep water
A = albedo of bottom assuming a Lambertian reflector
K_d = vertically averaged diffuse attenuation coefficients for downwelling irradiance
κ_C , κ_B = same as K_d but for upwelling irradiance from the water column and reflected from the bottom

Lee et al. then derive statistically averaged coefficients by calculating reflectances with a numerical model for a wide range of albedo and attenuation values (see [3], Appendix A). The same numerical model, HYDROLIGHT, is used in this study. It is a radiative transfer model that computes the water-leaving radiance as well as the spectral radiance distribution within the water as a function of depth, direction and wavelength [8]. The Version 4.3 software allows input of variable absorbing and scattering properties of the water column and bottom albedo, as well as time and geographic location to calculate solar irradiance representative of *in situ* data collection events. In addition to reflectance, the HYDROLIGHT output also provides the radiative transfer terms needed to convert subsurface reflectance to above-surface remote-sensing reflectance (*Rrs*) according to the equations in [9]. The "semianalytical model" used in this study is therefore the model detailed in [3] followed by this conversion to *Rrs*.

Running both these models requires values for absorption and backscattering coefficients of the water column. *In situ* observations (Moreton Bay, Queensland) and parameterizations of water quality presented in [4] provide those data inputs. Table 1 gives albedo and R_∞ values used in this study.

Table 1. Example values of albedo (A) and deep water reflectance (R_∞) used in this study. Albedos are from HYDROLIGHT 4.3 [8]; R_∞ is calculated using Moreton Bay data [4].

Wavelength (nm)	A(Coral Sand)	A(Brown Algae)	R_∞ (Min Concentration)	R_∞ (Max Concentration)
550	0.456	0.058	0.035	0.063
650	0.564	0.075	0.012	0.060

3. Reflectance Sensitivity Model Results

The sensitivity of remote-sensing reflectance to changes in depth forms the basis of all bathymetric algorithms. However, as indicated in equation 1, bottom albedo must also be accounted for if reflectance is to be inverted to obtain depth estimates. To explore the relative sensitivities of reflectance to depth and albedo, we take the partial differentials of equation 1 with respect to bottom albedo A and depth H,

$$\partial R(0,H)/\partial A = \exp(-(K_d + \kappa_B)H) \quad (2)$$

$$\partial R(0,H)/\partial H = (K_d + \kappa_C)R_\infty \exp(-(K_d + \kappa_C)H) - (K_d + \kappa_B)A\exp(-(K_d + \kappa_B)H) \quad (3)$$

We see that both the sensitivity to albedo and depth decrease with increasing depth. While the former is a simple function of depth (and independent of the magnitude of A itself), the sensitivity to depth is more complicated, being a function of deep water reflectance and bottom albedo as well as depth. Using the expressions in [3] and the data from Moreton Bay [4], we find very little difference in the magnitudes of κ_C and κ_B. Therefore the sign and magnitude of the sensitivity to changes in depth (equation 3) will depend mostly on the difference between R_∞ and A. From the values in table 1 we can expect sensitivity to depth to be large over sand bottoms ($A >> R_\infty$), but small over bottoms such as brown algae ($A \approx R_\infty$). Also sensitivity goes down with degrading water quality (larger R_∞).

Partial derivatives were also derived for the semianalytical model. In addition HYDROLIGHT was used to compute finite differences in reflectance due to small changes in depth. Runs were made with nominal H values of 1, 2, 3, 4, 5, and 8 m, and then repeated increasing H by 0.1 m. A Coral Sand bottom albedo was assumed throughout. The differences then represent the change in remote-sensing reflectance for a 0.1m change in depth. Comparison of the semianalytic and computational methods is shown in figure 1 where all *delta Rrs* are shown as positive. The good agreement with HYDROLIGHT indicates the robustness of the semianalytic model since the parameters were tuned with HYDROLIGHT outputs from a completely different dataset [3]. As a further check on how we were running the models, we repeated the HYDROLIGHT simulation for Moreton Bay carried out in [4]. Using a sand bottom albedo and 8m depth as they did, we found good agreement with their figure 5 for minimum concentration of water constituents. The optical properties for this case are used in all the calculations presented here. Also shown in figure 1 is the *NeDR* (noise-equivalent difference in reflectance) for the Hyperion hyperspectral imagery analyzed in [4], which provides a threshold value below which any *delta Rrs* will not be measureable due to system noise. The results indicate that changes of just 0.1m in depth are detectable in these coastal waters up to 5m deep, for some wavelengths and for the minimum concentrations of water constituents measured. Obviously larger changes in depth (i.e. degraded depth resolution) would be visible in even deeper waters.

Depth sensitivity calculations in figure 1 assume a uniform sand bottom with constant albedo, while equation 2 indicates that the reflectance will also change in response to variations in bottom albedo. Using results from the semianalytic model only, figure 2 shows the effect different bottom types has on reflectance sensitivity to depth. It also shows the change in remote sensing reflectance due to a 10% change of bottom albedo (equivalent to adding 20% brown algae to a sand footprint) which can be greater than the sensitivity to depth, even over sand, depending on wavelength. Both increased water depth and absorption (not shown) act to decrease the sensitivity to albedo changes.

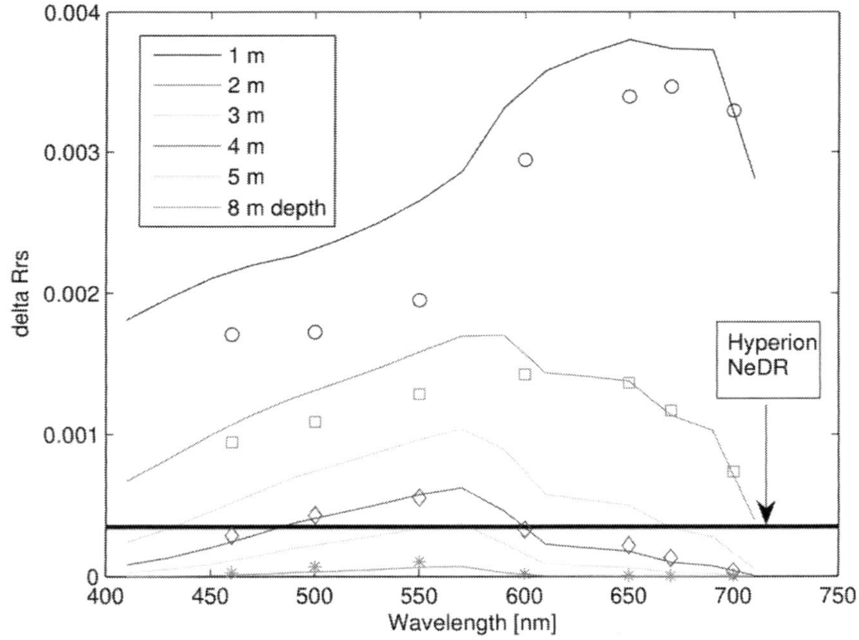

Figure 1. Semianalytical (symbols) and HYDROLIGHT (lines) model calculations of sensitivity to a 0.1m change in depth. Symbols are at calculation wavelengths.

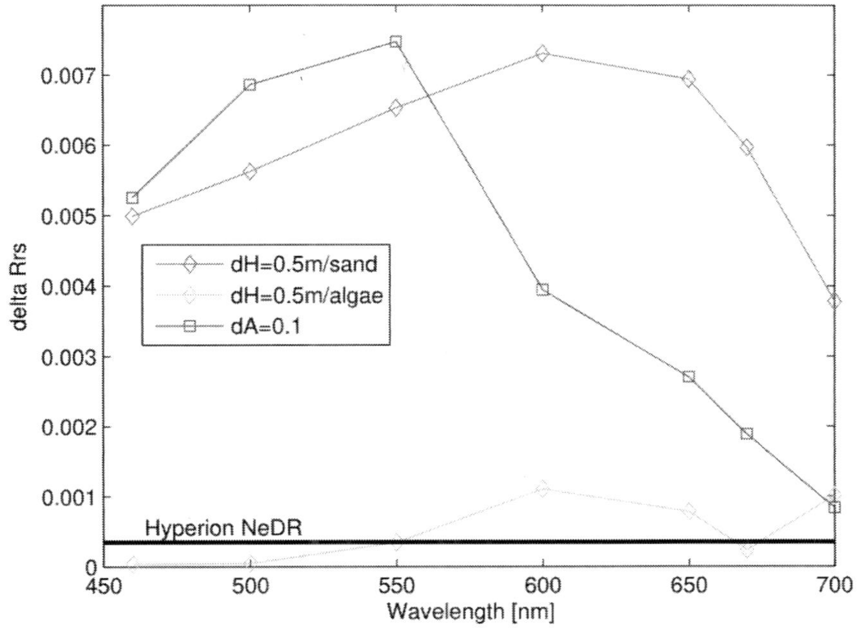

Figure 2. Red and green curves show change in reflectance due to change in depth in 2 m water for two extremes of bottom albedo; blue curve shows that sensitivity to just 10% change in albedo can be greater than sensitivity to 0.5m change in depth.

4. Impact on Bathymetry Algorithms

The curves in figure 2 illustrated how albedo variations can make the depth signal ambiguous. We now consider how this albedo sensitivity translates into errors in depth retrievals using standard bathymetry algorithms.

4.1. Definition of ratio algorithms

The basic algorithm to estimate depth from remote-sensing reflectance data is given by

$$H = \ln(R_b - Rinf) / K - \ln(Rrs - Rinf) / K \quad (4)$$

where the effective attenuation coefficient K, bottom reflectance R_b, and the deep water reflectance *Rinf* incorporate factors for transmission across the air-sea interface, atmospheric effects, and illumination (see [10]). Equation 4 is generally applied to the ratio of remote-sensing reflectance in two spectral bands to reduce the effects of measurement error and other uncertainties; hence the name ratio algorithm. H is now a linear function of $\ln(Rrs - Rinf)$ with the slope inversely proportional to the attenuation coefficient and the y-intercept determined by the bottom reflectance (i.e. albedo). A key step in obtaining a linear relationship for H is to subtract *Rinf* from the observed reflectance. Figure 3 illustrates this with the Moreton Bay data: for a given albedo depths are well separated along a straight line. The deep water correction, however, cannot be applied to all datasets since depth is undefined if *Rinf* is greater than the measured reflectance, as can occur if the albedo is low (algae or sea grass). An algorithm for this case has been developed [1] based on the ratio of the logs of measured blue and green reflectances. The performance of these two algorithms on the Moreton Bay data will be compared below.

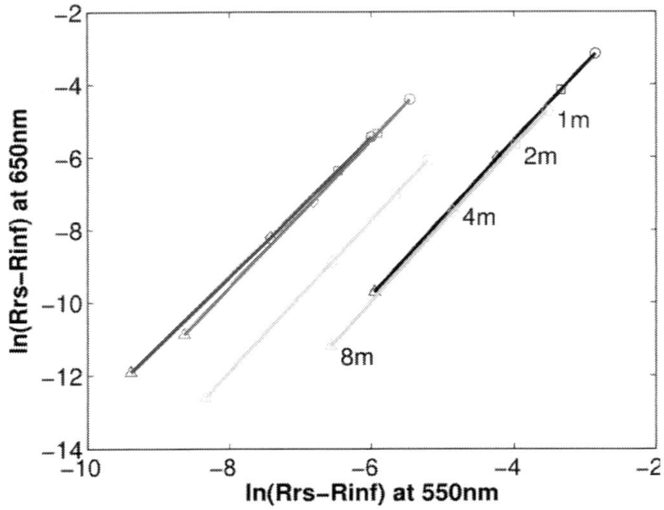

Figure 3. Correcting observations for deep water reflectance leads to linear relationship; also makes depth and albedo variation orthogonal. Key: blue = brown algae; red = red algae; cyan = sea grass; black = sand; green = green algae. Different symbols indicate depths of 1, 2, 4 and 8 m.

4.2. Effect of albedo sensitivity on depth retrieval

To examine how the *delta Rrs* due to albedo variation (figure 2) translates into an error bar on depth retrieval, we use the semianalytical model to calculate the remote-sensing reflectance at the three wavelengths used in the ratio algorithms described above for a mostly sand bottom (figure 4). The reflectances are then combined to form ratios as per the algorithm descriptions and plotted versus depth in figure 5. The line labelled "G/R algorithm" shows the linear algorithm ln($[Rrs - Rinf]_{550}$ / $[Rrs - Rinf]_{650}$), while the line "B/G algorithm" shows the uncorrected ratio ln(Rrs_{460})/ln(Rrs_{550}).

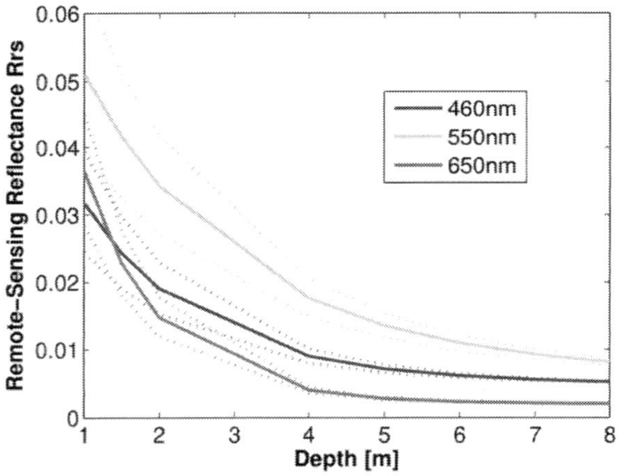

Figure 4. Solid line is for mixed albedo bottom; dotted lines indicate the +/-10% albedo variation.

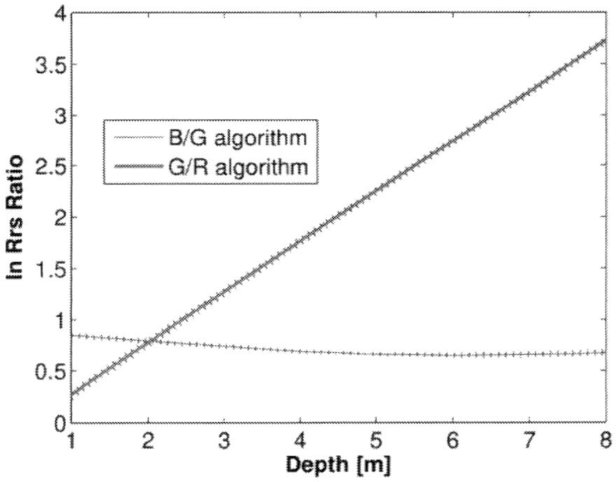

Figure 5. Depth retrieval plot indicates no skill for the uncorrected ratio algorithm (B/G).

The process of ratioing and taking the logarithms greatly reduces the effect of albedo sensitivity. For example, using 550nm alone results in a depth error of at least +/-0.5m at 2m depth (figure 4); the error with the G/R algorithm is approximately a tenth of that (figure 5). Table 2 presents depth errors for this algorithm. The errors are larger for dark bottoms (algae), but on average still approximately half that with the green channel alone (not shown). For both bottom types the B/G algorithm is nearly a horizontal line indicating almost no skill for depth retrieval. However this coastal dataset is very different from the tropical coral reef environment for which the algorithm was developed [1]. The errors presented in table 2 are meaningful only in the context of minimum detectability of the sensor and required depth resolution of the bathymetric mapping. For the Hyperion sensor deployed in the Moreton Bay experiment [4], a resolution of 0.1m is attained down to a depth of 3.5m for the most optimum conditions reported (minimum concentrations, sand bottom). This corresponds to the depth where the *delta Rrs* at 550 and 650 nm are still above *NeDR* in figure 1. Reaching the typical bottom depth of 8m would require degrading the resolution (the delta H) to 2m or more to increase the change-in-depth signal. The errors shown in table 2 then represent only 2 to 3% of the 2m resolution.

Table 2. Depth Errors (in m) for the G/R algorithm due to 10% (+/-) albedo variation

Bottom Type	Depth = 2m	Depth = 4m	Depth = 8m
Mostly Sand	0.0590 (+)	0.0503 (+)	0.0417 (+)
	0.0767 (-)	0.0699 (-)	0.0644 (-)
Mostly Brown Algae	0.2477 (+)	0.2552 (+)	0.2805 (+)
	1.4765 (-)	1.6515 (-)	2.2865 (-)

5. Conclusions

We have used a semianalytical reflectance model to derive (1) sensitivities to depth and bottom albedo, and (2) error bars on depth retrieval due to the sensitivity to albedo. We find that the use of a ratio algorithm (compared to using a single channel alone) reduces the effect of albedo sensitivity with resultant depth errors well below the detection limit of typical remote sensing systems. Such calculations can be a valuable companion to bathymetric observations to determine useful depth resolutions when variations due to tides, water quality, as well as changing bottom type are present. Future work should also consider the spatial as well as the radiometric resolution of the sensor.

Acknowledgments

The authors wish to acknowledge Roy Hughes of the Defence Science and Technology Organisation (DSTO) Maritime Operations Division for supplying the HYDROLIGHT software package.

References

[1] Stumpf R P, Holderied K and Sinclair M 2003 Determination of water depth with high-resolution satellite imagery over variable bottom types *Limnol. Oceanogr.* **48** 547-556

[2] Mobley C D et al 2005 Interpretation of hyperspectral remote-sensing imagery by spectrum matching and look-up tables *Appl. Optics* **44** 3576-3592

[3] Lee Z, Carder K L, Mobley C D, Steward R G and Patch J S 1999 Hyperspectral remote sensing for shallow waters: 2. Deriving bottom depths and water properties by optimization *Appl. Opt.* **38** 3831-3843

[4] Brando V E and Dekker A G 2003 Satellite hyperspectral remote sensing for estimating estuarine and coastal water quality *IEEE Trans. Geosci. Remote Sensing* **41** 1378-1387

[5] Wettle M and Brando V E 2006 SAMBUCA *CSIRO Land and Water Science Report* 22/06

[6] Brando V E, Anstee J M, Wettle M, Dekker A G, Phinn S R, and Roelfsema C 2009 A physics based retrieval and quality assessment of bathymetry from suboptimal hyperspectral data *Rem. Sens. Environm.* **113** 755-770

[7] Maritorena S, Morel A and Gentili B 1994 Diffuse reflectance of oceanic shallow waters: Influence of water depth and bottom albedo *Limnol. Oceanogr.* **39** 1689-1703

[8] Mobely C D and Sundman L K 2006 *Hydrolight 4.3 Users' Guide* (Bellevue, WA: Sequoia Scientific)

[9] Lee Z, Carder K L, Mobley C D, Steward R G and Patch J S 1998 Hyperspectral remote sensing for shallow waters: 1. A semianalytical model *Appl. Opt.* **37** 6329-6338

[10] Philpot W D 1989 Bathymetric mapping with passive mulispectral imagery *Applied Optics* **28** 1569-1578

Estimating 21st century changes in extreme sea levels around Western Australia

I D Haigh[1] **and C Pattiaratchi**
The Oceans Institute, University of Western Australia,
M470, 35 Stirling Highway, Crawley, WA 6009, Australia

E-mail: haigh@sese.uwa.edu.au pattiara@sese.uwa.edu.au

Abstract. Extreme sea levels are likely to increase in the future with an expected accelerated rise in mean sea level and through possible changes in storminess. Society is becoming more vulnerable to extreme sea levels due to considerable growth in human populations and economy at the coastal zone and this is particularly true for Western Australia, the fastest growing Australian state or region. This paper describes a novel approach used to estimate future changes in extreme sea level around the southwest coastline of Western Australia. Probabilities of extreme sea level for the present climate have been estimated using a 60 year hindcast of sea levels. The impact of climate change has been explored by adding a range of mean sea level rise projections to these probabilities. Estimates of possible future changes in recurrence intervals every decade over the 21st century are presented, showing that climate change has the potential to significantly reduce current average recurrence intervals and that the amount of reduction varies significantly around the coastline.

1. Introduction

Globally, mean sea levels (MSL) are rising [1] and the rate of rise is predicted to accelerate over the 21st century [2]. This will lead, along with possible changes in storminess (i.e. more intense cyclones [3]), to more extreme sea levels and hence an increased likelihood of coastal flooding and erosion. Society is becoming more vulnerable to extreme flooding events because of the growing human populations and economy in the coastal zone [4]. This is especially true for Western Australia (WA), the fastest growing Australian state or region for the last two years [5]. Some of the areas experiencing the largest population growth in the state are in flood prone areas (e.g. Rockingham, Mandurah, Bunbury and Busselton) and it has been estimated that between about 19,000 and 29,000 houses will be at risk of inundation in WA from a sea-level rise of 1.1 m, if flood defence measures are not upgraded. Hence there is an urgent need to provide accurate projections of possible future changes in extreme sea levels around WA to aid current coastal management and for future planning.

A simple estimate of future changes in extreme sea levels can be obtained by just adding projections of MSL to current recurrence intervals, estimated from existing sea level measurements [6] (we term this the 'MSL offset method'). However, there are problems with calculating current recurrence intervals from existing tide-gauge measurements. Firstly, the observational records vary in length at each site and have different periods of missing data. This can bias the calculation of current recurrence intervals. Secondly, the sea level observations are only available at discrete locations

[1] To whom any correspondence should be addressed.

around the coastline. The WA coastline is long (approximately 20,000 km) and characteristics of the astronomical tide and storm surges vary significantly around it. Therefore, a simple interpolation of current recurrence intervals, between the existing network of 28 tide-gauges in the state, is not likely to capture accurately the spatial variation in sea levels around the coastline.

An alternative method, first applied in 1998 [7], is to calculate recurrence intervals using multi-decadal hindcasts of sea level, created by driving hydrodynamic models, validated against observations, with meteorological fields. This has the advantage that maps of recurrence intervals can be estimated across the whole model domain. This method has recently been applied to estimate the frequency of sea level extremes and the impact of climate change in southeast Australia [8].

The aims of this paper are to: (1) briefly describe a 60 year hindcast of sea levels that has been created for the coastline of WA; and (2) use the results of this modelling exercise to provide an initial estimate of possible future changes in extreme sea level around the coastline of southwest Australia.

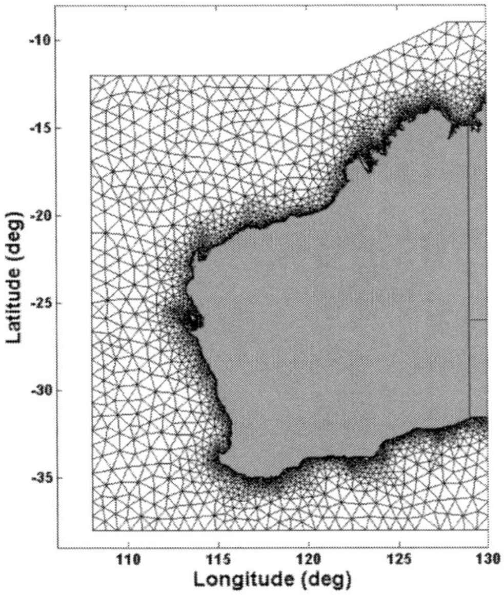

Figure 1. Model grid

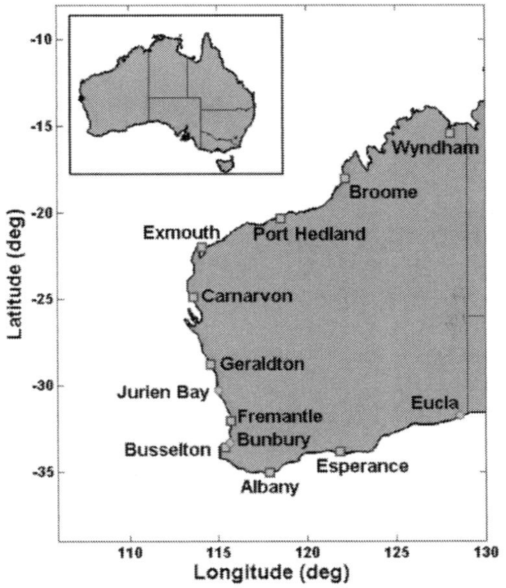

Figure 2. Tide-gauge locations (blue squares)

2. A 60 year hindcast of sea levels

Hydrodynamic numerical models have been used since the 1960's to improve understanding of sea level characteristics. Until recently, most of the past studies simulated sea levels over a particular storm event or for a few tidal cycles. It is only over the last decade that hydrodynamic models have been used to simulate multi-decadal hindcasts of sea level [9].

In this study, a depth-average tide-surge model has been created for the entire WA coastline using the Danish Hydraulic Institute's Mike21 FM (flexible mesh) suite of modelling tools. The Mike21 FM modelling system is based on the numerical solution of the incompressible Reynolds averaged Navier-Stokes equations invoking the assumptions of Boussinesq and hydostatic pressure [10]. The spatial discretization of the primitive equations, over the flexible mesh, is performed using a cell-centred finite volume method.

The model mesh, configured for WA, is shown in figure 1. It has a resolution of about 60 km at the open boundaries, increasing to 10 km along the entire coastline of WA. The bathymetric data, interpolated onto the model grid, was obtained from Geoscience Australia. In order to generate the astronomical tidal component of sea levels, the open model boundaries were driven with sea levels derived from 10 tidal constituents (M_2, S_2, N_2, K_2, K_1, O_1, P_1, Q_1, M_f, M_m), obtained from the Oregon State University Tidal Inversion Software [11]. To generate the surge component of sea levels, the model has been forced with atmospheric pressures and u and v components of wind, obtained from the US National Center for Environmental Prediction's (NCEP) global reanalysis [12]. These meteorological fields were available every 6 hours from 1949 to present and have a horizontal resolution of 2.5°.

The model has been validated against observations from 10 tide-gauges (Esperance, Albany, Busselton, Fremantle, Geraldton, Carnarvon, Exmouth, Port Hedland, Broome and Wyndham), the locations of which are shown in figure 2. The tide-gauge records were obtained from the WA Department of Transport. Validation was undertaken for the year 2003, as this was one of the stormiest years on record.

Two separate model simulations were undertaken. The model was first driven only with tidal forcing at the open boundary. In the second simulation the model was also forced with winds and surface air pressures. The first simulation provided only the astronomical tidal component of sea levels. The second simulation predicted the total sea level (i.e. tide + surge). By subtracting the second simulation from the first, the storm surge component of sea levels was obtained. The predicted total, tidal and surge levels were compared with the observed record and the derived tidal and storm surge components of the observed record (which were calculated by performing harmonic tidal analysis on the measured record at each site).

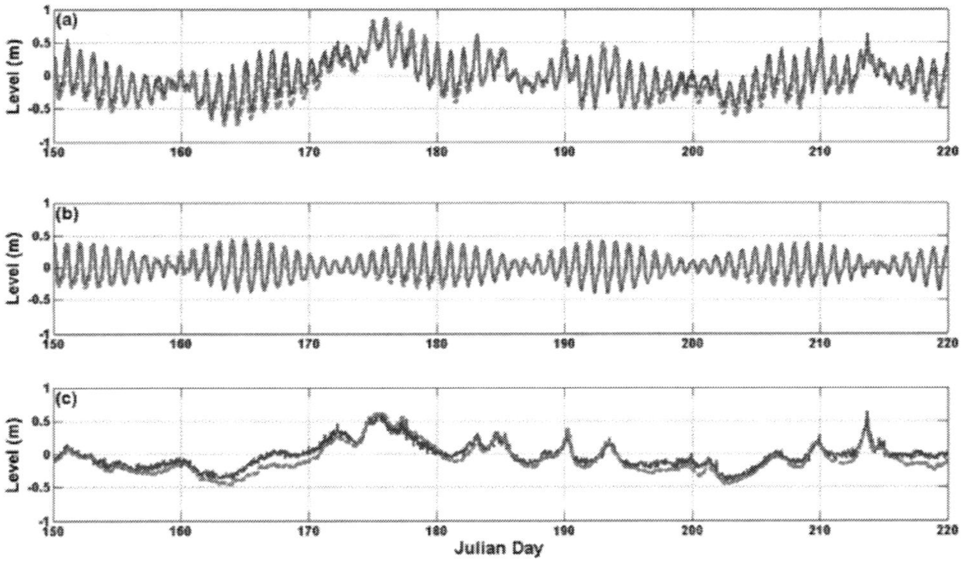

Figure 3. A comparison of the observed (solid blue) and model (dotted red) (a) total sea levels; (b) astronomical tidal levels; and (c) surge levels, for Fremantle during June and July 2003.

At each of the 10 sites, there is good agreement between the observed and modelled astronomical tidal component (root mean square errors are typically within 10 cm). However, the total sea levels and storm surge component of the sea levels only show good agreement (again root mean square errors are typically within 10 cm) for the southwest region of WA (Geraldton southwards). The poor agreement along the northwest part of the coastline is due the low spatial and temporal resolution of the NCEP air pressure and wind fields, which are not adequate to simulate the storm surge generated by tropical cyclones. A comparison of the measured and modelled total sea levels, tidal and surge levels is shown in figure 3 for Fremantle. Root mean square errors of the observed and modelled total, tidal and surge components of sea level are less than 5 cm at this site.

3. 21st century changes in extreme sea levels

Following the validation stage, the model was run for the 60 year period from 1949 to 2008. Results were saved every hour for each model grid cell. The 60 year simulated time-series of total sea levels were extracted for each model grid point around the coastline of southwest Australia (Eucla to Geraldton) and were then used to calculate current recurrence intervals. (At this stage in the study, we limit the analysis to the southwest region, as this is where there was good agreement between the model and observational dataset. Future work will address extending this analysis along the northwest shelf by using higher resolution wind and pressure fields, capable of more accurately predicting surges associated with cyclones in these regions.) The recurrence intervals were estimated by fitting an extreme value Gumbel distribution to the annual maximum sea level at each point [13]. The MSL offset method was then used with sea-level rise projections from the Intergovernmental Panel on Climate Change's (IPCC) Fourth Assessment Report (AR4) to provide an estimate of possible future changes in extreme sea level every decade from 1990 to 2100 around the southwest coast of WA.

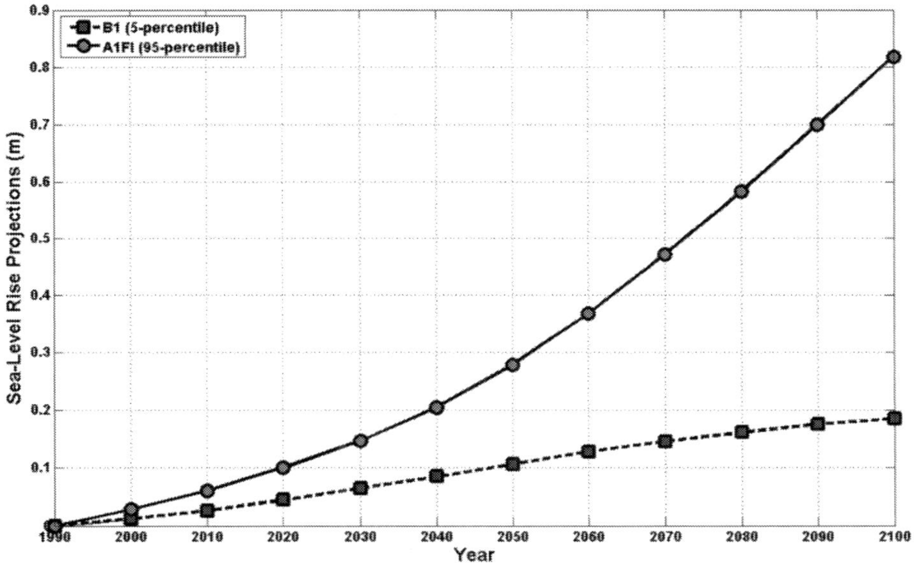

Figure 4. Lower and upper sea-level rise projections from the IPCC AR4. The lower projection corresponds to the 5th percentile of the low emissions scenario (B1) and the upper projection corresponds to the 95th percentile of the high emissions scenario (A1FI).

The IPCC's AR4 provided global projections of MSL rise at 2095 relative to 1990 for six SRES (Special Report on Emissions Scenarios) [2]. The sea-level rise projected by the models was between 18 and 79 cm for this period. The AR4 only provided projections of sea-level rise for 2095, unlike the IPCC's Third Assessment Report (TAR) which provided projections throughout the 21st century. Hunter [14] fitted the time series from the TAR to the AR4 projections to calculate sea-level rise projections at decadal resolutions for the different scenarios. We consider the 5th percentile of the B1 (low emission) and 95th percentile of the A1FI (high emission) sea-level rise scenarios (figure 4), to sample the range of change. The low scenario is approximately consistent with a continuation of the 17 cm of global change observed by tide-gauges over the 20th century [1], while the upper scenario represents a significant acceleration on 20th century rates of rise.

At each model grid point around the coastline of southwest Australia, and for each 10 year period from 1990 to 2100, the estimated current recurrence intervals have been increased by the lower and upper sea level projection for that specific decade. A cubic interpolation was then used, for each

10 year period, to find the recurrence interval associated with the level in 1990 that had a 100 year recurrence interval.

Results for the lower projection are shown in figure 5. Between Eucla and Albany, the level associated with a 100 year event in 1900 is predicted to occur about every 2 to 3 years by 2100, about a 50 fold increase in exceedence frequency. Between Albany and Bunbury, on the southwest corner of WA, the change in the recurrence interval becomes less dramatic. The level in 1990 associated with a 100 year event is predicted to occur about every 10 years by 2100, a 10 fold increase in exceedence frequency. Between Jurien Bay and Geraldton, there is a 100 fold increase in exceedence frequency. These results illustrate the dramatic changes in recurrence intervals of extreme high sea levels that can occur due simply to a small change in MSL and highlight the considerable spatial variation in the change around the coastline.

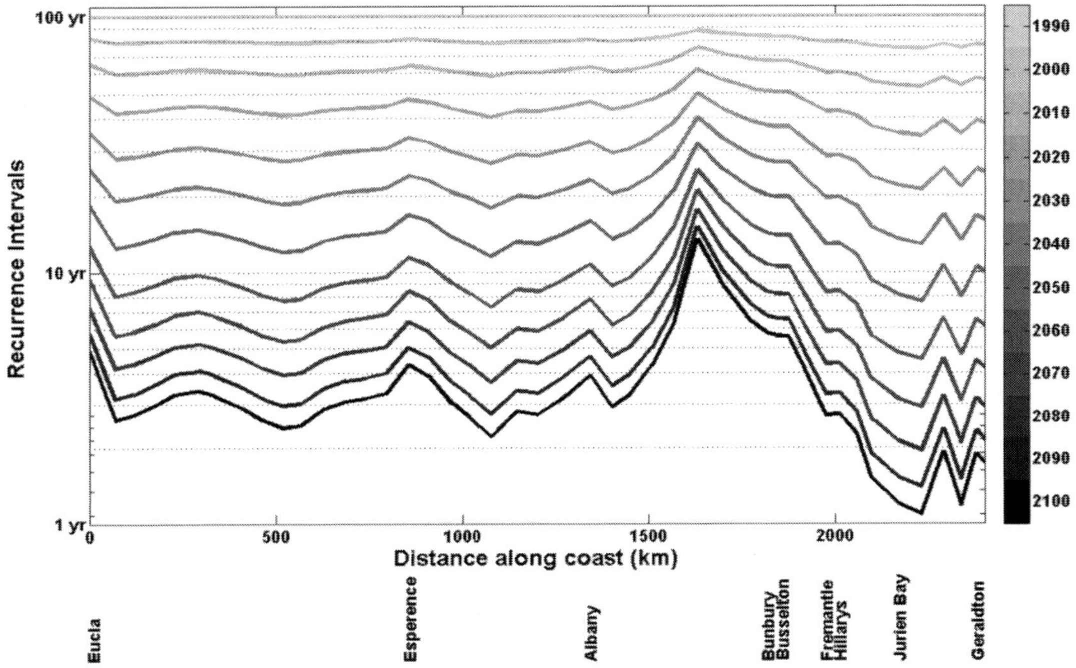

Figure 5. Recurrence intervals (years), every 10 years from 1990 to 2100, associated with a return level with a 100 year return period in 1990, for the lower sea-level rise projections.

For the upper sea-level rise projection (results not shown), the pattern of change around the coastline is similar to the lower projection, but the increase in exceedence frequency is significantly more pronounced. Between Eucla and Albany, the level associated with a 100 year event in 1990 is predicted to occur about every 2 to 3 years by 2050 (i.e. 50 years earlier than the lower projection). By 2100, the level associated with the 100 year event in 1990, is likely to occur at least once every spring/neap tidal cycle by 2100 along almost the entire southwest stretch of WA coastline (i.e. more than a 2,500 fold increase in exceedence frequency).

4. Conclusions

This paper has presented a 60 year hindcast of sea levels around Western Australia. This modelled sea level dataset has been used to map current recurrence intervals around the coastline of southwest Australia. Future changes in extreme sea levels throughout the 21st century have been estimated by increasing the current recurrence intervals by a low (18 cm) and upper (79 cm) sea-level rise projection. Results have shown that the predicted rise in sea level over the 21st century has the potential to significantly reduce current average recurrence intervals around southwest Australia and that the increase in exceedence frequency varies significantly around the coast. These estimates could increase further with enhanced storminess and this will be investigated in the next stage of the study.

Acknowledgments
We would like to thank the Department of Transport for providing the tide-gauge data. This study is being funded through the Western Australian Marine Science Institution.

References

[1] Bindoff N L *et al.* 2007 Observations: Oceanic climate change and sea level *In Climate Change 2007: The physical science basis. Contribution of working group I to the Fourth Assessment Report of the Intergovernmental Panel on Climate Change* ed S Solomon *et al.* (Cambridge University Press) chapter 5 pp 385-432
[2] Meehl G A *et al.* 2007 Global climate projections *In Climate Change 2007: The physical science basis. Contribution of working group I to the Fourth Assessment Report of the Intergovernmental Panel on Climate Change* ed S Solomon *et al.* (Cambridge University Press) chapter 10 pp 747-845
[3] Bender M A, Knutson T R, Tuleya R E, Sirutis J J, Vecchi G A, Garner S T and Held I M 2010 Modeled Impact of Anthropogenic Warming on the Frequency of Intense Atlantic Hurricanes *Science* **327** 454-58
[4] Nicholls R J, Wong P P, Burkett V R, Codignotto J O, Hay J E, McLean R F, Ragoonaden S, Woodroffe C D, 2007 Coastal systems and low-lying areas *In Climate change 2007: Impacts, adaptation and vulnerability. Contribution of working group II to the Fourth Assessment Report of the Intergovernmental Panel on Climate Change.* ed M L Parry *et al.* (Cambridge University Press) chapter 6 pp 315-356
[5] Department of climate change 2009 *Climate change risks to Australia's coast: A first pass national assessment* p 168
[6] Araújo I and Pugh D T 2008 Sea levels at Newlyn 1915-2005: Analysis of trends for future flooding risks *J. of Coastal Res.* **24** (sp3) 203-12
[7] Flather RA, Smith J A, Richards J D, Bell C and Blackman D L 1998 Direct estimates of extreme storm surge elevations from a 40-year numerical model simulation and from observations *The Global Atmosphere and Ocean Systems* **6** 165-76
[8] McInnes K L, Macadam I, Hubbert G D and O'Grady J D 2009 A modelling approach for estimating the frequency of sea level extremes and the impact of climate change in southeast Australia *Natural Hazards* **51** 115-37
[9] von Storch H and Woth K 2008 Storm surges, perspectives and options *Sustainability Science* **3** 33-4
[10] Danish Hydraulic Institute 2010 *Mike21 and Mike 3 flow model FM: Hydrodynamic and Transport Module Scientific Documentation* p 50
[11] http://www.oce.orst.edu/research/po/research/tide/otis.html
[12] Kalnay et al. 1996 The NCEP/NCAR 40-year reanalysis project *Bull. Amer. Meteor. Soc.* **77** 437-70
[13] Coles S 2001 *An introduction to statistical modelling of extreme values* (Springer) p 207
[14] Hunter J 2009 Estimating sea-level extremes under conditions of uncertain sea-level rise *Climatic Change* DOI: 10.1007/s10584-009-9671-6

IOP Conf. Series: Earth and Environmental Science **11** (2010) 012034 doi:10.1088/1755-1315/11/1/012034

Dynamics of separating western boundary currents in ocean models

Andrew E Kiss

School of Physical, Environmental and Mathematical Sciences, University of New South Wales at the Australian Defence Force Academy, Canberra ACT, Australia.

E-mail: a.kiss@adfa.edu.au

Abstract. Western boundary current (WBC) separation dynamics are investigated in a two-layer quasigeostrophic model of a wind-driven subtropical ocean gyre. The results suggest that the vorticity in the thin cyclonic layer next to the coast plays a key role in the WBC separation mechanism. It is found that the outflow from the western boundary current is a region where this cyclonic vorticity diminishes rapidly downstream, through the combined action of Laplacian viscosity and eddy vorticity flux divergence. These results are consistent with earlier work on laminar, unstratified WBC dynamics, and extend those findings to the more realistic case of turbulent, stratified flow.

1. Background and objectives

Western boundary currents (WBCs) are major currents found on the western side of all midlatitude oceans (examples include the East Australian Current and Gulf Stream). These currents separate abruptly from the coast and flow eastward as intense, meandering jets. Although many theories have been proposed, the dynamics underlying WBC separation remain unclear [1, 2, 3]. Accurate western boundary current (WBC) separation is difficult to achieve in ocean models, and separation often occurs much further downstream than in reality. Experience has shown that a horizontal resolution of 10 km or finer is usually a necessary condition for realistic separation [4, 5], but this empirical result is in need of a physical explanation.

Observations show that subtropical WBCs have strongly cyclonic shear in a layer ~ 30 km wide on their coastal side in which the relative vorticity ζ may exceed half the planetary vorticity f at that latitude [6, 7, 8, 9, 10]; this strong relative vorticity is likely to be dynamically significant: it makes the absolute vorticity $f+\zeta$ equal the to planetary vorticity about 25° further poleward. Elimination of this cyclonic vorticity was shown to be a key factor in separation of WBCs in an earlier study based on a laminar, unstratified model [11], suggesting that resolution of this thin layer might explain the 10 km threshold for realistic WBC separation in models. This study seeks to extend that laminar, unstratified result to the more realistic situation of the mean flow of a turbulent, stratified WBC.

2. Quasigeostrophic model

This study used the Quasi-Geostrophic Coupled Model (Q-GCM [12]) in ocean-only mode with two isopycnal layers. The basin is square, 2048 km on each side and uniformly 4 km deep. The upper-layer flow is driven by a steady vertical Ekman pumping velocity w at the surface due to an imposed wind stress curl. w is zero in the northern half-basin and negative (downward) in the southern half; it varies sinusoidally with latitude in the southern half, having its greatest magnitude ($-4.6 \times 10^{-6}\,\mathrm{m\,s^{-1}}$) at the central latitude of the southern half-basin, and falling off to zero at mid-basin and the southern boundary. This Ekman pumping drives a single anticyclonic (subtropical) gyre in the southern half and leaves the northern half largely stagnant, allowing the subtropical WBC to be studied in isolation, without collision with a subpolar WBC. The flow in this model is determined by the evolution of quasigeostrophic potential vorticity (PV)

$$q = f + \zeta - f_0 \frac{h}{H} \tag{1}$$

in each layer, where $f = f_0 + \beta(y - y_0)$ is the Coriolis parameter (planetary vorticity, with $f = f_0 = +10^{-4}\,\mathrm{s^{-1}}$ at $y_0 = 1024\,\mathrm{km}$ (mid-basin) and linear dependence on northward coordinate y via the constant $\beta = 2 \times 10^{-11}\,\mathrm{s^{-1}\,m^{-1}}$; since $0 \le y \le 2048\,\mathrm{km}$, this places the entire basin in the northern hemisphere). $\zeta = \mathbf{k} \cdot \nabla \times \mathbf{u}$ is the relative vorticity (where $\mathbf{u}$ is the horizontal velocity), H is the mean thickness of the layer and h is the variation in layer thickness due to vertical excursions of the interfaces between layers.

The time-mean potential vorticity equation for the top layer is

$$\underbrace{\bar{\mathbf{u}} \cdot \nabla \bar{q}}_{\substack{\text{downstream} \\ \text{change in } \bar{q}}} = \underbrace{\frac{f_0 w}{H}}_{\text{wind}} + \underbrace{A_2 \nabla^2 \bar{\zeta}}_{\substack{\text{Laplacian} \\ \text{viscosity}}} \underbrace{\underbrace{- \nabla \cdot \overline{(\mathbf{u}'\zeta')}}_{\substack{\text{eddy vorticity} \\ \text{flux divergence}}} \underbrace{+ \frac{f_0}{H} \nabla \cdot \overline{(\mathbf{u}'h')}}_{\substack{\text{eddy thickness} \\ \text{flux divergence}}}}_{\text{eddy PV flux divergence, } -\nabla \cdot \overline{(\mathbf{u}'q')}} \tag{2}$$

where A_2 is the kinematic Laplacian viscosity and all variables have been decomposed into time-averages (overbars) and anomalies (dashes). Numerous runs were undertaken, covering a wide variety of parameters. The subset reported here used two layers, of thickness H=1000 m (upper) and 3000 m (lower), with a density difference chosen to give a Rossby radius of 47.4 km. Results with Laplacian viscosity $A_2 = 40$, 100 or 400 $\mathrm{m^2\,s^{-1}}$ are reported here; this value controls the level of turbulence and also sets the WBC width via the Munk scale [13], which for these parameters is $\delta_M \equiv (A_2/\beta)^{1/3} = 12.6$, 17.1 or 27.1 km, respectively. This scale was well-resolved by using a uniform horizontal grid spacing of 2 km for $\delta_M = 12.6\,\mathrm{km}$ and 4 km for the other cases. No-slip lateral boundary conditions were used for the runs shown here.

3. WBC separation mechanism

As discussed by [11], a no-slip western boundary creates a WBC velocity profile in which the peak velocity is located some distance offshore, with cyclonic shear ($\bar{\zeta} > 0$) on the onshore side and anticyclonic shear ($\bar{\zeta} < 0$) on the offshore side, as is observed.[1] A realistically fast and narrow WBC has its $\bar{q}$ structure significantly altered by $\bar{\zeta}$, with increased $\bar{q}$ on the cyclonic (onshore) side and decreased $\bar{q}$ on the anticyclonic (offshore) side relative to the interior $\bar{q}$ at the same latitude. Outflow from the WBC must merge with the interior, where $\bar{\zeta}$ is insignificant; the WBC outflow must therefore change its $\bar{q}$ (i.e. eliminate any "excess" due to $\bar{\zeta}$) to match the interior. This requires loss of $\bar{q}$ downstream, i.e. that the right-hand side of equation (2) is strongly negative in the outflow from the cyclonic layer. Although the wind forcing term $f_0 w/H$ is negative, it is readily shown to be insignificant in the WBC vorticity balance. This implies that $A_2 \nabla^2 \bar{\zeta} - \nabla \cdot \overline{(\mathbf{u}'q')} \ll 0$ in the cyclonic outflow.

[11] shows that in a non-turbulent, unstratified WBC, loss of the excess cyclonic $\bar{q}$ requires strongly negative viscous dissipation ($A_2 \nabla^2 \bar{\zeta} \ll 0$), but since $\nabla^2 \bar{\zeta} \approx \frac{\partial^2 \bar{\zeta}}{\partial x^2}$ (due to boundary-layer

[1] For brevity, and without loss of generality, a northern hemisphere subtropical WBC is assumed.

scaling; x is the distance offshore), this implies an alteration in the cross-shore vorticity profile. $\frac{\partial^2 \bar{\zeta}}{\partial x^2} < 0$ can be obtained via relatively subtle profile changes, but to obtain $\frac{\partial^2 \bar{\zeta}}{\partial x^2} \ll 0$ the cyclonic vorticity maximum must move away from the coast in order to obtain a viscous flux of cyclonic vorticity into the boundary. The point at the coast where the offshore gradient of $\bar{\zeta}$ becomes positive marks the start of a region of "adverse" ageostrophic pressure gradient which strongly decelerates the WBC [14], implying offshore flow. The resulting dynamics prevent the WBC from following the boundary and steer it into the interior as a separated jet. A separated WBC jet is therefore the only flow structure which is consistent with the elimination of the excess potential vorticity of the cyclonic layer under realistically inertial conditions; it is not possible to have a broad and gradual outflow as is seen in the linear case [13].

We seek to extend these ideas to a turbulent, stratified WBC, in which the mean flow is subject to additional processes: eddy PV flux divergence $-\nabla \cdot \overline{(\mathbf{u}'q')}$ can change the $\bar{q}$ excess, perhaps eliminating the role of $A_2\nabla^2\bar{\zeta}$, and $\bar{q}$ now includes a layer thickness term $-f_0\bar{h}/H$, so mean flow driven across $\bar{h}$ contours might allow the $\bar{q}$ excess to be reduced without changing $\bar{\zeta}$.

4. Results

Figure 1 shows the time-average of a flow with relatively large viscosity ($A_2 = 400\,\mathrm{m^2s^{-1}}$). This results in a laminar and nearly-steady flow, a similar regime to that in [11] apart from the stratification and different geometry. The WBC flows rapidly northward before separating at mid-basin to form a meandering jet. The cyclonic layer is visible as the red region against the coast in the WBC. Figure 2 shows a "side-on" view of $\bar{q}(x,y)$; the rightmost red profile is along the western boundary and the others are spaced every 4 km further east. The black diagonal line is the $\bar{q}$ profile in the interior. It is clear that the cyclonic layer is relatively thin, and that $\bar{q}$ in this layer far exceeds that in the interior at the same latitude (very close to the coast, $\bar{q}$ exceeds even the northernmost interior value). This excess PV must therefore be lost to allow this fluid to re-enter the interior. The close-up of the WBC in figure 3 shows the terms $A_2\nabla^2\bar{\zeta}$ and $-\nabla \cdot \overline{(\mathbf{u}'q')}$ in equation (2); downstream $\bar{q}$ loss from the cyclonic layer outflow (the blue region at roughly $800 < y < 900\,\mathrm{km}$) is dominated by the Laplacian viscosity. This term becomes strongly negative via a change in the offshore $\bar{\zeta}$ profile whereby the maximum moves offshore (figure 4), which implies an adverse ageostrophic pressure gradient [14]. These dynamics and structures are the same as in [11], despite the presence of stratification. Further results indicate that the loss of $\bar{q}$ is due to a loss of $\bar{\zeta}$ rather than an increase in $\bar{h}$ (reflecting the close alignment of streamlines and $\bar{h}$ contours), so the separation dynamics are not significantly altered by the stratification. Experiments (not shown) with partial-slip boundaries also reinforce the importance of the cyclonic layer to separation: as the amount of boundary slip increases, the cyclonic vorticity weakens and so the PV excess is reduced. When this excess is sufficiently small, the mean WBC adopts a very different geometry in which it leaves the coast as a broad, gradual outflow rather than a narrow jet. With a free-slip coast (no cyclonic layer at all) the mean WBC "overshoots" all the way to the northern boundary before turning back and heading east along the midline.

When A_2 is reduced to $100\,\mathrm{m^2s^{-1}}$ the flow becomes more strongly time-dependent. In this case both $\nabla^2\bar{\zeta}$ and $-\nabla \cdot \overline{(\mathbf{u}'q')}$ contribute to $\bar{q}$ loss, but the $\bar{\zeta}$ maximum still leaves the coast at separation (figures 5 and 6). A further reduction to $A_2 = 40\,\mathrm{m^2s^{-1}}$ yields a strongly turbulent flow in which downstream loss of $\bar{q}$ from the cyclonic layer is due almost entirely to eddy PV flux divergence (figures 7 and 8). Interestingly, the $\bar{\zeta}$ profile still shows similar changes (particularly the maximum moving offshore), despite the unimportance of the Laplacian term relative to $-\nabla \cdot \overline{(\mathbf{u}'q')}$. Equation (2) shows $-\nabla \cdot \overline{(\mathbf{u}'q')} = -\nabla \cdot \overline{(\mathbf{u}'\zeta')} + \frac{f_0}{H}\nabla \cdot \overline{(\mathbf{u}'h')}$; a comparison of these terms reveals that the second term is negligible, indicating that barotropic instability is more important than baroclinic instability in this region. A working hypothesis (still under investigation) is that $-\nabla \cdot \overline{(\mathbf{u}'\zeta')}$ acts as a down-gradient diffusion of $\bar{\zeta}$ in this region, so obtaining

IOP Conf. Series: Earth and Environmental Science **11** (2010) 012034 doi:10.1088/1755-1315/11/1/012034

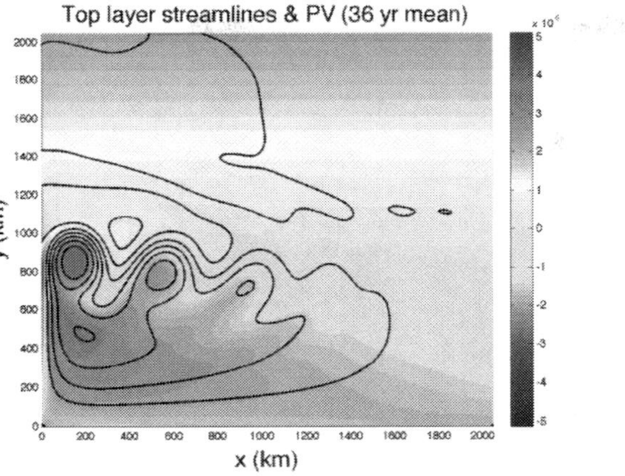

Figure 1. Streamlines coloured by $\bar{q}$ ($m^{-1} s^{-1}$) for a 36-year mean of the flow in the top layer with $A_2 = 400\,m^2s^{-1}$ and 4 km resolution.

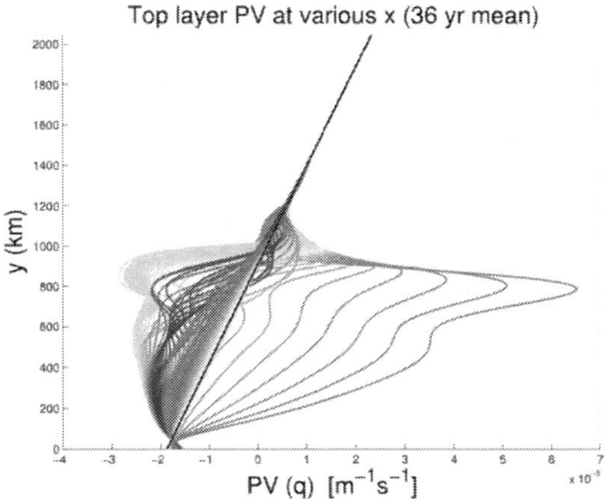

Figure 2. $\bar{q}$ profiles vs. y at various x for the flow in figure 1.

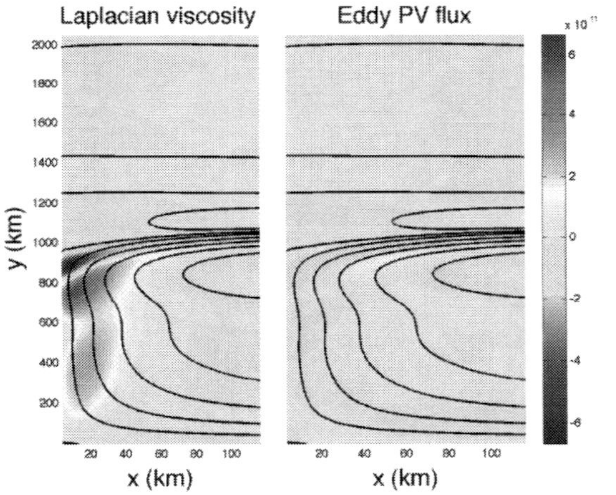

Figure 3. Closeup of time-averaged Laplacian viscosity $A_2\nabla^2\bar{\zeta}$ (left) and eddy PV flux divergence $-\nabla\cdot\overline{(\mathbf{u}'q')}$ (right) against the western boundary for the flow in figure 1, with mean streamlines overlaid.

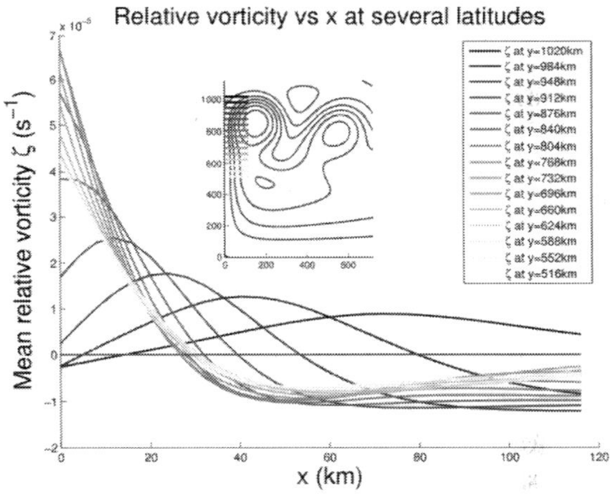

Figure 4. Closeup of mean relative vorticity transects in the WBC for the flow in figure 1. Inset: transect locations and lengths relative to mean streamlines.

$-\nabla\cdot\overline{(\mathbf{u}'\zeta')} \ll 0$ requires changes to the vorticity structure (i.e. the $\bar{\zeta}$ maximum moving offshore, producing separation of the WBC) which are essentially the same as for Laplacian viscosity.

Experiments (not shown) were also conducted with antisymmetric double-gyre wind forcing (i.e. with the northern half of the basin also driven by a wind stress curl, equal and opposite to that used here in the southern half) and no-slip boundaries. This produces a subpolar gyre with a southward-flowing WBC, in addition to the subtropical gyre shown here. With $A_2 = 400\,m^2s^{-1}$, both WBCs separate about 100 km before mid-basin, and then join to form an eastward jet. The jet trajectory is more zonal than that shown in figure 1, but in each gyre the relative importance and spatial structure of the terms in equation 2 close to the coast were nearly identical to those

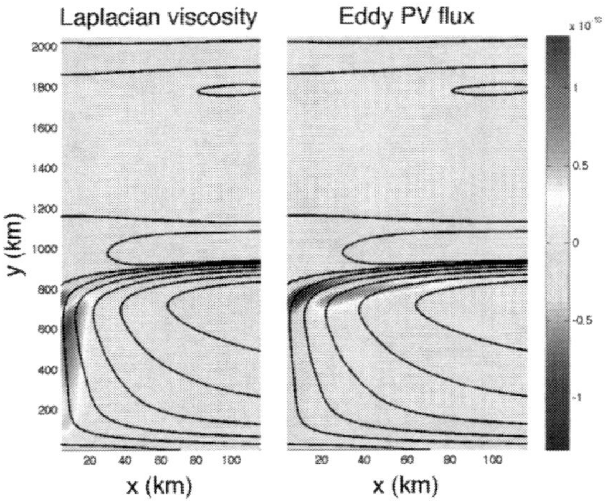

Figure 5. Closeup of 36-year time-average of Laplacian viscosity $A_2\nabla^2\bar{\zeta}$ (left) and eddy PV flux divergence $-\nabla\cdot\overline{(\mathbf{u}'q')}$ (right) against the western boundary for flow in the top layer with $A_2 = 100\,\mathrm{m^2s^{-1}}$ and 4 km resolution.

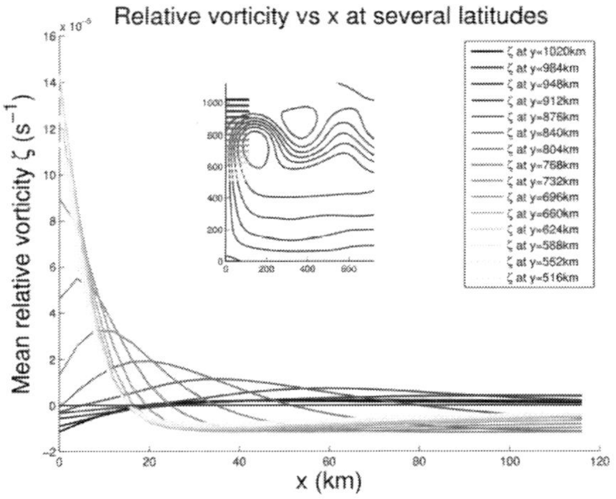

Figure 6. Mean relative vorticity transects in the WBC for the flow in figure 5. Inset: transect locations and lengths relative to mean streamlines.

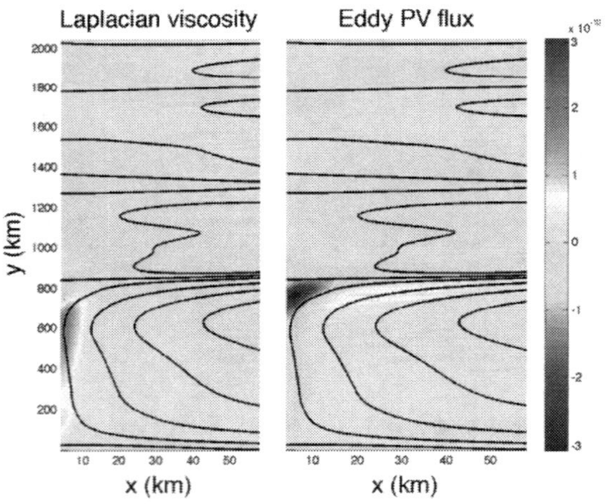

Figure 7. Closeup of 26-year time-average of Laplacian viscosity $A_2\nabla^2\bar{\zeta}$ (left) and eddy PV flux divergence $-\nabla\cdot\overline{(\mathbf{u}'q')}$ (right) against the western boundary for flow in the top layer with $A_2 = 40\,\mathrm{m^2s^{-1}}$ and 2 km resolution.

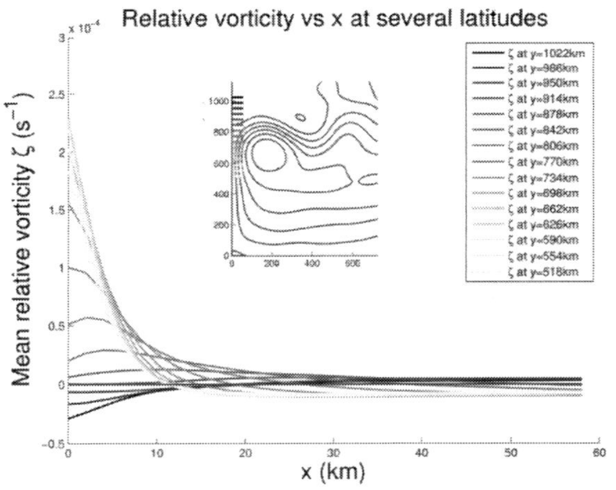

Figure 8. Mean relative vorticity transects in the WBC for the flow in figure 7. Inset: transect locations and lengths relative to mean streamlines.

shown here, and involved the vorticity maximum moving offshore. This indicates that vorticity flux to the boundary is more important than flux between the WBCs post-separation, so the presence of a second WBC is unimportant to the separation process. Similar results were also found for two-gyre flow with $A_2 = 40\,\mathrm{m^2s^{-1}}$. However, under free-slip boundary conditions the presence of a counter-flowing WBC dramatically alters the flow by preventing overshooting.

5. Conclusions

Separation of the mean WBC in a turbulent two-layer quasigeostrophic model with no-slip boundaries involves very high mean potential vorticity $\bar{q}$ in a thin, strongly cyclonic layer on the onshore side of the WBC, and abrupt loss of this $\bar{q}$ excess via viscosity and/or divergence of eddy potential vorticity flux in the WBC outflow, involving the vorticity maximum moving offshore, i.e. a separated flow. This mechanism is essentially the same as that identified previously in a non-turbulent, unstratified model [11] because stratification has negligible effect on downstream $\bar{q}$ variation, and eddy vorticity flux replaces viscosity as the main cause of $\bar{q}$ loss as turbulence increases. The deliberately idealised geometry and forcing of this process study make it difficult to apply to actual WBCs. Further work is planned to better understand the separation process and to extend these ideas into more realistic situations including varied coastline geometry and non-quasigeostrophic dynamics.

Acknowledgments

This research was undertaken on the NCI National Facility in Canberra, Australia, which is supported by the Australian Commonwealth Government. I thank Andy Hogg for his help with Q-GCM.

References

[1] Dengg J, Beckmann A and Gerdes R 1996 *The Warmwatersphere of the North Atlantic Ocean* ed Krauss W (Berlin, Stuttgart: Gebrüder Borntraeger) chap 9, pp 253–290
[2] McWilliams J C 1996 *Ann. Rev. Fluid Mech.* **28** 215–248
[3] Chassignet E and Marshall D 2008 *Ocean Modeling in an Eddying Regime* (*Geophysical Monograph Series* vol 177) ed Hecht M and Hasumi H (American Geophysical Union) URL `http://ocean.fsu.edu/faculty/chassignet/recentpubs.html`
[4] Bryan F O, Hecht M W and Smith R D 2007 *Ocean Modelling* **16** 141–159
[5] Hurlburt H E and Hogan P J 2000 *Dynamics of Atmospheres and Oceans* **32** 283–329 URL `http://www.sciencedirect.com/science/article/B6VCR-40SFH3K-9/2/1ee51e74f36d0e07035c287b0d169062`
[6] Logoutov O, Sutyrin G and Watts D R 2001 *Journal of Physical Oceanography* **31** 637–644
[7] Rossby T and Zhang H M 2001 *Journal of Marine Research* **59** 949–975
[8] Leaman K D, Johns E and Rossby T 1989 *Journal of Physical Oceanography* **19** 36–51
[9] Roughan M, Oke P R and Middleton J H 2003 *Journal of Physical Oceanography* **33** 2551–2564 URL `<GotoISI>://000187327400006`
[10] Chelton D, Schlax M, Freilich M and Milliff R 2004 *Science* **303** 978–983
[11] Kiss A E 2002 *J. Mar. Res.* **60** 779–803
[12] Hogg A M, Dewar W K, Killworth P D and Blundell J R 2003 *Monthly Weather Review* **131** 2261–2278 URL `<GotoISI>://000185353100004`
[13] Munk W H 1950 *J. Meteorology* **7** 79–93
[14] Tritton D J 1988 *Physical Fluid Dynamics* 2nd ed (Oxford Science Publications)

Modeling tides and vertical tidal mixing: A reality check

Robin Robertson
School of Physical, Environmental, and Mathematical Sciences,
University of New South Wales @ Australian Defence Force Academy
Canberra, ACT 2600 Australia
Email: r.robertson@adfa.edu.au

Abstract. Recently, there has been a great interest in the tidal contribution to vertical mixing in the ocean. In models, vertical mixing is estimated using parameterization of the sub-grid scale processes. Estimates of the vertical mixing varied widely depending on which vertical mixing parameterization was used. This study investigated the performance of ten different vertical mixing parameterizations in a terrain-following ocean model when simulating internal tides. The vertical mixing parameterization was found to have minor effects on the velocity fields at the tidal frequencies, but large effects on the estimates of vertical diffusivity of temperature. Although there was no definitive best performer for the vertical mixing parameterization, several parameterizations were eliminated based on comparison of the vertical diffusivity estimates with observations. The best performers were the new generic coefficients for the generic length scale schemes and *Mellor-Yamada's* 2.5 level closure scheme.

1. Introduction

Half of the ocean's vertical mixing is attributed to tides [1]. Regional models can simulate tides and the tidal mixing, but how well they do it and the best performer of the many vertical mixing parameterizations is unknown. The latter is very important, since the mixing estimate depends on the vertical mixing parameterization. An earlier evaluation found a wide range of estimates of vertical diffusivities for the identical region and tidal forcing dependent on the vertical parameterizations (figure 1), even though the tidal velocity fields showed only minor differences (figure 2). Evaluation of the model performance, specifically how well the models replicate the tidal dynamics, baroclinic tidal fields, and vertical diffusivities, is needed to appropriately apply model results. The goals of this study were to evaluate the performance of ten different vertical diffusivity parameterizations in a terrain-following, hydrostatic model in simulating baroclinic tides and vertical diffusivities and to determine which one of them performs best in replicating the internal tides, energy exchanges between constituents, and the dissipation or removal of energy at high frequencies. These tidal simulations were performed for the Fieberling Guyot region, which has as extensive set of observations available for evaluation of the model results. This extends an earlier model evaluation on resolution and parameterizations [2], investigating effects of vertical mixing on the higher frequency motions.

2. Model description

Terrain-following models, such as the Regional Ocean Model System (ROMS) have been successfully used to model barotropic and baroclinic tides [2]. ROMS was used in this study due to both its broad use in the community and the multiple vertical mixing parameterizations available. ROMS is

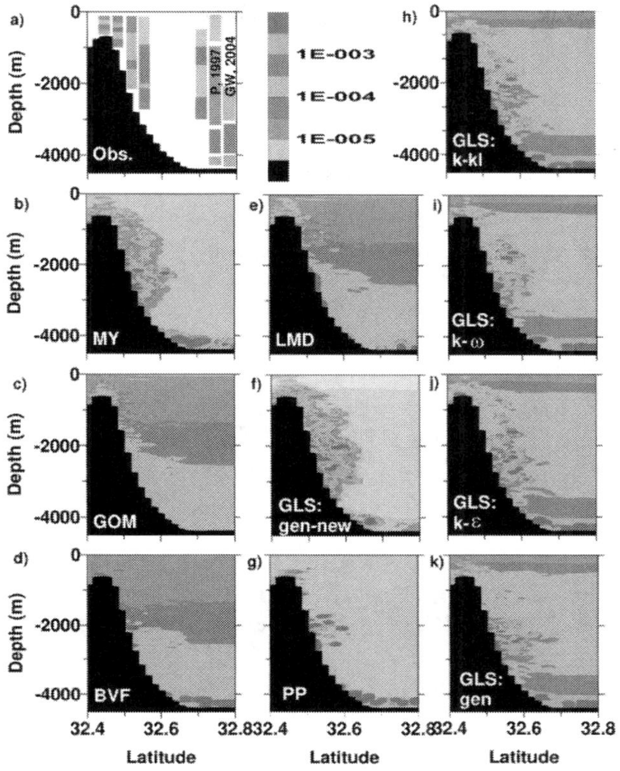

Figure 1. A portion of the north-south transect of the vertical temperature diffusivity averaged over two semi-diurnal tidal cycles at 127° 41.0' W for different vertical mixing parameterizations: a) observations from *Kunze and Toole* [10] and *Toole et al.* [13], b) *Mellor-Yamada* (MY), c) Generic Ocean Mixing (GOM), d) *Brunt-Väisäla* frequency (BVF), e) *Large-McWilliams-Doney* (LMD), f) new generic coefficients in the generic length scale (GLS:gen-new), g) *Pacanowski-Philander* (PP), h) the generic length scale version of the *k-kl* parameterization (GLS: *k-kl*), i) the generic length scale version of the κ-ω parameterization (GLS: κ-ω), j) the generic length scale version of the κ-ε parameterization (GLS: κ-ε), and k) the generic length scale version for a generic parameterization (GLS: gen). In a), GW, 2004 refers to the values from *Ganachaud and Wunsch* [12] and P, 1997 the observations of *Polzin et al.* [13].

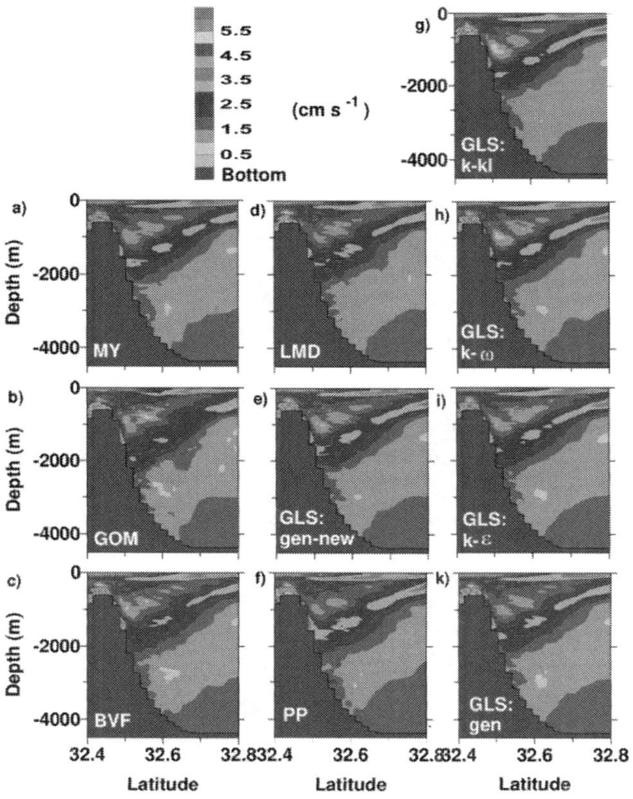

Figure 2. A portion of a north-south transect of the major axes of the M_2 tidal ellipses at 127° 41.0' W for different vertical mixing parameterizations: a) *Mellor-Yamada* (MY) [6], b) Generic Ocean Mixing (GOM), c) *Brunt-Väisäla* frequency (BVF) [5], d) *Large-McWilliams-Doney* (LMD) [7], e) new generic coefficients in the generic length scale (GLS: gen new) [8], f) *Pacanowski-Philander* (PP) [4], g) the generic length scale version of the *k-kl* parameterization (GLS: *k-kl)* [8], h) the generic length scale version of the κ-ω parameterization (GLS: κ-ω) [8], i) the generic length scale version of the κ-ε parameterization (GLS: κ-ε [8], and j) the generic length scale version for a generic parameterization (GLS: gen) [8].

described fully on their website (http://www.myroms.org/index.php). The model domain (figure 3), tidal forcing, and other operational considerations are given in [2]. The resolution was 2 km in the horizontal directions with 60 vertical levels. *T_Tide* routines [3] were used to calculate the major axes for four constituents, M_2, S_2, K_1, and O_1.

3. Vertical mixing parameterizations

Ten different parameterizations were investigated for their effects on both the velocity fields and the vertical diffusivity: 1) *Pacanowski-Philander* (*PP*) turbulence closure scheme [4], 2) an N^1 based scheme (*BVF*) [5], 3) *Mellor-Yamada* (*MY*) 2.5 level turbulence closure scheme [6], 4) *Large-McWilliams-Doney* (*LMD*) *Kpp* profile scheme [7], 5) Generic Ocean Model (GOM) scheme developed by the ROMS team (http://www.myroms.org/index.php), and 6) five versions of the generic length scale scheme (*GLS*) [8], κ-κl, κ-ω, κ-ε, and two generic versions (*GLS:k-kl* m=1, n=1, and p=0.; *GLS:*κ-ε: m=1.5, n=-1, and p=3; *GLS:*κ-ω: m=0.5, n=-1, and p=-1). The generic version of the *GLS* scheme has two sets of coefficients associated with it, the old (m=1, n=-0.67, and p=1) and the new ones (m=1, n=-0.67, and p=2). No rationale was given for this change in coefficients (http://www.myroms.org/index.php). A brief description of each of these parameterizations, including their basic assumptions and drawbacks, is given in *Robertson* [2].

ROMS has shown reasonable agreement with observations in various regions and reproduced observed spectra both in background regions and in regions of active internal tide generation [9]. However, figure 1 clearly shows a wide range of diffusivities between the different schemes and several of them are inconsistent with the observations (figure 1a).

It should be noted that ROMS estimates vertical diffusivities of momentum, K_ν, temperature, K_T, and salinity, K_S. Observations generally determine a diapycnal diffusivity of density, K_ρ, which is based on a calculated dissipation from velocity shear measurements. Ignoring cabbelling effects and assuming the observed diapycnal diffusivity is vertical, K_T is equivalent to K_ρ.

4. Observations

Observational data was collected from ten moorings in the area to investigate tidal interactions with rough topography [10]. The data set is also described in [2]. The observational major axes of the tidal ellipses for the four constituents were determined using the *T_Tide* routines [3].

5. Simulations

Semi-diurnal, M_2 and S_2, internal tides were generated over the crown and upper flanks of the guyot (M_2 shown in figure 4a) in agreement with linear internal wave theory. The internal tides propagated away from the guyot following internal wave characteristics (white dashed lines in figure 4a).

Linear internal wave theory predicts a primarily barotropic response for the diurnal constituents since this site is poleward of the diurnal critical latitudes (latitude where the inertial frequency equals the tidal frequency). The model response basically followed this theory with weak baroclinic tides generated over the rim and upper flank of the guyot trapped at the bottom at ~500-800 m (figure 4b). The diurnal response was weaker than the semi-diurnal with peak values of 6 cm s^{-1} and 8 cm s^{-1}, respectively.

The mean velocities were depth-dependent (figure 4c) with a general pattern of a series of toroids around the guyot centered at ~ 500, 1000, and 2000 m depth. Toroid diameter increased with increasing depth (figure 4c) and the direction alternated with depth. The toroids are attributed primarily to generation through tidal rectification.

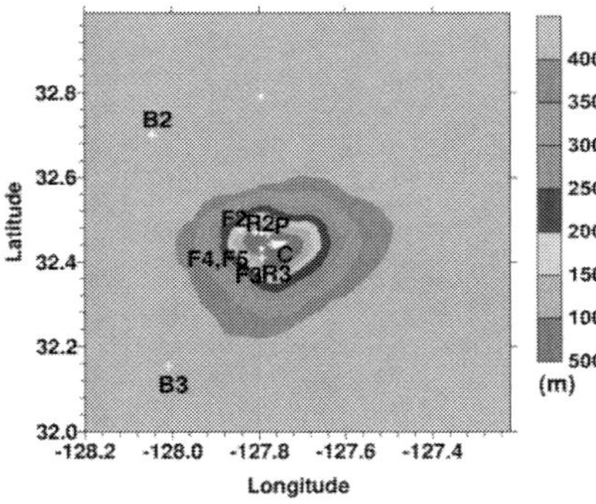

Figure 3. Bathymetry over the domain contoured from the data provided by *Eriksen* (personal communication). The locations of the ten moorings are shown as crosses and labeled.

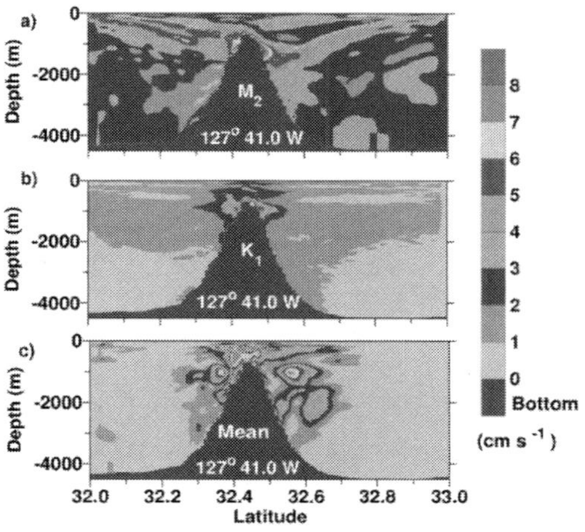

Figure 4. North-south transects of the major axes of the tidal ellipses at 127° 41.0' W for a) M_2, b) K_1, and c) the mean (magnitude). Internal wave characteristics for M_2 are shown as white dashed lines in a) and were calculated using the hydrography.

Table 1. The rms differences for the major axes between the observations and the model results for the ten different vertical mixing parameterizations along with their integrated vertical diffusivities of temperature over the domain.

Vertical Mixing Parameterization	Rms M_2 Major Axis (cm s^{-1})	Rms S_2 Major Axis (cm s^{-1})	Rms K_1 Major Axis (cm s^{-1})	Rms O_1 Major Axis (cm s^{-1})	Integrated Vertical Diffusivity (m^2 s^{-1})
Mellor-Yamada 2.5	3.7	0.9	6.5	2.1	1.4×10^{-3}
Generic Ocean Mixing	3.8	0.9	6.5	2.1	9.0×10^{-4}
Brunt-Väisälä Frequency	3.8	0.9	6.4	2.2	1.3×10^{-2}
Large-McWilliams-Doney	3.9	0.9	6.5	2.1	1.1×10^{-3}
Pacanowski-Philander	3.7	0.9	6.5	2.1	8.0×10^{-5}
Generic Length Scale k-kl	3.8	0.9	6.5	2.1	3.8×10^{-3}
Generic Length Scale k-ω	3.8	0.9	6.5	2.1	3.7×10^{-3}
Generic Length Scale k-ε	3.8	0.9	6.5	2.1	1.9×10^{-3}
Generic Length Scale generic old	3.7	0.9	6.5	2.2	1.8×10^{-3}
Generic Length Scale generic new	3.8	0.9	6.5	2.2	3.7×10^{-4}

6. Comparison

6.1. Tidal constituents
Performance of the ten vertical mixing parameterizations was nearly uniform at the tidal constituent frequencies. Only minor differences appear in the major axes of the M_2 tidal ellipses (figure 2) and the rms differences between the model estimates of the major axes and the observational major axes are essentially equivalent between the different vertical mixing parameterizations for all four tidal constituents (table 1).

6.2. Spectral
Spectral comparisons evaluate not only the energy at the tidal frequencies, but also energy at other frequencies and the transfer of energy between frequencies. Spectra at three levels in the water column for both the East-West and North-South velocities are similar (figure 5) between the different vertical mixing parameterizations with very few points exceeding the 95% confidence levels (black dashed lines in figure 5). The exception is *GOM*, which generally exceeds the spectral values for frequencies greater than 0.1 cph. All of the spectra are red and roughly follow the slope of the *Garrett-Munk* spectra [11] (not shown). The spectra shown are within the region of internal tidal activity, but spectra in a background region are similar (not shown).

6.3. Diffusivity
Comparison of the diffusivities for the ten parameterizations (figure 1b-k) with the observed diffusivities (figure 1a) show that several of the vertical mixing schemes fall short. *GOM*, *BVF*, and *LMD* fail to predict the diffusivities due to shears along the internal wave beam. Except for the new generic coefficients, the *GLS* schemes all add high diffusivities in the lower water column, which are not supported by the observations.

6.4. Integrated diffusivity
The average diffusivity in the ocean varies, but is estimated to be ~$1 x 10^{-4}$ m^2 s^{-1} as an average value including active and inactive mixing regions [12, 13]. Although differences may occur between the integrated diffusivity over the domain and the ocean's average diffusivity due to the domain having a higher percentage of rough topography compared to the global ocean, the integrated diffusivities should be similar. Most of the integrated diffusivities for the vertical mixing parameterizations ranged from roughly 1-40 $x 10^{-4}$ m^2 s^{-1} (table 1). There were outliers, such as the *BVF*, whose value was too high over 1 $x 10^{-2}$ m^2 s^{-1} and it did not perform sufficiently according to this criterion.

7. Summary
The performance of ten different vertical mixing schemes for the Regional Ocean Model System (ROMS) was evaluated using data from Fieberling Guyot for comparison. Four different criteria were used for the evaluation: comparison of major axes of tidal ellipses against observations, spectral comparison of velocities, comparison of vertical diffusivity estimate from the model against the observed values and the integrated vertical diffusivity over the domain. These criteria indicated that *Mellor-Yamada* 2.5 level turbulence closure and the new generic version of the Generic Length Scale are the best performers.

References
[1] Egbert G D and Ray R 2000 *Nature* **405** 775
[2] Robertson R 2006 *Ocean Dynamics* doi 10.1007/s10236-006-0062
[3] Pawlowicz R, Beardsley R and Lentz S 2002 *Computers and Geoscience* **28** 929
[4] Pacanowski R and Philander G 1981 *J. Phys. Oceanogr.* **11** 1442-1451

[5] Gargett A E and Holloway G 1984 *J. Mar. Res.* **42** 15-27
[6] Mellor G L and Yamada T 1982 *Rev. of Geophys. and Space Physics* **20** 851-875
[7] Large W G and Gent P R 1999 *J. Phys. Oceanogr.* **29** 449-464
[8] Umlauf L and Burchard H 2003 *J. Mar. Res.* **61** 235-265
[9] Robertson R 2010 accepted by *Dynamics of Atmospheres and Oceans* February 2010
[10] Brink K H 1991 *Ann. Rev. Fluid Mech.* **23** 399
Brink K H 1995 *J. Geophys. Res.* **100** 10817
Eriksen C C 1991 *J. Geophys. Res.* **96** 15227
Eriksen C C 1998 *J Geophys Res* **103** 2977
Kunze E and Toole J M 1997 *J. Phys. Oceanogr.* **28** 811
Kunze E and Toole J M 1995 in Topographic Effects in the Ocean. *Proceedings of Aha Huliko'a, Hawaiian Winter Workshop* ed. by P Müller and D Henderson, School of Ocean and Earth Sci. And Technol, Univ. of Hawaii, Honolulu 15
Toole J M, Schmitt R W and Polzin K L 1997 *J. Geophys. Res.* **102** 947
[11] Garrett C and Munk W H 1975 *J. Geophys. Res.* **3** 225-264
[12] Polzin K L, Toole J M, Ledwell J R and Schmitt RW 1997 *Science* **276** 93
[13] Ganachaud A and Wunsch C 2004 *Nature* **410** 240

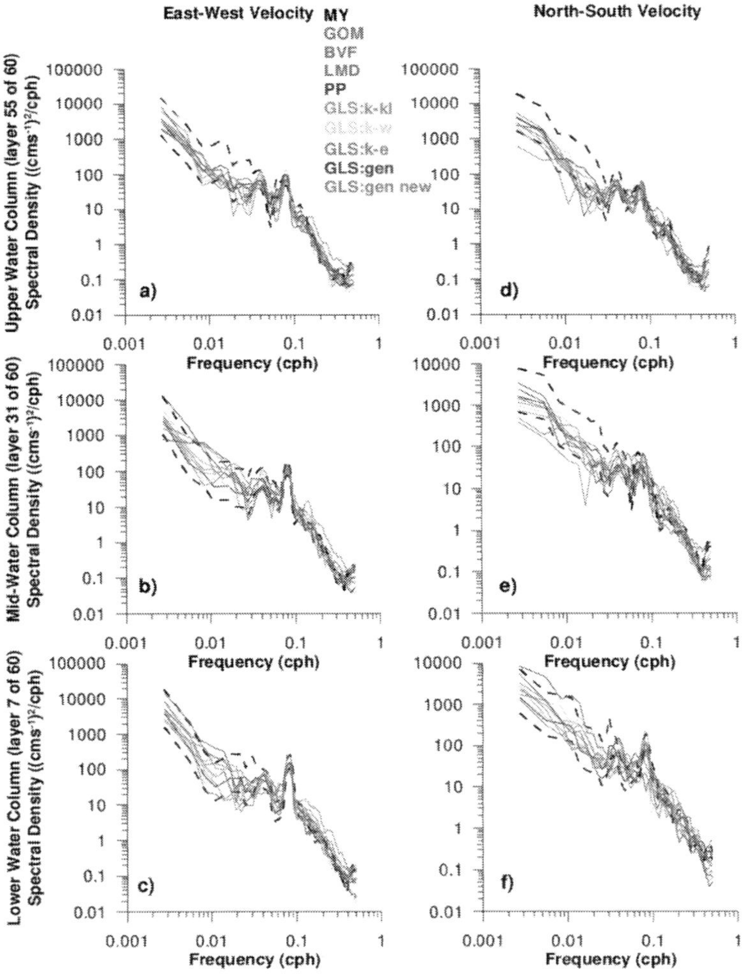

Figure 5. East-West (left panels: a-c) and North-South (right panels: d-f) velocity spectra from three levels, near surface (upper: a, d), mid-water column (middle: b, e), and near bottom (lower: c, f) for each of the ten vertical mixing parameterizations, color coded as indicated. The heavy dashed black lines indicate the 95% confidence limits for the MY25 spectra.

An examination of Southwest Pacific explosive cyclones, 1989 to 2009

M T Black[1], A B Pezza[1] and P Kreft[2]

[1] The Department of Earth Sciences, The University of Melbourne, Victoria 3010, Australia
[2] Meteorological Service of New Zealand Limited, Kelburn, Wellington 6012, New Zealand

Email: m.black4@pgrad.unimelb.edu.au

Abstract. This study has assembled a climatology of Southwest Pacific explosively developing cyclones, based on the European Centre for Medium-Range Weather Forecasts' ERA-Interim reanalysis data, over the 21-year period from 1989 to 2009. The recently developed 'combined explosive' expression, a refinement of the 'relative explosive' criterion, was used to identify cyclones deemed explosive with respect to both the drop in central pressure and the climatological pressure gradient. Over the period of analysis, 47 explosive cyclones were identified within the Southwest Pacific, equating to an average of 2.2 explosive events per year. Seasonally, explosive cyclones are most frequent during the winter months, while least frequent during the summer. Two case explosive systems are briefly considered, with their corresponding measures of intensity and scale placed into climatological perspective.

1. Introduction

Cyclonic systems experiencing explosive development, also known as meteorological 'bombs', are characterised by a reduction of central pressure at the critical rate of at least 24 hPa in 24 hours at a latitude of 60 degrees [1]. This critical rate would have to change with latitude ϕ, as modulated by the Coriolis parameter ($f = 2\Omega\sin\phi$; $\Omega = 7.29\times10^{-5}$ rad s^{-1}), to be geostrophically equivalent (1). For example, a 14 hPa pressure fall over 24 hours is required at a latitude of 30 degrees, while 28 hPa is required at the pole.

$$\text{Explosive Development Critical Rate} = 24\frac{\sin\phi}{\sin 60}\ \text{hPa (24 hours)}^{-1} \qquad (1)$$

Explosive cyclones are of concern to Australia and New Zealand (NZ), as general features associated with these systems include destructive winds, flood rains and coastal storm surges. Notable previous explosive events in the Southwest Pacific region (SWP) have resulted in extensive damage and the loss of life [2,3,4]. On the 23rd July 2008, an exceptional cyclonic system described as "no ordinary storm" [5] formed over the Coral Sea, exhibiting a rapid deepening as it moved towards NZ. This system had an observed 24 hour pressure drop of 28 hPa averaged at 30°S, which is double that required to satisfy explosive classification. With record low surface pressures and winds gusting to 130-160 km h^{-1}, the system struck the NZ North Island as one of the most severe storms in decades. Flooding was observed in Northland, Auckland and Coromandel, significant wind damage in Northland, Auckland, Taranaki and the central plateau, and a number of weather-related fatalities were recorded. This 'Auckland bomb' was followed by a second explosive event in a matter of days, called here the 'Nelson bomb', giving little chance for NZ to recover.

While there has been a large amount of research exploring explosive cyclogenesis, the focus has been predominantly on Northern Hemisphere (NH) events. Explosive systems are often poorly forecast (e.g. [3]), hence there is scope for further understanding of these systems in order to enable improved planning and community preparedness.

A global climatology of explosive cyclones was assembled by Lim and Simmonds [6] for the 21-year period 1979-1999, using the Melbourne University automated cyclone finding and tracking scheme in conjunction with the National Centres for Environmental Prediction – Department of Energy reanalysis-2 data. Each year, Lim and Simmonds identified an average of 26 explosive events in the Southern Hemisphere (SH) and almost twice as many in the NH. Explosive cyclones were found to be predominant during winter in both hemispheres, although the degree of seasonality was much less in the SH. With respect to the SH, wintertime explosive cyclone density maxima were found to the south of Australia (45°-60°S, 90°-180°E) and in the western Pacific Ocean near 30°S, while coastal southeastern Australia and the Tasman Sea also hosted many explosive developments; the summertime density maxima was found over Western Australia.

A regional climatology of explosive cyclones originating in the south-eastern Tasman Sea was achieved by Leslie *et al.* [7] for the 15-year period 1990-2004, based on the examination of official archived mean sea level pressure (MSLP) analyses. Leslie *et al.* identified the occurrence of explosive cyclones to be unevenly distributed throughout the year, with a major peak experienced in June-July, a secondary peak in November-December, and minima for January-March and September-October.

This current study aims to develop a regional climatology of explosive systems within the SWP, in a similar fashion to Leslie *et al.*, albeit using an automated tracking scheme and a recently refined expression of explosive development. The ability of the tracking scheme to describe the location and physical features of explosive cyclone centres throughout their existence, at 6-hourly intervals, provides a means of conducting detailed case studies and placing systems within a climatological perspective.

2. Data and methodology

The landmark study by Sanders and Gyakum [1] initially characterised explosive cyclones by a reduction of central pressure at the rate of 24 hPa in 24 hours, relative to 60 degrees of latitude. Subsequently, Lim and Simmonds [6] considered that a more useful indicator of rapid development could be found in analysing the reduction in central pressure in relation to the background of climatological pressure. Therefore, their definition of explosive cyclones took into consideration the spatial changes of climatological pressure along the cyclone path. While this latter approach may be more appropriate for characterising explosive development in the SH, given the strong meridional climatological pressure gradients there, the nature of this expression results in a few systems being erroneously described as explosive. Specifically, the movement of cyclones into regions of higher climatological pressure may result in the system sufficing the 'relative explosive' criterion, despite experiencing negligible increases in intensity or falling central pressure (e.g. [8]). In recognising this, Allen *et al.* [8] introduced the 'combined explosive' expression to describe systems that simultaneously satisfy the two previous definitions of explosive development: that is, systems that experience a fall in both the relative and actual central pressure at the rate of at least 24 hPa in 24 hours, relative to 60 degrees of latitude. Accordingly, the 'combined' expression accounts for most systems given by Lim and Simmonds, while rejecting cases that were deemed explosive simply due to migration into regions of higher climatological pressure. In this sense, the 'combined explosive' expression is used in this study.

The use of global reanalysis data, in conjunction with an automated cyclone finding and tracking scheme, provides a means of assembling a climatology of explosive cyclones. The European Centre for Medium-Range Weather Forecasts' ERA-Interim digital MSLP reanalysis [9] was used in this study, spanning the period from 1 January 1989 to present (30 November 2009). Formed with a spectral resolution of T255L60 and providing fields on 1.5° x 1.5° latitude-longitude grid, ERA-Interim is the most recently available reanalysis of the state of the atmosphere.

Operating on the single input variable MSLP, the objective and automated Melbourne University cyclone finding and tracking scheme [10,11] was used to identify and track SH cyclones by way of a two-step process. In the first step, the 'low-finding' routine determined the location and physical features of cyclone centres from the digital MSLP fields at each 6-hourly time interval; this information was then used for 'tracking' the cyclones through space and time. All resulting cyclone tracks that satisfied the 'combined explosive' expression within the SWP (defined here to encapsulate 20-50°S, 140-190°E) constituted the population of explosive systems considered in this climatology.

3. A climatology of Southwest Pacific explosive cyclones

Over the period from 1 January 1989 – 30 November 2009, 47 cyclones experienced explosive development within the Southwest Pacific region. On average, 2.2 explosive events occurred each year, although considerable inter-annual variability exists across the 21 years of analysis (figure 1a). For some years, as many as 5-6 explosive events were identified within the SWP (e.g. 1990, 2008 and 2009), while during other years no such events occurred (e.g. 1994 and 2001). The collated monthly frequencies of explosive events (figure 1b) reveal a pronounced peak of 25 events during the winter months (June-August), secondary peaks of 9 events during both autumn (March-May) and spring (September-November), and a minimum of 4 events during summer (December-February), confirming that explosive cyclones are predominantly wintertime events in the SWP.

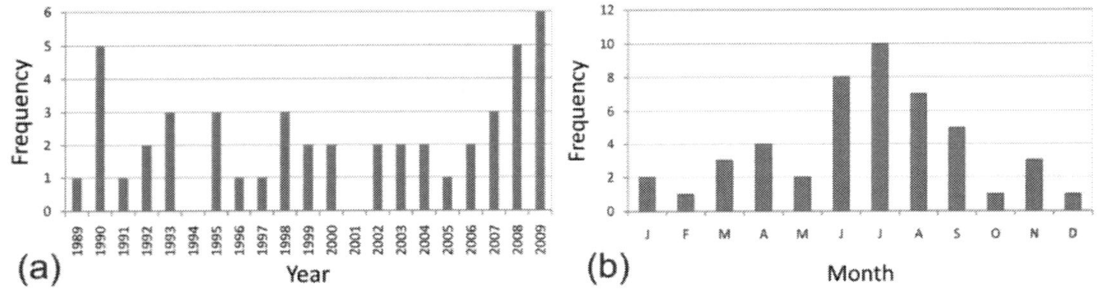

Figure 1. Time series of (a) annual and (b) monthly explosive cyclone occurrence in the Southwest Pacific region (20-50°S, 140-190°E) for the period 1 January 1989 – 30 November 2009.

The geographical distribution of SWP explosive cyclones is displayed in figure 2. Importantly, these cyclone trajectories provide a potentially valuable source of information for understanding the nature of these events. For example, the cyclone tracks in figure 2 reveal a southeasterly direction of motion consistent with Sutcliffe's Rules [12], with systems having formed along the eastern Australian coastline or over the Coral Sea. The identified cyclogenesis along the eastern coastline of Australia highlights the possibility of type-2 East Coast Lows, which often experience rapid deepening [13].

An important feature to note from figure 2 is that a cyclone may persist for an extended period of time following the explosive development. The temporal lengths of the cyclones' lifecycle (from cyclogenesis to cyclolysis) are displayed in figure 3; this graph highlights that SWP explosive cyclones retained their cyclonic features for an average of about 8 days, with all systems existing for at least 4 days and one system as long as 16 days. Consequently, the cyclones experiencing explosive development within the SWP often move large distances from the study region, with some systems observed as far away as the Drake Passage and the Ross Sea (figure 2). Despite this long temporal existence, the onset of explosive development occurred for all identified cyclones within the first few days following cyclogenesis (figure 3). In fact, of the 47 identified explosive cyclones, 28 systems had experienced explosive deepening within 24 hours of cyclogenesis, while the remaining 19 systems had done so within 96 hours.

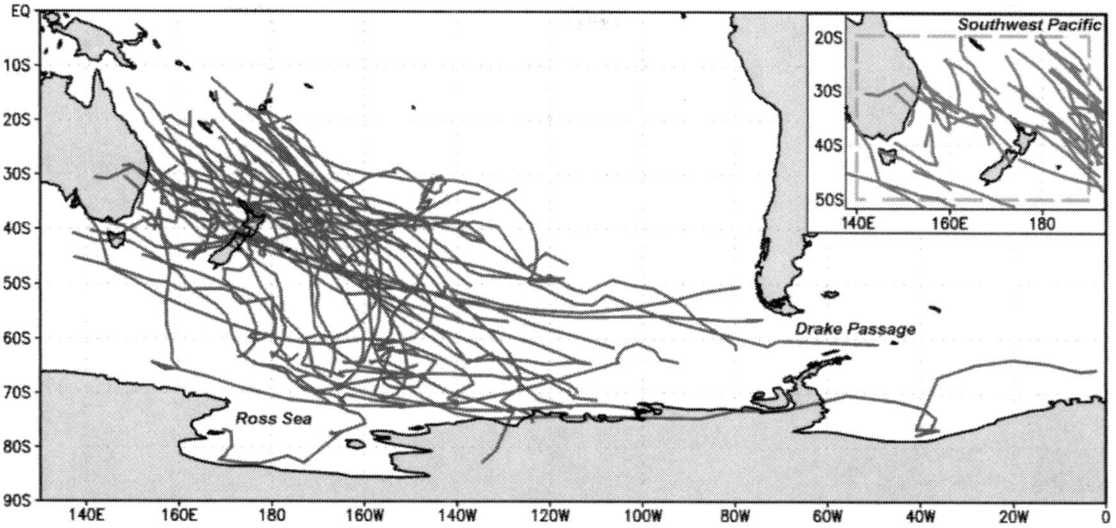

Figure 2. Inset: Tracks of the 47 explosive cyclones, plotted with 6-h segments over the 24 hours of explosive development, as identified for the period 1989 to 2009 within the Southwest Pacific region (20-50°S, 140-190°E; dashed region). Main image: Corresponding complete cyclone tracks, plotted with 6-hourly segments over their total period of existence (cyclogenesis to cyclolysis).

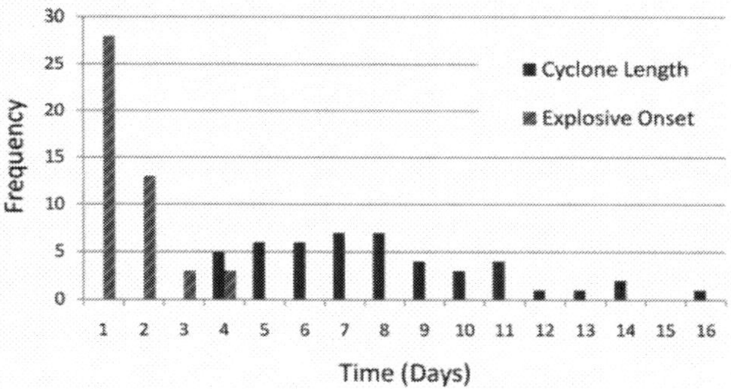

Figure 3. Temporal lengths (in days) of the identified Southwest Pacific explosive cyclone lifecycles (from cyclogenesis to cyclolysis), accompanied by the time (in days) following cyclogenesis within which explosive development occurred.

Given that the tracking system used in this study identifies and tracks all cyclone systems (both explosive and non-explosive), it was possible to calculate the proportion of explosive cyclones in the SWP between 1989 and 2009 (table 1). Only 0.57% of all cyclone systems identified experienced explosive development. Seasonally, the highest percentage of cyclones experiencing explosive development occurred during winter (1.29%), while the lowest occurred during summer (0.17%). Accordingly, the occurrence of explosive development within the SWP is rare. Following this point, it is important to note that the current study does not capture all important storms within the SWP, rather the most extreme ones based on the pressure criteria.

4. Consideration of two case explosive cyclones

A final point of interest in this study is to place the Auckland and Nelson 'bombs' into a basic climatological perspective. The intensity and scale of the two case systems was examined in relation to the mean values for all explosive systems identified within the SWP. The Laplacian of pressure ($\nabla^2 p$) averaged over a calculated radial distance from the cyclone centre (2° latitude in this study) provides one measure of the intensity of a cyclone, given the relationship with relative cyclonic geostrophic vorticity (ξ):

$$\xi = \frac{1}{\rho f}\nabla^2 p \tag{2}$$

where ρ is the air density and f is the Coriolis parameter.

Table 1. The proportion of cyclones systems occurring within the Southwest Pacific region (SWP) that experience explosive development over the 21-year period from 1989 to 2009.

	Total	Summer (DJF)	Autumn (MAM)	Winter (JJA)	Spring (SON)
SWP cyclones	8258	2293	1820	1935	2210
SWP explosive cyclones	47	4	9	25	9
Percentage of all SWP cyclones experiencing explosive development	0.57 %	0.17 %	0.49%	1.29%	0.41%

The temporal evolution of the $\nabla^2 p$ following the commencement of explosive development (figure 4a) reveals a rapid increase in cyclone intensity, with maximum values achieved in around 30 hours. The $\nabla^2 p$ values were significantly higher (at the 95% confidence level) for the 'Auckland bomb' and significantly lower for the 'Nelson bomb' than the mean for all SWP explosive cyclones. Indeed, the peak $\nabla^2 p$ value for the 'Auckland bomb' (5.15 hPa ($^{\circ}$ lat)$^{-2}$) was the second highest value recorded for all SWP explosive cyclones.

The radius of SWP explosive cyclones, taken here to be the weighted mean distance from the cyclone center to points where $\nabla^2 p$ is zero at the 'edge' of the cyclone, provides one measure of the spatial extent of the systems. On average, SWP cyclones experience a slight increase in radius following the onset of explosive development (figure 4b), from 5 to 6 degrees of latitude. The 'Auckland bomb' achieved radius values less than or equal to the SWP mean, while the 'Nelson bomb' recorded the largest radius of all SWP explosive cyclones (9.18 degrees of latitude). This understanding of the complementary measures of intensity and scale provide a more complete picture of the characteristics of the two case systems: the 'Auckland bomb' was a comparatively smaller and more intense system than the 'Nelson bomb'.

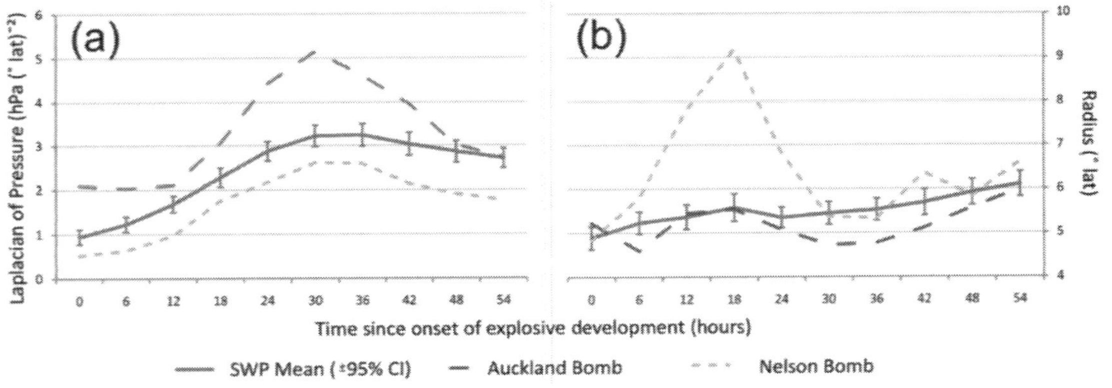

Figure 4. Evolution of the mean (a) area-averaged Laplacian of pressure over a 2° latitude radial distance from the cyclone centre, and (b) system radius, of the 47 identified Southwest Pacific explosive cyclones (with 95% confidence intervals) following the onset of explosive development. The corresponding temporal values for the 'Auckland' and 'Nelson' case explosive systems are also shown.

5. Conclusion and further work

A comprehensive climatology of Southwest Pacific explosive cyclones for the 21-year period from 1989 to 2009 has been presented here, using an automated cyclone tracking scheme and a recently refined expression of explosive development. A total of 47 explosively developing cyclones were identified within the region of interest, with events being most frequent during the winter months. In total, only 0.57% of all cyclones identified within the Southwest Pacific over the period of analysis experienced explosive development, indicating that such events in that region are rare. Of the systems experiencing explosive development, most exhibited a southeasterly trajectory and maintained their cyclonic identity for an extended period of time. Following the onset of rapid deepening, SWP explosive cyclones experience a mean increase in intensity and spatial extent.

This study is one constituent of a Masters thesis examining Southern Hemisphere explosive cyclones. Future research will explore the environmental features associated with the development of explosive cyclones, as well as the connections between these cyclones and indices of the large-scale circulation.

References

[1] Sanders F and Gyakum J R 1980 Synoptic-dynamic climatology of the bomb *Mon. Weather Rev.* **108** 1589-606
[2] Hill H W 1980 A case of intense cyclogenesis in the Tasman Sea *New Zeal. J. Sci.* **23** 111-27
[3] Buckley B W and Leslie L M 2000 The Australian Boxing Day storm of 1998 – synoptic description and numerical simulations *Weather Forecast* **15** 543-58
[4] Revell M J and Gorman R M 2003 The "Wahine storm": evaluation of a numerical forecast of a severe wind and wave event for the New Zealand coast *New Zeal. J. Mar. Fresh.* **37** 251-66
[5] National Institute of Water and Atmospheric Research 2008 *Natural Hazards 2008* (Wellington) p 9
[6] Lim E P and Simmonds I 2002 Explosive cyclone development in the Southern Hemisphere and a comparison with Northern Hemisphere events *Mon. Weather Rev.* **130** 2188-209
[7] Leslie L M, Laplastrier M, Buckley B W and Qi L 2005 Climatology of meteorological "bombs" in the New Zealand region *Meteorol. Atmos. Phys.* **89** 207-14
[8] Allen J T, Pezza A B and Black M T 2010 Explosive cyclogenesis: a global climatology comparing multiple reanalyses *submitted J. Climate*
[9] Berrisford P, Dee D, Fielding K, Fuentes M, Kallberg P, Kobayshi S and Uppala S 2009 The ERA-Interim archive – Version 1.0 *ERA report series* (Reading, United Kingdom-European Centre for Medium Range Weather Forecasts) 20pp
[10] Simmonds I and Murray R J 1999 Southern extratropical cyclone behaviour in ECMWF analyses during the FROST special observing period *Weather Forecast* **14** 878-91
[11] Simmonds I, Murray R J and Leighton R M 1999 A refinement of cyclone tracking methods with data from FROST *Aust. Meteorol. Mag.* **special edition** 35-40
[12] Sutcliffe R C 1947 A contribution to the problem of development *Quart. J. Roy. Meteorol. Soc.* **73** 370-83
[13] Holland G J, Lynch A H and Leslie L M 1987 Australian east-coast cyclones. 1. Synoptic overview and case-study *Mon. Weather Rev.* **12** 3024-36

Rainfall declines over Queensland from 1951-2007 and links to the Subtropical Ridge and the SAM

D A Cottrill[1] and J Ribbe[2]
[1]Bureau of Meteorology, 700 Collins St, Docklands, Melbourne, Victoria, Australia.
[2]University of Southern Queensland, Toowoomba, Queensland, Australia.

E-mail: a.cottrill@bom.gov.au

Abstract. Much of southern and eastern Australia including Queensland have experienced rainfall declines over recent decades affecting agricultural production and accelerating water infrastructure development. Rainfall declines from southern Australia have now been directly related to changes in the Southern Annular Mode (SAM) and the subtropical ridge. In southern and coastal Queensland, the rainfall declines have occurred mostly in the austral summer and autumn. Observations from this region reveal the rainfall decline is correlated to an increase in the mean sea level pressure (MSLP) at many stations. The largest increases in MSLP are over southeast Queensland and coastal regions, where some of the largest rainfall declines occur. This study indicates the subtropical ridge as one of the main factors in the rainfall decline over this region. SAM is also likely to be important, although its seasonal influence, apart from winter, is harder to determine.

1. Introduction

Rainfall declines over much of Queensland began in ~1951 and although there have been about four La Niña events since 1996, the rainfall declines have now intensified and cover most of the state [1,2]. Rainfall has decreased over coastal as well as inland regions of Queensland and also during the La Niña events, although useful rains have occurred over the last three years or so in some areas. Negative decadal trends in annual rainfall (up to 200 mm decade^{-1}) have occurred along the Queensland coast and over half of Queensland is now affected by decreasing rainfall of 20 mm decade^{-1} or greater (figure 1). If the rainfall trends from 1951–2007 are compared to the climatological mean (1961–1990), over half of inland Queensland, as well as coastal areas have had a significant rainfall decline of 20% or higher, with some areas, such as north of Mackay, a 50% or higher decrease in rainfall (figure 2).

Although the El Niño-Southern Oscillation (ENSO) and tropical sea surface temperatures are important factors on the year to year rainfall variability in Queensland [3-5], other factors on decadal time scales must be underpinning the long-term rainfall decline. Over southern Australia, some of the long-term rainfall declines have been attributed to increases in the mean sea level pressure (MSLP) and to changes in the Southern Annular Mode (SAM), and are less likely related to changes in ENSO or the Indian Ocean Dipole (IOD)[6,7]. However, it has been argued that the key driver of major droughts over Victoria is the IOD [8]. These studies and others, indicate the causes of the rainfall declines (and associated droughts) in Australia are likely to be complex and vary between regions.

IOP Conf. Series: Earth and Environmental Science **11** (2010) 012037 doi:10.1088/1755-1315/11/1/012037

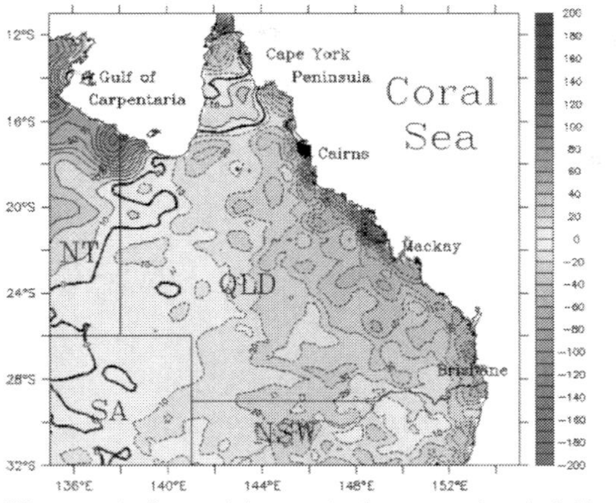

Figure 1. Decadal trends in annual rainfall (mm decade $^{-1}$) over Queensland from 1951–2007. Gridded data (0.25° by 0.25°) from the Australian Bureau of Meteorology.

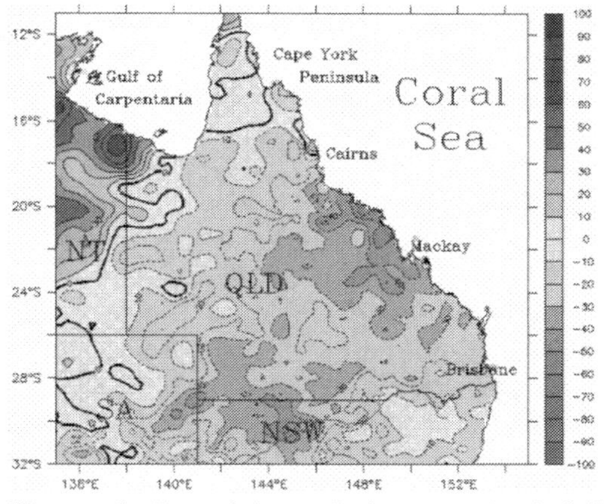

Figure 2. Decadal trends in annual rainfall over Queensland (1951–2007) as a percentage of the annual mean (1961–1990).

2. Data and Method

Monthly rainfall and MSLP data were obtained from the Australian Bureau of Meteorology (BoM) from 25 stations from across the Queensland and adjacent regions including Lord Howe and Norfolk Islands. Although several stations have reliable records starting in the 1880s and 1890s, no significant rainfall or MSLP changes occur prior to ~1950, and so the focus of this study is on the period from 1951–2007. Stations were selected based on the continuity of rainfall and MSLP data and the location to provide statewide coverage. The monthly MSLP data consist of 9am and 3pm readings and both have been averaged to calculate the mean MSLP value over the time series at each station. Linear regression analysis is used to calculate the change in MSLP from the start to the end of each time series for seasonal and annual values. The decadal trend in seasonal and annual MSLP is calculated from the mean MSLP change. All time series of the MSLP exhibit considerable year to year variability and shows an increase from the start to the end of each time series.

A correlation analysis between the seasonal and annual rainfall and MSLP across the Queensland region was completed to investigate the relationship between lower rainfall and higher surface pressure. Since many of the time series start in a different year, the level of significance (using the 5% level) varies slightly for each and these values are shown in the corresponding tables (tables 1 and 2). Only results from ten stations are shown in the tables to provide broad spatial coverage.

A correlation analysis between the MSLP from Queensland and the SAM from 1957–2007 using data from the Southern Hemisphere Annular Mode Index located at http://www.antarctica.ac.uk/met/gjma/sam.html (and links therein) is completed to indicate any relationships to the SAM, such as over southern Australia. This SAM dataset was selected as it has been corrected by a number of observational based stations around the South Polar Region and the mid-latitudes [9].

A correlation analysis between an ensemble mean of the annual MSLP from fourteen stations (identified by having an annual correlation value above the 5% level to the SAM) from the Queensland region and Norfolk Island for the period 1957–2007 highlights MSLP changes and the links to the subtropical ridge. Norfolk Island is located in the north Tasman Sea, east of southern Queensland and changes in the MSLP in the austral spring, summer and autumn should be unaffected by the Australian land mass and continental heating. The correlation between the annual MSLP from fourteen stations in an ensemble mean and the SAM index is provided to show the overall teleconnections between the two regions.

3. Changes in the MSLP across the Queensland Region

Contoured plots of the decadal trend in seasonal and annual MSLP from the Queensland region are shown in figure 3. In DJF, the increase in MSLP is highest over southeast Queensland (SE QLD)(> 0.3 hPa decade^{-1}) and the north Tasman Sea, with the largest values at Toowoomba, Norfolk Island, Lord Howe Island, Burketown and Longreach. A small decrease in MSLP occurs over far southwestern Queensland around Thargomindah. In MAM, all of Queensland is covered by an increase in MSLP, the highest values (>0.5 hPa decade^{-1}) over SE QLD at Toowoomba and at Norfolk and Lord Howe Islands. The coastal regions south of Rockhampton and parts of the Cape York Peninsula also show moderate increases in MSLP (>0.3 hPa decade^{-1}). In JJA, the increase in MSLP is slightly stronger over Queensland than in MAM, with a larger region covered by an increase in MSLP of 0.3 hPa decade^{-1} or higher. The largest increase (>0.5 hPa decade^{-1}) occurs at Toowoomba and Norfolk and Lord Howe Islands. In SON, the increase in MSLP is lower over most of Queensland compared to other seasons, with the largest increases (>0.3 hPa decade^{-1}) over SE QLD at Toowoomba and the Tasman Sea (Norfolk and Lord Howe Islands). A small decrease in MSLP occurs over southwest Queensland and the Mount Isa region. The strongest increases in seasonal MSLP occur in MAM and JJA. The decadal trend in the annual MSLP shows nearly the entire region is covered by small to moderate increase (>0.2 hPa decade^{-1}) in MSLP, strongest over SE QLD and the north Tasman Sea (southeast of Brisbane and south of the Coral Sea).

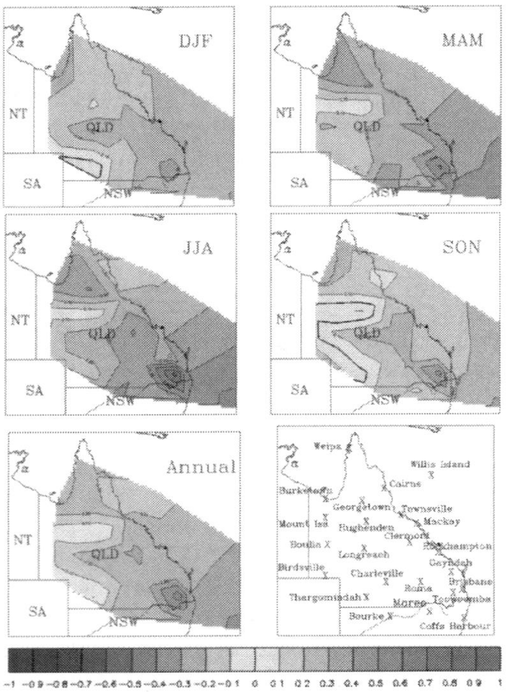

Figure 3. Contoured seasonal and annual decadal MSLP trends (hPa decade^{-1}) from BoM observations over the Queensland region (1951–2007) in DJF, MAM, JJA, SON and annual periods. Regions shaded in red show increasing MSLP and blue decreasing MSLP. Bold contour line is zero MSLP change. Station locations - lower right plot.

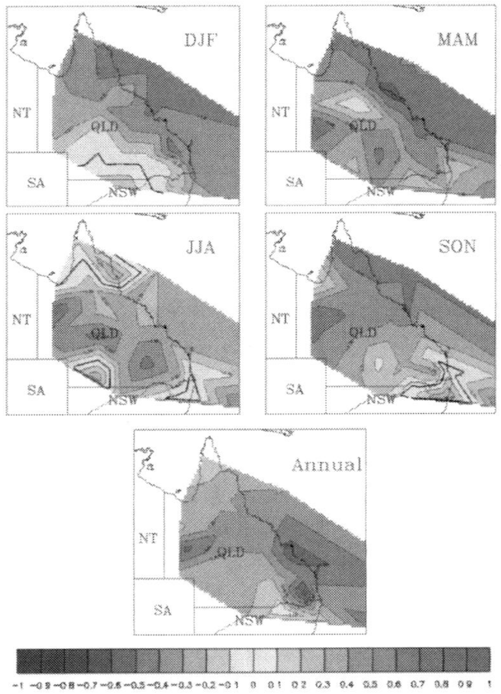

Figure 4. Contoured seasonal and annual correlation between rainfall and MSLP from BoM observations over the Queensland region (1951–2007) in DJF, MAM, JJA, SON and annual periods. Regions shaded in red show positive correlation and blue negative correlation. Bold contour line is zero correlation.

4. Correlation between MSLP and Rainfall

A summary of the correlation values between rainfall and MSLP over the Queensland region from ten selected stations are shown in table 1. In DJF, the locations with the highest negative correlation are Gayndah (-0.50), Cairns (-0.49), Willis Island (-0.49) and Mackay (-0.43). Eleven out of 25 stations have correlation values which are statistically significant above the 5% level of significance. In MAM, the highest negative correlation values occur at Townsville (-0.54), Norfolk Island (-0.50), Boulia (-0.50) and Mackay (-0.49). Sixteen stations have correlations which are statistically significant (above the 5% level) and this is the largest number of stations for all seasons. In JJA, four stations have negative correlations between -0.4 and -0.6 including Lord Howe Island (-0.51), Mount Isa (-0.51), Roma (-0.45) and Longreach (-0.43). Eight stations have correlation values above the 5% level. In SON, the highest negative correlations occur at Weipa (-0.50), Lord Howe Island (-0.46), Willis Island (-0.43) and Boulia (-0.43). Thirteen out of 25 stations have correlation values above the 5% level of significance. The annual correlation is highest (negative) at Toowoomba (r=-0.65) and r^2 indicates ~42% of the rainfall decrease can be explained by an increase in MSLP. Other large annual negative correlations are at Rockhampton (-0.56), Boulia (-0.54) and Lord Howe Island (-0.45). Twelve out of the 25 stations have correlation values which are statistically significant above the 5% level.

Regionally (figure 4), in DJF, the highest negative correlation values (>-0.3) are located over eastern, central and northern coastal Queensland. A small region over the southwest border of Queensland with New South Wales (NSW) shows a small positive correlation (>0.1). In MAM, the negative correlation covers the entire Queensland region, with the highest values (>-0.3) along the east coast regions and southern and western Queensland. In JJA, the negative correlation covers most of Queensland and is highest in western and south-central Queensland, except for a small positive correlation over southwest parts and the Cape York Peninsula. In SON, the negative correlation covers nearly the entire Queensland region, highest over northern Queensland. A small positive correlation occurs in northeast NSW. The highest correlations occur in MAM, followed by SON and DJF. The annual correlation shows the entire Queensland region is covered by negative values, with the largest values (>-0.4) over western, central and coastal SE QLD. We have shown there is a significant relationship between the increasing MSLP and lower rainfall over Queensland in all seasons, generally strongest in coastal areas and SE QLD.

Table 1. Seasonal and annual correlation between the MSLP and rainfall from a selection of ten stations across the Queensland region and the north Tasman Sea.

Location	Level of Significance at 5%	Seasonal Correlation				Annual Correlation
		DJF	MAM	JJA	SON	
1. Toowoomba	0.35	-0.12	-0.36	-0.24	-0.24	-0.65
2. Brisbane Aero	0.27	-0.35	-0.13	0.01	0.06	-0.36
3. Norfolk Island	0.27	-0.35	-0.50	-0.33	-0.42	-0.32
4. Lord Howe Island	0.27	-0.38	-0.44	-0.51	-0.46	-0.45
5. Burketown	0.33	-0.28	-0.32	0.04	-0.01	-0.24
6. Longreach	0.33	-0.07	-0.29	-0.43	-0.29	-0.29
7. Rockhampton	0.27	-0.36	-0.45	-0.27	-0.22	-0.56
8. Mackay PO	0.29	-0.43	-0.49	-0.13	-0.41	-0.41
9. Townsville	0.27	-0.23	-0.54	-0.25	-0.36	-0.31
10. Cairns	0.27	-0.49	-0.42	0.29	-0.43	-0.31

5. Correlation between MSLP and the SAM

A summary of the correlation values between MSLP and the SAM index from ten selected stations over the Queensland region are shown in table 2. In DJF, only three stations, Townville (0.32), Cairns (0.29) and Gayndah PO (0.29) have significant correlation values above the statistically significant level of 5%. In MAM, no stations have correlations statistically significant at the 5% level of significance. In JJA, nineteen stations have correlation values greater than the statistically significant level of 5%, with Coffs Harbour (0.46), Lord Howe Island and Burketown (0.44), Bourke (0.43) and Toowoomba (0.41) above 0.40. This indicates nineteen stations show an increase in the MSLP that may be related to the decrease in MSLP in the South Polar Region, with higher MSLP in the mid–latitudes. In SON, only one station (Coffs Harbour) has a correlation value above the statistically significant level of 5%. The annual correlation between the MSLP and the SAM index shows fourteen stations have correlation values greater than the statistically significant level of 5%, with the highest values at Bourke, Coffs Harbour and Lord Howe Island (0.46). Thus the signal from the SAM is strong enough to be significant in the annual correlation, even though only the austral winter season has a significant correlation at a number of stations.

Table 2. Seasonal and annual correlation between MSLP and the SAM from a selection of ten stations across the Queensland region and the north Tasman Sea.

Location	Level of Significance at 5% Level	Seasonal Correlation				Annual Correlation
		DJF	MAM	JJA	SON	
1. Toowoomba	0.35	0.21	0.07	0.41	0.08	0.43
2. Brisbane Aero	0.27	0.26	0.12	0.39	0.20	0.43
3. Norfolk Island	0.27	0.24	0.18	0.39	0.09	0.44
4. Lord Howe Island	0.27	0.19	0.18	0.44	0.14	0.46
5. Burketown	0.33	0.29	0.08	0.44	-0.08	0.26
6. Longreach	0.33	0.28	0.03	0.28	0.04	0.32
7. Rockhampton	0.27	0.21	0.03	0.34	0.15	0.40
8. Mackay PO	0.29	0.22	-0.02	0.32	0.04	0.34
9. Townsville	0.27	0.32	-0.03	0.37	0.03	0.34
10. Cairns	0.27	0.29	-0.10	0.33	0.07	0.25

6. Relationship of MSLP to the Subtropical Ridge and the SAM

A time series of the annual MSLP at Norfolk Island and an ensemble mean using fourteen stations from the Queensland region is shown in figure 5. This shows there has been an increase in MSLP by ~2 hPa in both regions, with a strong correlation of 0.89 (r^2 =79%). The correlations from the ensemble mean for the austral summer is 0.78, autumn (0.67), winter (0.88) and spring (0.80) and indicates all seasons are well correlated to the changes in the MSLP at Norfolk Island. Seasonal global MSLP data shows the strongest MSLP changes are in the Tasman Sea [1] and this indicates the subtropical ridge is the most likely cause of the increase in MSLP over the Queensland region.

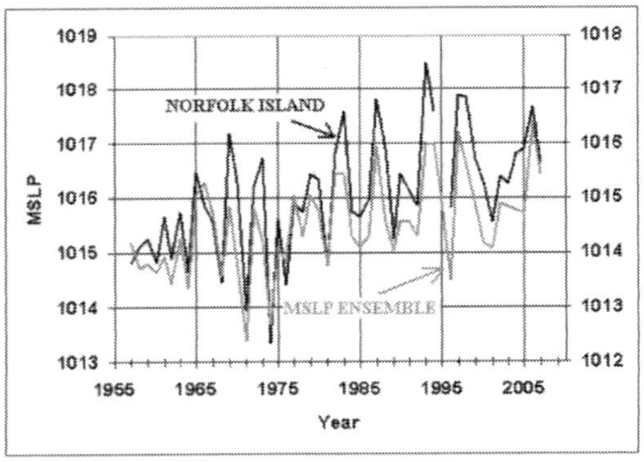

Figure 5. Time series of the annual MSLP at Norfolk Island (black) and ensemble mean (purple) from the Queensland region for 1957–2007. Correlation = 0.89.

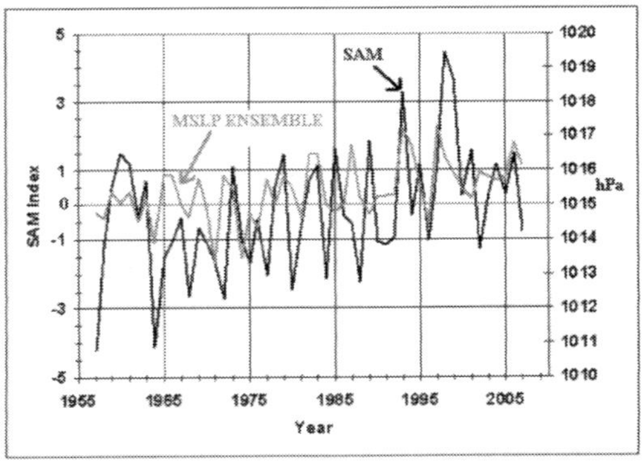

Figure 6. Time series of the annual SAM mode index (black) and the annual MSLP ensemble (purple) over the Queensland region for 1957-2007. Correlation = 0.44.

A time series of the annual SAM index and an ensemble mean of annual MSLP are shown in figure 6. The SAM index shows considerable variability from year to year and varies between ~2–4 hPa. A long–term increase in the MSLP in the SAM index from negative to positive values is evident. The MSLP ensemble from the Queensland region has a lower year to year variability (~1–2 hPa) but has many peaks and troughs in phase with the SAM index. The MSLP ensemble also shows an increase in the MSLP over the 50 year period from 1015 hPa to 1016 hPa. Similar peaks between the SAM index and the MSLP ensemble occurred in 1963, 1965/66, 1969, 1982/1983, 1993 and 2006 and troughs in 1962, 1964, 1968 and 1996 respectively. Several other periods (1971/72, 1980/81 and 1997/98) show the MSLP ensemble from the Queensland region leads or lags the SAM index peak or trough by one year. From 1984 to 1990, the changes in the SAM index and the MSLP ensemble are distinctly opposite to one another, for reasons which remain unclear. The correlation between the SAM index and the ensemble mean for the entire period (1957–2007) is 0.44, indicating ~19% of the variance in the MSLP can be explained at these stations by changes in the annual SAM index.

7. Conclusions

Although ENSO plays an important role in year to year rainfall variability over the Queensland region, it is unlikely to be playing a major role in the rainfall declines. This study shows the subtropical ridge and to a lesser degree the SAM are important factors in the rainfall declines over southern Queensland, well north of previous studies from southern Australia [7] and New Zealand [10]. Increases in MSLP are significant across much of the Queensland region, especially SE QLD and the coastal regions, with the largest increases in austral autumn and winter, indicating the subtropical ridge is driving some of the MSLP changes over the region. The strongest impacts from the SAM on MSLP were shown to be in the austral winter and for the annual period, with lower impacts in the other seasons. Although we have not investigated direct links to the descending region of the Hadley circulation over the Australian region, changes during this time may have enhanced the drying trend through increased subsidence in association with the subtropical ridge [12]. We have demonstrated changes in the MSLP have directly altered rainfall in the Queensland region, however further studies are required to identify the significant synoptic features which have led to the rainfall decline over the Queensland region, which are clearly different from those in southern regions of Australia.

In this paper, we have shown the rainfall declines have been accompanied by an increase in the mean sea level pressure (MSLP) over the region. The increases in the MSLP are directly linked to changes in the behaviour of the subtropical ridge in the southeast Australian region and at least partially due to the positive polarity of the SAM. Since the changes in the SAM are directly related to ozone depletion and rising greenhouse gases [6,11], the changes in rainfall over much of Queensland are could be at least partly anthropogenic in origin and may represent climate change that has developed and persisted in rainfall trends for now over half a century.

References

[1] Cottrill D A M 2009 *PhD thesis*: University of Southern Queensland
[2] Gallant A J E, Hennessy K J and Risbey J 2007 *Aust. Met. Mag.* **56** 223-39
[3] Cai W, Whetton P H and Pittock A B 2001 *Climate Dyn.* **17** 421-32
[4] Stone R C and Auliciems A 1992 *Int. J. Climatol.* **12** 6 625-36
[5] Murphy B F and Ribbe J 2004 *Int. J. Climatol.* **24** 703-21
[6] Cai W and Cowan T 2006 *Geophys. Res. Lett.*, **33** L24708
[7] Nicholls N 2009 *Climate Dyn.* Doi:10.1007/s00382-0090527-6
[8] Ummenhofer C C, England M H, McIntosh P C, Meyers G A, Pook M J, Risbey J S, Gupta A S and Taschetto A S 2009 *Geophys. Res. Lett.*, **36** L04706
[9] Marshall G J 2003 *J. Climate* **16** 4134-43
[10] Ummenhofer C C and England M H 2007 *J. Climate* **20** 5418-40
[11] Crook G D, Gillet N P and Keely S P E 2008 *Geophys. Res. Lett.*, **35** L07806
[12] Seidel D J, Fu Q, Randel W J and Reichler T J 2007 *Nat. Geosci.* **1** 24-7

Assessing rainfall trends and remote drivers in regional climate change projections: the demanding test case of Tasmania

MR Grose[1], S Corney[1], J Bennett[2], CJ White[1], GK Holz[1] and NL Bindoff[1,3]
[1] Climate Futures for Tasmania, Antarctic Climate and Ecosystems Co-operative Research Centre, University of Tasmania, Private Bag 80, Hobart, TAS 7001, Australia
[2] Hydro Tasmania Consulting, 89 Cambridge Park Drive, Cambridge, TAS 7170, Australia
[3] Centre for Australian Weather and Climate Research, Aspendale, VIC 3195, Australia

E-mail: Michael.Grose@acecrc.org.au

Abstract. Understanding and modelling Tasmanian rainfall variability and making future projections of Tasmanian rainfall are challenging tasks. Tasmania has spatially and temporally complex rainfall patterns. Rainfall variability is influenced by a complex suite of remote drivers and these influences vary by season. The *Climate Futures for Tasmania* high-resolution model simulations project small changes to annual rainfall averaged over Tasmania, but larger changes to the spatial patterns and seasonality of rainfall. A case study of changes to summer rainfall under a high greenhouse gas emission scenario is shown here. The projected summer decrease in rainfall in the western rainfall region is consistent with the southerly movement and intensification of the subtropical ridge as well as an enhancement of the high phase of the Southern Annular Mode. The increase along the east coastal strip is consistent with an increase in blocking in the Tasman Sea as well as an increase in sea surface temperature, relative humidity and convective rainfall. We propose that projections of rainfall for places like Tasmania are strengthened through dynamical downscaling and also the analysis of the rainfall mechanisms within the model at all length scales.

1. Introduction

Tasmania is an island with a temperate maritime climate that lies between 39.5 and 44°S in the Roaring 40s belt of westerly airflow. The interaction of the westerlies with rugged topography leads to a strong pattern of rainfall variability across Tasmania, with over 3000 mm falling per annum near the west coast and approximately 600 mm per annum in the central midlands. The westerlies are weakest in summer. Westerly flow can be interrupted in any season by blocking highs in the Tasman Sea [1] and these blocks are associated with cutoff lows that can cause high rainfall events in the northeast [2]. The seasonality of rainfall varies in different regions of the state, with a distinct seasonal cycle in the western region, and no significant cycle in the east.

Tasmanian rainfall variability is influenced by various remote drivers, including the El Niño Southern Oscillation (ENSO), Southern Annular Mode (SAM), Indian Ocean Dipole (IOD) and atmospheric blocking in the Tasman Sea. The strength of the influence of these drivers varies across the state and in different seasons. In summer, rainfall variability shows the greatest correlation with an index of SAM on the west coast and an index of blocking on the east coast [3].

2. Climate modelling: methods

Output from six global climate models (GCMs) were dynamically downscaled to a fine resolution appropriate to Tasmania using the CSIRO Conformal Cubic Atmospheric Model (CCAM) [4]. This dynamical downscaling method takes output from a host GCM and uses it as forcing into a stretched grid global atmospheric model. The result is a fine-scale dynamical model over the area of interest. To achieve a desired final resolution of 0.1°, a two stage downscaling process was implemented. The first stage (intermediate model) downscaled host GCM outputs to a grid with the high-resolution face of the grid covering all of Australia at a resolution of approximately 0.5°. The second stage placed the high-resolution face over Tasmania and the Bass Strait islands at an approximate resolution of 0.1°.

The representation of the average annual total rainfall for the baseline climatology period of 1961-1990 is improved through both of the downscaling stages (figure 1). A typical GCM resolution (left panel) only has a few grid cells that cover the state, suggesting the average state-wide annual total precipitation is approximately 750 mm. The 0.5° resolution model (centre panel) shows greater spatial structure and an average total of approximately 1000 mm. The final 0.1° resolution model (right panel) has a finely resolved spatial pattern of rainfall and an average total of 1385 mm, closely resembling the observed spatial patterns and total of 1390 mm measured by the Bureau of Meteorology [5]. It was concluded that the high-resolution 0.1° dynamical downscaling process had the ability to model the present climate of Tasmania with a high degree of fidelity, including seasonality, spatial variance and relationships between the different climate variables [6].

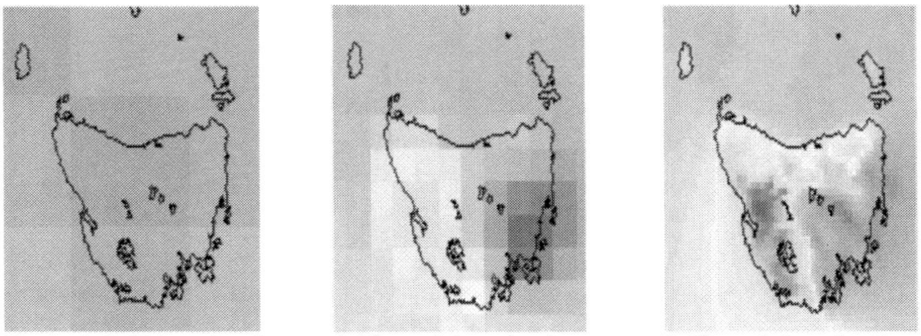

Typical GCM projection   0.5° model projection   0.1° model projection

Figure 1. Average annual precipitation totals for Tasmania projected on typical GCM, 0.5° and 0.1° grids. Precipitation scaled from 0-3000 mm per annum

Six GCMs were selected based on an assessment of their reproduction of present-day climate of eastern Australia [7]. The GCMs selected were CSIRO-Mk3.5 (Australia), GFDL-CM2.0 and GFDL-CM2.1 (USA), ECHAM5/MPI-OM (Germany), UKMO-HadCM3 (UK), and MIROC3.2(medres) (Japan). Two SRES emission scenarios were used to provide a range of likely projected futures, from a low (B1) to a high (A2) emission of greenhouse gases. For the purposes of this brief paper the six-model mean for all analyses is presented. Rainfall changes over Tasmania are examined in the finest scale model outputs (0.1° grid) and changes to large-scale drivers are examined in the intermediate resolution model simulations (0.5° grid).

3. Results

The six-model mean of the projections shows a change of less than 5% in total annual rainfall under either emission scenario and larger changes of up to 15% on a seasonal basis. Individual districts within Tasmania show much larger changes than the statewide average. It should be noted that the

Tasmanian rainfall average is dominated by the high rainfall region of the west coast. The projections under the two scenarios show a similar spatial pattern of change, but there is an enhancement of the magnitude of the change in the higher emission scenario is observed compared to the lower. Summer rainfall under the A2 scenario shows a steadily emerging pattern of increased rainfall in the east and northeast areas of Tasmania and a decrease in summer rainfall on the west and southwest regions (figure 2).

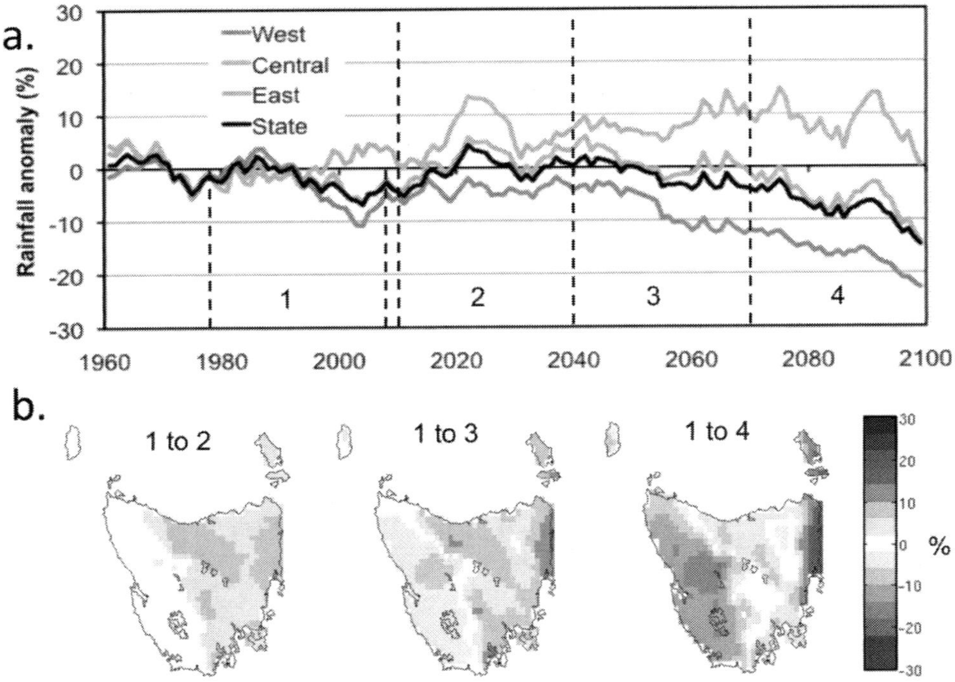

Figure 2. Projection of six-model-mean Tasmanian summer rainfall for the A2 scenario, a) 11-year moving average timeseries of proportional rainfall anomaly from 1961-1990 baseline for Tasmania and three Bureau of Meteorology forecast zones (East = 'East Coast', Central = 'Central Plateau & Upper Derwent Valley', West = 'West & South Coast & Highlands'), b) map of proportional difference in summer rainfall between the periods marked on the timeseries plot (1 = 1978-2007, 2 = 2010-2039, 3 = 2040-2069, 4 = 2070-2099)

The six-model mean of selected indices of relevance to the climate variability of Tasmania is shown in figure 3. An index of the summer intensity of the subtropical ridge (STR) [8] is seen to increase by 1 hPa over the projections. An index of the position of the STR [9] shows a southerly shift of >1 degree in summer. The blocking index at 140 °E [10] shows a steady increase over the period and the regional Antarctic Oscillation Index [11] shows a steady intensification of the Summer SAM. These all appear as a continuation of the current direction of trend as indicated by anomaly values from NCEP Reanalysis 1 [12] also shown on the plots.

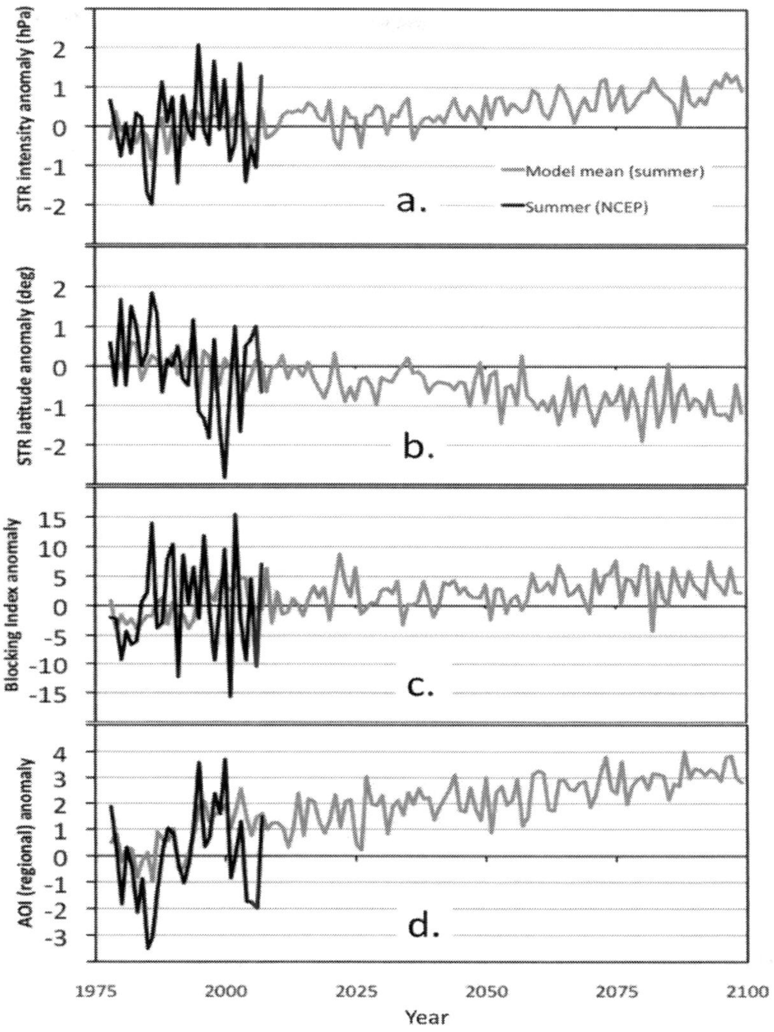

Figure 3. Timeseries of climate indices anomaly from 1978-2007 mean calculated from NCEP Reanalysis and the six-model-mean, a) summer central pressure (intensity) of the subtropical ridge across the Australian zone, b) summer latitude anomaly of the subtropical ridge, c) summer blocking index anomaly at 140 °E, d) summer regional Antarctic Oscillation Index anomaly

Changes to summer mean sea level pressure (MSLP), zonal wind and sea surface temperature (SST) are shown in figure 4. Maps indicate the modelled mean for the recent period (1978-2007) and the change between the recent period and the end of the century (2070-2099). There is an increase in summer MSLP in the region of Tasmania. The increase is greater over New Zealand to the east and in the Indian Ocean to the west. Zonal wind is strongly positive (westerly wind) in the latitude of Tasmania. The average strength of the westerlies is seen to decrease by approximately 1 ms^{-1} in the region of Tasmania in these projections. Summer SST shows a current gradient from >17 °C in the northeast to <15 °C in the southwest. This gradient is seen to increase in these projections, with an increase in summer SST of >3 °C on the east coast and <2.3 °C on the west coast. This is likely to be due to a strengthening or increased extension of the East Australia Current.

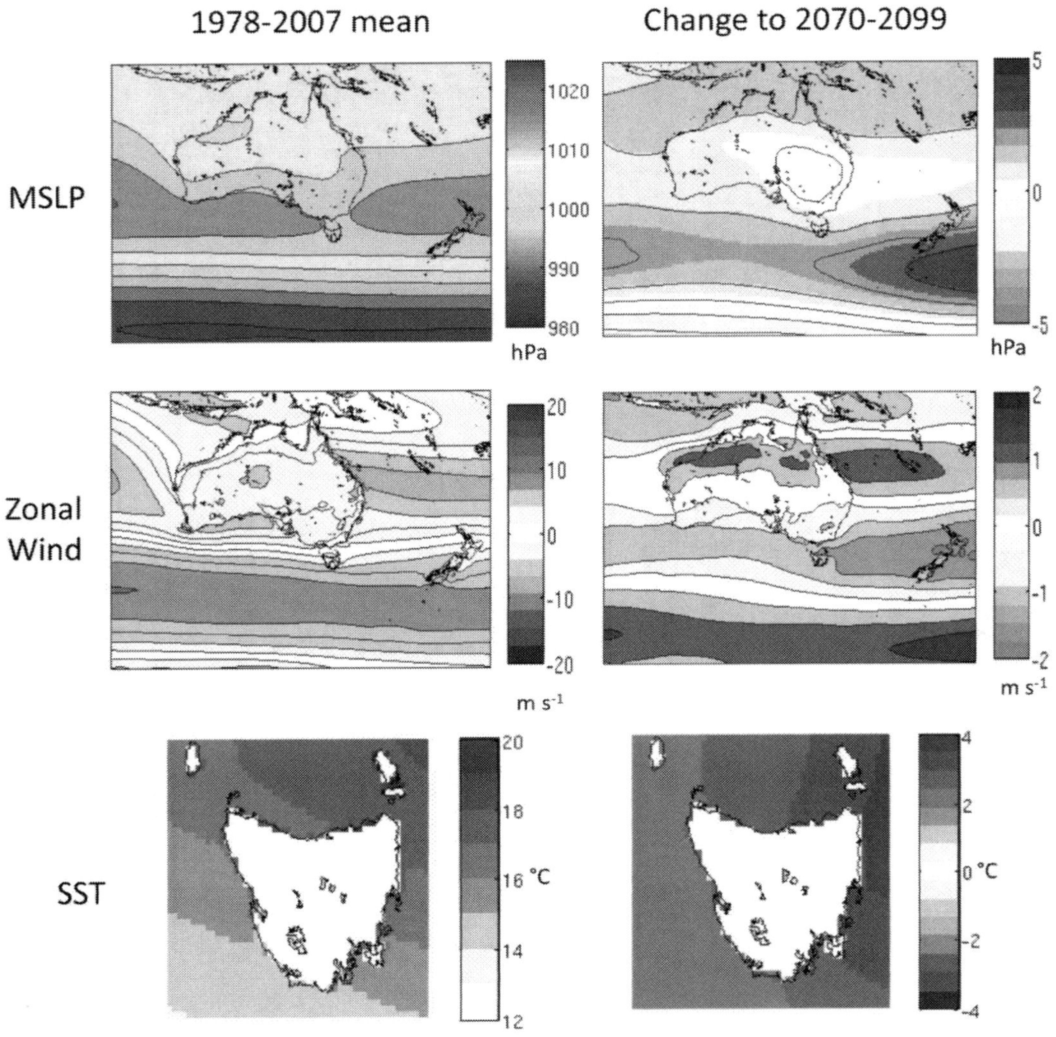

Figure 4. Maps showing the average mean sea level pressure (MSLP), zonal wind and sea surface temperature (SST) present in the models in summer 1978-2007, and the change in these means to 2070-2099

There is an east-west pattern in variables related to rainfall and heat balance (not shown). Over the eastern strip there is an increase in cloud cover, a reduction in solar radiation and an increase in relative humidity whereas the western district experiences a decrease in cloud cover, an increase in solar radiation and a decrease in relative humidity. There is an increase in convective available potential energy (CAPE) over the entire state, but the increase is greater over the eastern region. Similarly, there is a general increase in the proportion of rainfall that falls through convective processes and this change is greater on the eastern side. There is an increase in the maximum precipitation rate over the eastern strip and a decrease over the west.

4. Discussion and Conclusion

The changes to mean summer rainfall shown in figure 2 can be related coherently to the changes in the climate drivers shown in figures 3 and 4. The decrease in rainfall on the west coast is consistent with the weakening and deflection of westerly systems at the latitude of Tasmania indicated by the intensification and southerly movement of the STR and a decrease in the dominant zonal wind. The increase in the SAM index is also consistent with a rainfall decline. In contrast, the rainfall increase seen in the northeast strip is consistent with the increase in blocking index and a change to the mean zonal wind, which implies an increase in northeasterly rain-bearing systems including cutoff lows. The increase is also consistent with an increase in SST, CAPE, convective processes and rainfall intensity indicating an increase in dynamic precipitation events.

Tasmanian rainfall distribution is reliant on a complex suite of mechanisms and drivers at all scales, and an interaction of systems with complex topography. Fine scale model simulations appear more able to account for all the relevant drivers at all scales, including the interaction of air with topography than the corresponding coarse scale GCMs. Also, the projections of rainfall can be used with more confidence when the drivers behind rainfall changes are examined, and the projected changes are plausible and are consistent with our knowledge of climate dynamics.

Acknowledgements

This work was supported by the Australian Government's Cooperative Research Centre Programme through the Antarctic Climate and Ecosystems Cooperative Research Centre (ACE CRC). Climate Futures for Tasmania is possible with support through funding and research of a consortium of state and national partners.

References

[1] Sturman A and Tapper N 1996 The weather and climate of Australia and New Zealand. (Melbourne, Vic: Oxford University Press)
[2] Pook MJ, Risbey JS and McIntosh PC *IOP Conf. Ser. : Earth. Environ. Sci.* this volume
[3] Risbey JS, Pook MJ, McIntosh PC, Wheeler MC and Hendon HH 2009 *Mon. Weath. Rev.* **137** 3233-3253
[4] McGregor JL 2005 *CCAM: geometric aspects and dynamical formulation Tech. Paper 70* (Melbourne, Vic: CSIRO Marine and Atmospheric)
[5] Jones DA, Wang W and Fawcett R 2009 *Aust. Met. & Oceanog.* J. **58** 233-248
[6] Corney S, Katzfey JJ, McGregor JL, Grose MR, Bennett J, White CJ, Holz GK and Bindoff NL *In Press* ACE CRC report, Hobart Tasmania
[7] Smith IN and Chandler E 2009 *Clim. Change*. DOI 10.1007/s10584-009-9757-1
[8] Larsen SH and Nicholls N 2009 *Geophys. Res. Lett.* **36** L08708
[9] Drosdowsky W 2005 *Int. J. of Climatol.* **25** 1291-1299
[10] Pook MJ and Gibson TT 1999 *Aust. Met. Mag.* **48** 51-60
[11] Meneghini B, Simmonds I and Smith IN 2007 *Int. J. of Climatol.* **27** 109-121
[12] Kalnay E *et al.* 1996 *Bull. Am. Met. Soc.* **77** 437-471

IOP Conf. Series: Earth and Environmental Science 11 (2010) 012039 doi:10.1088/1755-1315/11/1/012039

Scatterometer-derived Southern Ocean mesoscale cyclone activity over the last decade

D Irving[1,2], I Simmonds[1], K Keay[1]

[1] School of Earth Sciences, The University of Melbourne, Victoria, Australia

[2] CSIRO Marine and Atmospheric Research, Aspendale, Victoria, Australia

Email: damien.irving@csiro.au

Abstract. Mesoscale cyclones (or mesocyclones) play an important role in the weather and climate of the Southern Hemisphere (SH) mid-to-high latitudes. However, the relatively small size and short lifetime of these systems, combined with the lack of available conventional data in this region, means that we have a poor understanding of their climatological characteristics. In this study, the University of Melbourne cyclone finding algorithm was applied to relatively high resolution scatterometer-derived surface pressure fields in order to produce a decade-long (1999-2008) climatology of mesocyclone activity over the Southern Ocean. The frequency of mesocyclone activity was found to be highest just to the north of the sea-ice zone, reaching a maximum over the Amundsen and Bellingshausen Seas (ABS) and off the coast of Wilkes Land, while the southern Indian Ocean was associated with systems of the largest depth, intensity and mean surface fluxes of momentum and mechanical energy. Maximum wind speed data indicated that SH polar lows may be more frequent than the current literature suggests, while strong positive trends identified in mesocyclone frequency over the ABS may represent a contributing factor to the rapid warming observed in that region over recent years. Partial correlation analyses indicated a link between mesocyclone frequency and the Southern Annular Mode.

1. Introduction

The sparsely populated and observed mid-to-high southern latitudes represent the region on Earth where we probably have the poorest understanding of the atmospheric system. This is particularly true on the mesoscale, where a lack of conventional data combined with the small size and relatively short lifetime of mesoscale cyclones (or 'mesocyclones') presents a problem for the detection and diagnosis of these systems. With a diameter of less than 1000 km and a lifetime often less than 24 hours, mesocyclones can manifest as insignificant, minor vortices with only weak surface circulation and no distinct cloud signature, through to the very active disturbances known as polar lows, which have been known to be associated with winds of hurricane force and heavy snowfall [1].

The importance of Southern Hemisphere (SH) mesocyclones in the general circulation of the atmosphere and in the coupled atmosphere-ocean-ice system is still unclear [2]. However, because they are so numerous [3], mesocyclones may make a substantial contribution to the poleward transport of heat and moisture, to the maintenance of the global atmospheric momentum budget, and to fluxes of momentum, mechanical energy and sensible and latent heat at the air-sea interface [3]. One factor contributing to our relatively poor understanding of the importance of these systems is our limited knowledge of their climatological characteristics [1]. Early research efforts in this field were observational in nature, with researchers making use of satellite imagery to locate cloud signatures indicative of mesocyclone activity [e.g. 4-5]. However, the manual identification of mesocyclones is a very time-consuming process and hence these observational studies could typically only focus on regions of suspected high activity around the Antarctic coastline for periods of no more than 12 months. In addition, there are reliability issues associated with such a subjective methodology, and not all mesocyclones are associated with a distinctive cloud signature [6-7].

In more recent times, the advent of various automated cyclone finding and tracking algorithms [e.g. 8-10] has freed researchers from many of the issues associated with manual techniques. As such, a number of studies have applied these algorithms to reanalysis pressure fields in order to produce climatologies of SH cyclone activity that span the last three to four decades of the twentieth century [e.g. 11-13]. However, while these studies have provided a wealth of information regarding synoptic-scale cyclone activity, the relatively coarse resolution of current global reanalysis products (which vary in resolution from 2.5° × 2.5° to 1.125° × 1.125°) means that they are able to capture only a limited percentage of mesoscale activity [14]. It is therefore fair to say that our climatological understanding of SH mesocyclone activity has progressed only modestly from the observational studies of the 1990s. As such, the focus of this paper was to make use of relatively high resolution scatterometer-derived surface pressure fields, in conjunction with an automated cyclone finding algorithm, to assemble a decade-long climatology of mesocyclone activity for the extratropical SH.

2. Data and methods

The polar orbiting National Aeronautics and Space Administration Quick Scatterometer (QuikSCAT) satellite has been providing estimates of the surface wind over the ice-free global oceans since mid-1999 (<http://winds.jpl.nasa.gov/missions/quikscat/index.cfm>), and the Department of Atmospheric Sciences at the University of Washington have made available to the public 0.5° × 0.5° oceanic swath-based (as opposed to products of greater spatial coverage) surface pressure fields derived from these data (hereinafter referred to as the UWQS pressure fields [15-16]). At 1600 km the effective swath-width of the QuikSCAT satellite is far greater than its predecessors and more than sufficient for the study of mesocyclone activity (although, it should be noted that this width is still too narrow to capture the full extent of synoptic activity). Hence, the UWQS pressure fields were used in conjunction with the University of Melbourne (UM) cyclone finding algorithm [10, 17] in order to determine climatological mesocyclone statistics. For this analysis, mesocyclones were defined as all systems of diameter less than 1000 km, which is consistent with previous work in the area [e.g. 6, 18]). The swath-based nature of the UWQS pressure fields meant that cyclone tracking could not be performed in an automated fashion, and hence the statistics presented can be thought of as the culmination of tens of thousands of unbiased, instantaneous snapshots of mesocyclone activity.

3. Results and discussion

In total, 82,907 cyclonic systems were identified poleward of 30°S from the UWQS pressure fields over the period September 1999 to November 2008, which represents an average of 0.87 systems per field/swath. The radius of these systems ranged from 0.41-11.08 °lat (46-1230 km), however since mesocyclones were defined to have an upper radial bound of 4.5 °lat (500 km) the largest 15% were excluded, leaving a final total of 70,354 mesocyclones (0.74 per field) for further analysis.

3.1. Mesocyclone frequency

The system density statistic reported by the UM cyclone finding software represents the proportion of the time that a given grid point was under cyclonic influence [10]. The annual and seasonal geographic distributions of this statistic showed an arc of highest values located 5-10 °lat north of the sea ice (or missing data) zone, extending from the south of Africa eastwards to Drake Passage (figure 1), which was generally consistent with previous reanalysis-based synoptic climatologies [e.g. 12-13, 19-20].

Seasonal area-weighted average system densities calculated over the entire study domain revealed autumn and spring as the most and least active seasons respectively.

The annual (figure 2) and seasonal trends observed in system density showed increases situated around the Antarctic Peninsula, with declines located throughout much of the remaining study domain. Again, these patterns were generally consistent with the aforementioned reanalysis-based climatologies, however the strong increasing trends seen over the Amundsen and Bellingshausen Seas (ABS) have not been identified in studies heretofore. These were of particular interest given that the mechanisms responsible for the rapid rise in mean near-surface air temperature observed to the west of the Antarctic Peninsula over the past half century are yet to be fully comprehended [21-22]. Since mesocyclone activity has been shown to yield an increase in sensible heat flux from the ocean to the lower atmosphere [23], an increased frequency of mesocyclone activity may represent a contributing factor to this rapid warming.

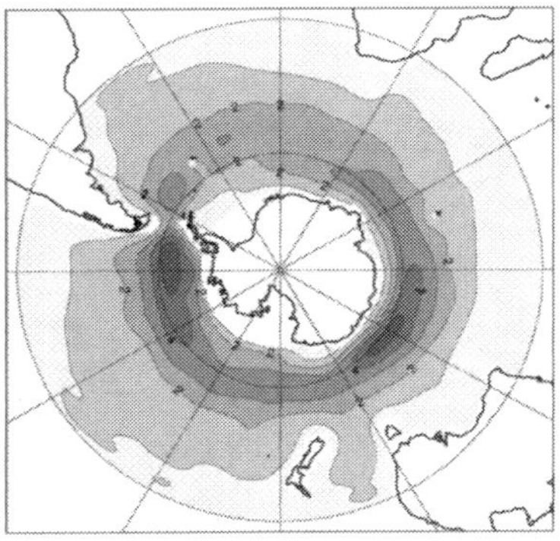

Figure 1. Annual mean system density (represented as the mean number of mesocyclones found in a 10^3 (°lat)^2 area per analysis). The contour interval is 1×10^{-3} (°lat)$^{-2}$.

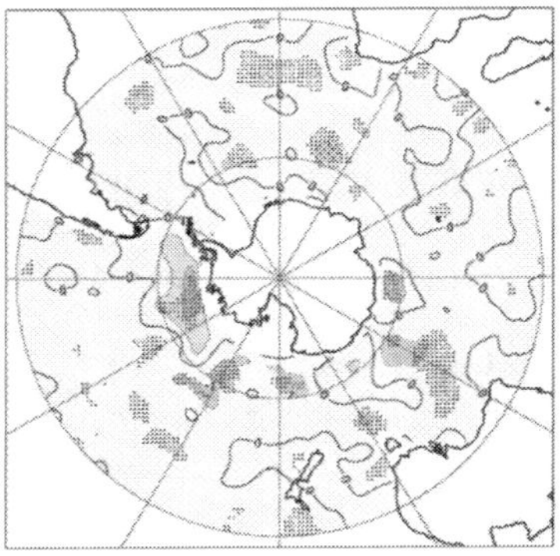

Figure 2. Annual trend (slope of the least squares line of best fit) in system density over the entire study period (contour interval is 1 system per 10^3 (°lat)^2 decade^{-1}). Regions over which the trend differs significantly from zero (at the 90% level) are indicated by stippling, while trends that differ from zero by greater than one contour interval are indicated by successively darker shades of blue (negative trend) or red (positive trend).

Seasonal Antarctic Oscillation Index (AOI) data [24] were used as a measure of the Southern Annual Mode (SAM), and broadly speaking a hemispheric, quasi-annular pattern of partial correlation was observed between the AOI and mesocyclone system density (figure 3). Mesocyclone activity is also thought to be associated with the El Niño Southern Oscillation (ENSO) and the Trans Polar Index (TPI) [25], and hence the calculation of *partial* correlations [26] allowed for the removal of the influence of these additional modes of climate variability. Since a positive phase of the SAM (i.e. positive AOI values) would be expected to be associated with a southward shift in the storm track [e.g. 27], the observed pattern of negative values from 45-60°S and regions of positive correlation poleward of this latitude band was suggestive of a sound physical association. Partial correlation fields were also calculated for ENSO and the TPI, however widespread regions of statistical significance were rare, and spatial patterns consistent with known physical mechanisms were not apparent.

3.2. Mesocyclone size and strength

In quantifying the size of each system it identifies, the UM cyclone finding algorithm calculates a 'radius' statistic, while the strength of a system is captured by the 'depth' and 'intensity' [11-12]. The spatial structures seen in the annual and seasonal mean depth and intensity fields were very similar, and hence only the depth statistic is presented in figure 4. As in previous reanalysis-based climatologies, which have used the UM definition for cyclone depth and intensity [e.g. 12-13] as well as alternative definitions such as those that involve a relative vorticity criterion [e.g. 28], the southern Indian and eastern South Atlantic Oceans were associated with systems of the greatest strength, with winter (summer) returning the largest (smallest) area-weighted average depth and intensity. The South Pacific Ocean was associated with mesocyclones of greater radius than the South Atlantic or Indian Oceans (figure 5), while the distribution of observed radii peaked at approximately 3.2-3.6 °lat (~350-400 km) and showed very little seasonality.

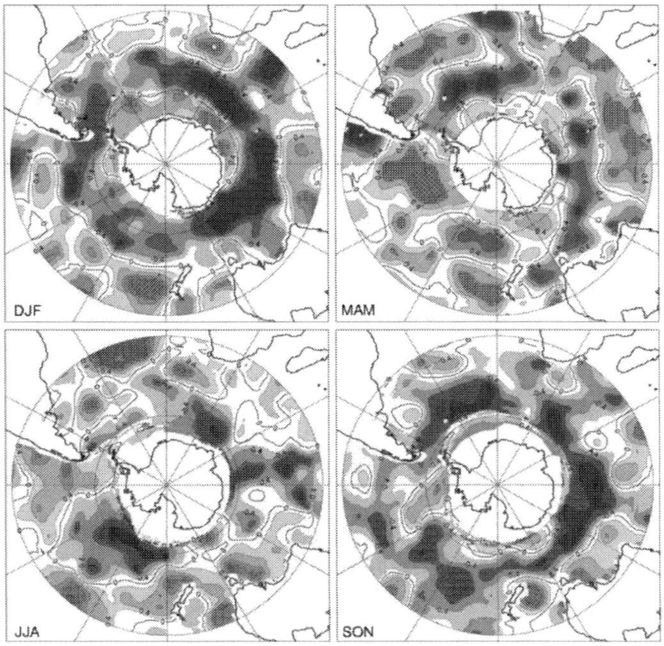

Figure 3. Seasonal partial correlation between system density and the AOI over the entire study period (contour interval is 0.2). Negative contours are dashed and the zero contour is highlighted in bold. Regions over which the correlation differs significantly from zero (at the 90% confidence level) are indicated by stippling, while correlations that differ from zero by greater than one contour interval are indicated by successively darker shades of blue (negative correlation) or red (positive correlation).

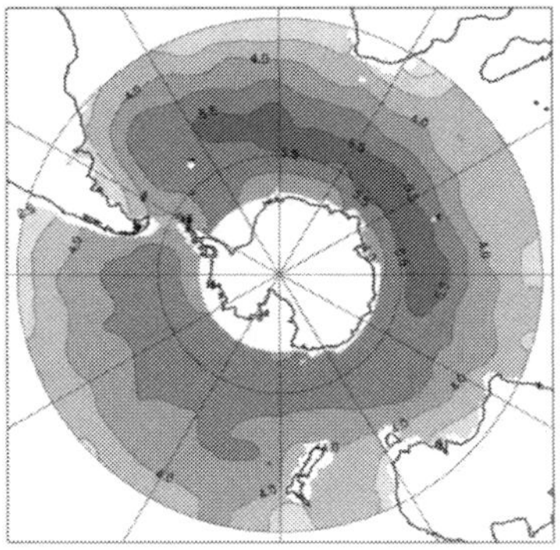

Figure 4. Annual mean mesocyclone depth (hPa). The contour interval is 0.75 hPa.

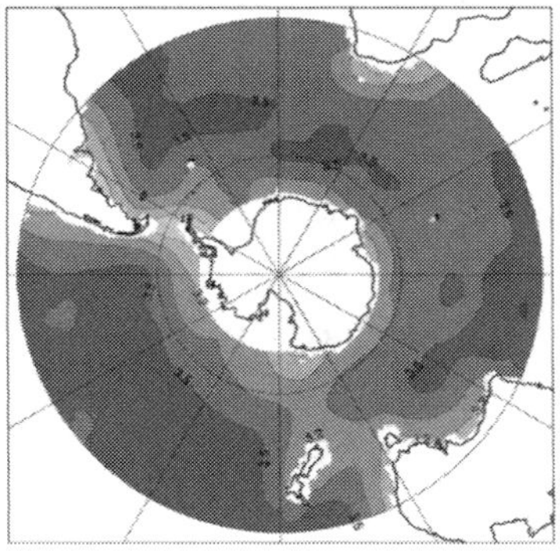

Figure 5. Annual mean mesocyclone radius (°lat). The contour interval is 0.15 °lat.

3.3. Air-sea interaction

The instantaneous scalar wind stress (or momentum flux, τ) and the mechanical energy flux (w) acting at the ocean surface can be parameterized as

$$\tau = \rho C_D u^2 \qquad w = \rho C_D u^3$$

[e.g. 29-30] where ρ is the atmospheric density, u the wind speed at some (near surface) reference level, and C_D the drag coefficient with respect to that level. While these expressions assume a constant drag coefficient and stationary ocean, which can lead to errors [31], they were considered a useful approximation for the purposes of this study. In applying these expressions ρ and C_D were approximated as 1.2 kg m^{-3} and 1.4×10^{-3} [32] respectively, and the mean u^2 and u^3 over the domain of each mesocyclone (taken here to be twice the system radius) was determined from QuikSCAT data. By using the mesocyclone frequency data presented earlier, it was then possible to create spatial fields that estimated the mean momentum and mechanical energy flux at any given time instant, associated with mesocyclone activity (i.e. the mesocyclone flux per analysis). Since both statistics showed a very similar spatial distribution, only the momentum flux per analysis is presented in figure 6. The highest annual mean values were located over the southern Indian Ocean and this maximum was most pronounced during winter, consistent with a recent model-based study [3]. By making use of the QuikSCAT data to calculate relevant spatial integrals, it was also possible to estimate that mesocyclones are responsible for 17% (20%) of the total climatological momentum (mechanical energy) flux over the entire Southern Ocean.

3.4. Maximum surface wind speed

The geographic distribution of the annual mean mesocyclone maximum wind speed is shown in figure 7, and is of particular relevance with respect to the classification of polar lows. This severe sub-category of mesocyclones is typically defined as those systems occurring poleward of the polar front (which varies in location between 60-70°S [e.g. 33]) with a diameter <1000 km and maximum wind speed >15 m s^{-1}. While SH polar lows are considered relatively rare in comparison to their Northern Hemisphere counterparts [1], we observed that 79% of all mesocyclones identified poleward of 60°S had a maximum wind speed of >15 m s^{-1}. This may suggest that SH polar lows are more frequent than the current literature suggests.

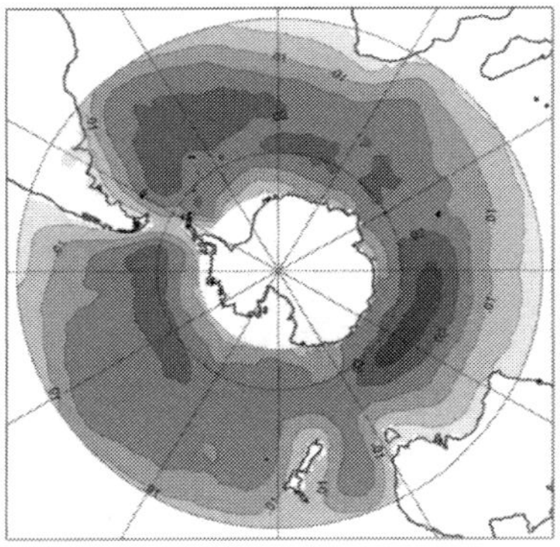

Figure 6. Annual mean mesocyclone momentum flux per analysis (N m^{-2}). The contour interval is 0.005 N m^{-2}.

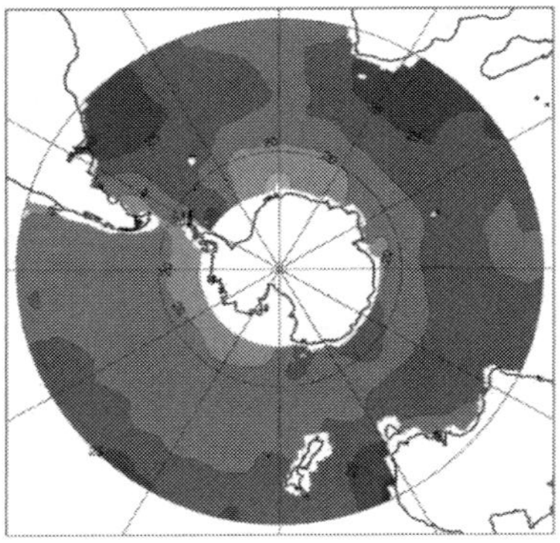

Figure 7. Annual mean mesocyclone maximum surface wind speed (m s^{-1}). The contour interval is 2.5 m s^{-1}

4. Conclusions

The amount of mesoscale detail contained in the UWQS pressure fields is arguably superior to that of any other data currently available for the climatological investigation of cyclonic activity in the southern extratropics. By using these fields in conjunction with the UM cyclone finding algorithm, a decade-long climatology of mesocyclone activity over the ice-free regions of the Southern Ocean was compiled.

The spatial patterns of mean mesocyclone radius, depth and intensity identified from the UWQS pressure fields, in addition to patterns of mesocyclone frequency and their associated trends, showed a close resemblance to those reported by existing synoptic-scale cyclone climatologies based on reanalysis data for the last three to four decades of the twentieth century. This finding suggests that synoptic and mesoscale cyclone activity are intimately linked, which is of particular interest considering that the dynamical processes responsible for the formation and intensification of mesoscale systems (particularly polar lows) is an area of ongoing, active research [1]. However, one feature not noted in existing synoptic climatologies was the strong positive trends in mesocyclone frequency observed over the ABS, which may represent a contributing factor to the rapid rise in mean near-surface air temperature observed to the west of the Antarctic Peninsula in recent years.

Analysis of QuikSCAT surface wind data indicated that mesocyclone activity makes a substantial contribution (~20%) to the total climatological fluxes of momentum and mechanical energy into the Southern Ocean, and that the occurrence of polar lows in the high southern latitudes may be more frequent than the current literature suggests. In addition, partial correlation analyses indicated a link between the frequency of mesocyclone activity and the SAM.

References

[1] Rasmussen E A and Turner J 2003 *Polar Lows: Mesoscale Weather Systems in the Polar Regions* (New York: Cambridge University Press)
[2] King J C and Turner J 1997 *Antarctic Meteorology and Climatology* (New York: Cambridge University Press)
[3] Yuan X, Patoux J and Li C 2009 *J. Geophys. Res.* **114** D04106 doi:10.1029/2008JD010874
[4] Carrasco J F and Bromwich D H 1992 *Antarct. J.* **27** 289-91
[5] Turner J and Thomas J P 1994 *Int. J. Climatol.* **14** 871-94
[6] Carleton A M 1995 *Int. J. Remote Sens.* **16** 2457-85
[7] Heinemann G and Klein T 2003 *Mon. Wea. Rev.* **131** 302-16
[8] Hodges K I 1995 *Mon.Wea. Rev.* **123** 3458-65
[9] König W, Sausen R and Sielmann F 1993 *J. Climate* **6** 2217-31
[10] Murray R J and Simmonds I 1991 *Aust. Meteor. Mag.* **39** 167-80
[11] Lim E -P and Simmonds I 2007 *J. Climate* **20** 2675-90
[12] Simmonds I and Keay K 2000 *J. Climate* **13** 873-85
[13] Simmonds I, Keay K and Lim E -P 2003 *Mon. Wea. Rev.* **131** 272-88
[14] Condron A, Bigg G R and Renfrew I A 2006 *Mon. Wea. Rev.* **134** 1518-33
[15] Patoux J, Foster R C and Brown R A 2003 *J. Appl. Meteor.* **42** 813-26
[16] Patoux J, Yuan X and Li C 2009 *J. Geophys. Res.* **114** D04105 doi:10.1029/2008JD010873
[17] Simmonds I, Murray R J and Leighton R M 1999 *Aust. Met. Mag.* **Special Edition** 35-49
[18] Turner J and Thomas J P 1994 *Int. J. Climatol.* **14** 871-94
[19] Pezza A B, Simmonds I and Renwick J A 2007 *Int. J. Climatol.* **27** 1403-19
[20] Wernli H and Schwierz C 2006 *J. Atmos. Sci.* **63** 2486-507
[21] Turner J, Colwell S R, Marshall G J, Lachlan-Cope T A, Carleton A M, Jones P D, Lagun V, Reid P A and Iagovkina S 2005 *Int. J. Climatol.* **25** 279-94
[22] Vaughan D G, Marshall G J, Connolley W M, King J C, Mulvaney R 2001 *Science* **293** 1777-9
[23] Lubin D, Wittenmyer R A, Bromwich D H and Marshall G J 2008 *Geophys. Res. Lett.* **35** L02808 doi:02810.01029/02007GL032170
[24] Marshall G J 2003 *J. Climate* **16** 4134-43
[25] Claud C, Carleton A M, Duchiron B and Terray P 2009 *Climate Dyn.* **33** 383-408
[26] Spiegel M R and Stephens L J 1999 *Schaum's Outline of Theory and Problems of Statistics, 3rd ed.* (New York: McGraw-Hill)
[27] Rao V B, Do Carmo A M C and Franchito S H 2003 *Int. J. Climatol.* **23** 1537-45
[28] Sinclair M R 1994 *Mon. Wea. Rev* **122** 2239-2256
[29] Simmonds I 1985 *J. Geophys. Res.* **90** 5637-5660
[30] Simmonds I, Rafter A, Cowan T, Watkins A B and Keay K 2005 *Bound.-Layer Meteor.* **117** 149-77
[31] Kara A B, Wallcraft A J, Metzger E J, Hurlburt H and Fairall C W 2007 *J. Climate* **20** 5856-64
[32] Chelton D B, Schlax M G, Freilich M H and Milliff R F 2004 *Science* **303** 978-83
[33] Streten N A and Pike D J 1980 *Aust. Meteor. Mag.* **28** 201-16